ANIMAL WASTE UTILIZATION

Effective Use of Manure as a Soil Resource

Edited by
J.L. Hatfield
B.A. Stewart

CRC Press
Taylor & Francis Group
Boca Raton London New York

CRC Press is an imprint of the
Taylor & Francis Group, an **Informa** business

First published 2002 by Lewis Publishers

Published 2019 by CRC Press
Taylor & Francis Group
6000 Broken Sound Parkway NW, Suite 300
Boca Raton, FL 33487-2742

© 2002 by Taylor & Francis Group, LLC
CRC Press is an imprint of Taylor & Francis Group, an Informa business

First issued in paperback 2019

No claim to original U.S. Government works

ISBN-13: 978-0-367-44804-2 (pbk)
ISBN-13: 978-1-57504-068-4 (hbk)

Visit the Taylor & Francis Web site at
http://www.taylorandfrancis.com

and the CRC Press Web site at
http://www.crcpress.com

Library of Congress Card Number 97-30973

Library of Congress Cataloging-in-Publication Data

Animal waste utilization : effective use of manure as a soil resource / edited by J.L.
Hatfield, B.A. Stewart.
    p. cm.
    Includes bibliographical references and index.
    ISBN 1-57504-068-9
    1. Farm manure—Congresses. I. Hatfield, Jerry L. II. Stewart, B.A. (Bobby
    2. Biology—molecular. I. McLachlan, Alan. II. Title.
    Alton), 1932-
S655.A57 1998
631.6′61   97-30973
                                                 CIP

Preface

Utilization of animal manure as a soil resource is a concept that was practiced widely before the advent of commercially available fertilizers and the increase in the size of farm and livestock operations. Throughout the world there is an increasing concern about the generation of animal manure in volumes that could potentially pose environmental problems and inefficient use in agricultural systems. There is an increasing social dilemma over the use of manure because of the odor problems and costs of application and handling of manure compared to commercial fertilizers. These are only a few of the emerging concerns about the use of manure.

Manure is often considered a waste and its decomposition is referred to as waste disposal rather than resource utilization. This attitude toward manure has led to much of the current misunderstanding of how we could use this resource to supply crop nutrients and increase soil organic matter. If one looks through the history of agricultural research, it is easy to see that our current understanding of manure is based on research conducted in the late 1960s with a few studies in the 1970s. Much of that research focused on the supplying of crop nutrients and not on the environmental consequences of surface runoff of phosphorus or leaching of excess nitrate-nitrogen through the root zone. We have also changed the primary tillage practices, and much of the manure application is onto land in which there is a requirement for a crop residue cover. This residue cover requirement limits the incorporation of manure and there is little equipment technology available to help the producer through these problems.

We have a research base on which to draw initial answers about the effective use of manure; however, these have not been summarized in any treatise for use by a range of audiences. In 1994 a workshop was held on the Effective Use of Manure as a Soil Resource as part of the National Soil Tilth Laboratory's series on Long-Term Soil Management. The workshop was held with the goal of bringing together researchers who had developed much of the current knowledge base on manure use and handling and of drawing inferences from their research and understanding of the problem to provide a base that could be used to develop solutions for the problems of today and tomorrow. The chapters contained within this volume include one on the attitudes of farmers about the use of manure by Pete Nowak and his co-workers and one on the economics issues surrounding manure usage by Lynn Forster. We are fortunate to have their expertise available to us as we try to develop new programs for manure utilization.

The chapters on swine, dairy, and poultry manure show examples of current problems and the limitations of technology specific to a given livestock industry. These authors provide a basis for improved understanding of manure generation and utilization as a soil resource. Manure is often considered to be a cropland resource; however, application to rangeland and grass pasture is often practiced over a wider range of climates and manure types. Use of manure on grazing lands helps to define the potential uses on this type of system. Environmental concerns from the use of manure are often associated with ground and surface water quality. This chapter details the impacts of nitrate-nitrogen and phosphorus movement from different manure sources and the potential environmental impacts. To help develop an improved management tool for manure, the final chapter describes the use of system engineering principles to help develop manure management and utilization scenarios.

This volume is intended to help promote interest in the use of manure; however, it also captures our current knowledge base so that we can develop effective research programs that build upon this existing knowledge base. It is imperative that we continue to develop solutions that can be readily adopted by the user community and that when adopted, instill confidence in the user and society that the agricultural community is interested in efficient production, a high quality environment, and being good neighbors. It is our desire that this book serve as an initial step in that process.

J.L. Hatfield
B.A. Stewart

Contents

About the Editors:

Dr. J.L. Hatfield has been the Laboratory Director of the United States Department of Agriculture Agricultural Research Service, National Soil Tilth Laboratory in Ames, Iowa since 1989. He has been with the USDA-ARS since 1983, previously as the research leader of the Plant Stress and Water Conservation Unit in Lubbock, Texas. After receiving his Ph.D., Dr. Hatfield served on the faculty at the University of California, Davis, from 1975 through 1983. Dr. Hayfield received his Ph.D. from Iowa State University in 1975, a M.S. from the University of Kentucky in 1972, and a B.S. from Kansas State University in 1971. He is a Fellow in the American Society of Agronomy, Crop Science Society of America, and Soil Science Society of America. He served as editor of the *Agronomy Journal* from 1989 through 1995. Dr. Hatfield is the author or co-author of more than 225 articles and book chapters. He is the co-editor of *Biometeorology and Integrated Pest Management* and five volumes of *Advances in Soil Science*. He began the Long-Term Soil Management Workshops in 1991, of which this volume and other volumes of Advances in Soil Science are derived, as a means of evaluating the current state of knowledge regarding soil management and basic soil processes. He has an active research program in soil-plant-atmosphere interactions with emphasis on the energy exchanges as the soil surface under different tillage and crop residue management methods and the estimation of the evapotranspiration.

Dr. B.A. Stewart is a Distinguished Professor of Soil Science, and Director of the Dryland Agriculture Institute at West Texas A&M University, Canyon, Texas. Prior to joining West Texas A&M University in 1993, he was Director of the USDA Conservation and Production Research Laboratory, Bushland, Texas. Dr. Stewart is past president of the Soil Science Society of America, and was a member of the 1990-1993 Committee of Long Range Soil and Water Policy, National Research Council, National Academy of Sciences. He is a Fellow of the Soil Science Society of America, American Society of Agronomy, Soil and Water Conservation Society, a recipient of the USDA Superior Service Award, and a recipient of the Hugh Hammond Bennett Award by the Soil and Water Conservation Society.

Contributors

D.R. Bouldin, Department of Soil, Crop and Atmospheric Sciences, Cornell University, Ithaca, NY 14853, USA

A. Breeuwsma, Agricultural Research Department, The Winand Staring Research Centre, Marijkeweg 11/22, NL-6700 AC Wageningen, The Netherlands

Michael C. Brumm, University of Nebraska, Northeast Research and Extension Center, Concord, NE 68728, USA

T.C. Daniel, Department of Agronomy, University of Arkansas, Fayetteville, AR 72701, USA

Donald L. Day, Agricultural Engineering Department, University of Illinois at Urbana-Champaign, Urbana, IL 61801, USA

D. Lynn Forster, Agricultural Economics Department, The Ohio State University, Columbus, OH 43210, USA

Ted L. Funk, Agricultural Engineering Department, University of Illinois at Urbana-Champaign, Urbana, IL 61801, USA

D.L. Karlen, U.S. Department of Agriculture, Agricultural Research Service, National Soil Tilth Laboratory, Ames, IA 50011, USA

S.D. Klausner, Department of Soil, Crop and Atmospheric Sciences, Cornell University, Ithaca, NY 14853, USA

Fred Madison, Department of Soil Science, College of Agricultural and Life Sciences and University of Wisconsin Extension, Madison, WI 53706, USA

A.P. Mallarino, Agronomy Department, Iowa State University, Ames, IA 50011, USA

J.J. Meisinger, U.S. Department of Agriculture, Agricultural Research Service, Environmental Chemistry Laboratory, BARC-West, Beltsville, MD 20705, USA

Philip A. Moore, Jr., U.S. Department of Agriculture, Agricultural Research Service, PPPSRU, University of Arkansas, Fayetteville, AR 72701, USA

Pete Nowak, Department of Rural Sociology, College of Agricultural and Life Sciences and University of Wisconsin Extension, Madison, WI 53706, USA

William A. Phillips, U.S. Department of Agriculture, Agricultural Research Service, Grazinglands Research Laboratory, El Reno, OK 73036, USA

J.R. Russell, Animal Science Department, Iowa State University, Ames, IA 50011, USA

J.S. Schepers, U.S. Department of Agriculture, Agricultural Research Service, Soil and Water Conservation Unit, University of Nebraska, Lincoln, NE 68583, USA

Andrew Sharpley, U.S. Department of Agriculture, Agricultural Research Service, Pasture Systems and Watershed Research Lab., Curtin Road, University Park, PA 16802-3702, USA

Robin Shepard, Environmental Resources Center, College of Agricultural and Life Sciences and University of Wisconsin Extension, Madison, WI 53706, USA

J.T. Sims, Department of Plant Science, University of Delaware, Newark, DE 19717-1303, USA

John M. Sweeten, Texas Agricultural Experiment Station, The Texas A&M University System, Agricultural Research and Extension Center, Amarillo, TX 79106, USA

Farmers and Manure Management: A Critical Analysis

P. Nowak, R. Shepard, and F. Madison

I. Introduction

Manure management, the focus of this paper, is the use of animal manures in a way that is appropriate to the capabilities and goals of the farm firm while enhancing soil and water quality, crop nutrition, and farm profits. While it is possible to provide a general definition for manure management, the same cannot be said of the farms with this responsibility. The role of manure within a farm situation is diverse in form and occurrence in that the farms that generate manure vary from feedlots, dairy and beef farms, horse operations, and poultry operations to open-range ranches. The form, nutrient content, and handling procedures associated with animal manure in these situations vary dramatically. The agronomic and environmental context in which this manure is introduced also varies in terms of assimilative capacity and vulnerability to degradation. Finally, there is also significant variation in the extent the market and institutional context recognizes and supports animal manure as a crop nutrient source or promotes alternative, commercial crop nutrients. Two implications result from these overlapping patterns of diversity.

First, there will be no single technological solution to the current mismanagement of animal manures. As noted, the composition, form, prevailing management patterns, and physical setting for manure preclude any universal solution based on new technologies. While any one new technology may have adequate applicability, it is unlikely to be employed on a universal or even widespread basis. This is due to the aforementioned diversity and the fact that the operators of the farms and ranches responsible for managing manures are also diverse in terms of managerial skills, economic objectives, access to supporting programs, and ability and willingness to adopt various manure management technologies.

Second, changing patterns and consequences of manure management are predicated on the ongoing process of changing human behavior. This is the fundamental principle of manure management. Manure management from the farmer's[1] perspective is not an end objective. Manure management is an ongoing, evolving process for the livestock farmer. While analysis of manure management is often dominated by discussion of why changes are needed due to environmental degradation, or technical investigations of what remedial technologies and practices should be employed, the fact remains that behavioral change is the only criterion for measuring success in the area of manure management. Any assessment of a manure management program will ultimately have to be based on the extent the program has induced behavioral change with targeted livestock and poultry managers.

A consequence of these overlapping patterns of diversity is that any attempt to change farmer behavior by uniformly promoting a "one-size-fits-all" remedial program based on some mix of financial, educational, or regulatory efforts will be ineffective. The premise of this paper is that manure management as defined above is not possible either through seeking a quick "technical fix" or through reliance on "shotgunning" uniform policy tools at diverse farm audiences operating in diverse settings. Instead, the complexities in the physical and engineering dimensions of manure management need to be matched by understanding the complexities in what farmers are actually doing and why it is being done relative to manure management. Moreover, this complexity needs to be specified within exact physical, technological, and farm system contexts. "Bringing the farmer in" to establish a behavioral foundation will be the basis for sound manure management. While there is a role for technological and programmatic innovation, these creative efforts must be guided by an understanding of the farmer's current situation. Technological and programmatic innovation in manure management cannot continue to blindly accept untested assumptions, repeat glib generalizations, or base efforts on political platitudes when it comes to the behavior of livestock farmers. The behavior of livestock farmers relative to patterns of manure management and mismanagement is a research question and must be addressed as such.

This paper has two functions. The starting point must be an understanding of current patterns of manure management and mismanagement. One cannot explain why farmers do not use manures more efficiently until one first examines how

[1] Farmer will be used in a generic sense to refer to landusers who manage livestock and poultry.

manures are currently being used. This issue will be examined by reporting on research exploring the extent and accuracy of manure management within Wisconsin. Data on the extent and accuracy in crediting manure, total nutrients applied in the production of corn, efficacy of storage structures, and a dimension of manure distribution will be presented. While there are significant limitations in generalizing the results beyond upper-Midwest dairy-livestock systems, they do present many research issues to be explored in other settings under different types of livestock systems.

The second function of this paper is to provide a better understanding of the farmer's situation relative to manure management. As noted, the objectives of manure management are going to be achieved by changing the behaviors of farmers responsible for managing this on-farm resource. However, if managing manures in an economically and environmentally sound fashion is the "right" thing to do, then why are not more farmers doing it? Policy analysts, program managers, agricultural researchers, farm organizations, environmentalists, and equipment manufacturers all have answers to this question. All these explanations contain some validity. However, the perspective of the most important group — farmers responsible for actually managing this manure — is often lacking from this discussion. Consequently, the second portion of this paper will present a number of reasons from the perspective of the livestock farmer on why they do not manage manures according to various technical and policy recommendations.

II. Methods

Data were collected from 1,179 Wisconsin farmers. A standardized survey instrument was used between 1990 and 1994 to assess current agronomic and manure management behavior in eight different geographic locations. The survey instrument focused on commercial fertilizer, manure storage and application issues, crop rotations, pesticide selection, operator knowledge of management practices, and preferred sources of management information.

A flexible instrument was designed so it could be employed in personal interviews, group meetings, or mail surveys. The format of the assessment instrument was the outcome of an interdisciplinary process. Questions in the assessment were based on relevant research, University of Wisconsin Extension bulletins, fact sheets, and publications. Questions were peer reviewed for technical accuracy by a multidisciplinary group of university specialists and researchers.

The instrument was printed using high quality graphics, color, easily understood language (i.e., farmer friendly tone to the writing style), and a variety of question styles that include Likert scales, multiple choice answers, and fill-in-the-blank numerical responses. Pretest versions were modified to enhance the validity and reliability of responses. On average, 50 questions have been used across the eight collection points. Additional modifications continue to be made to focus on selected issues and to make the instrument applicable to special geographic areas and types of production systems.

Table 1. Method and location of survey with response rate

Wisconsin geographic location	Project type	Delivery method	Number of responses	Response rate (%)
South	Watershed	Face-to-face	208	88
Southwest	Watershed	Mail	139	77
East	County-wide	Small groups	214	76
Northeast	Watershed	Mail	101	75
Central	County-wide	Mail	227	80
Central	Watershed	Face-to-face	45	90
West	County-wide	Face-to-face	195	77
North	County-wide	Face-to-face	50	86
Totals			1,179	80

In four of the eight data collection locations, face-to-face interviews were used. Free well water tests (nitrate-nitrogen and bacteria) were offered as an incentive to complete the questionnaire in one of these locations. Three other locations used mail delivery techniques following a modified Dillman approach to survey research (Dillman, 1978). The remaining uses of the assessment were based on a series of group meetings where the instrument was administered to participants. All respondents were screened on two criteria: 1) they operated at least 16 ha of land, and 2) they had at least 15 dairy or beef cattle. The average survey response rate for these eight different data collection locations was 80% (Table 1) with a range between 75% and 90%.

Each farmer was asked to identify the form and rate of nutrients applied to a representative field. The research strategy used a representative field rather than collecting detailed information on multiple fields due to logistical costs. Data were analyzed to determine mean rates of nitrogen, phosphorus, and potassium application on this representative field within the farm operation. It was decided after pretesting and talking with farmers that the most productive field in corn during the year of the interview would be the representative field. The representative nature of this field was assessed by asking whether the nutrient rates used on this field were higher, lower, or the same as on other corn fields in production the year of the assessment. This field was judged to be representative as 80% of the farmers did not differentiate between corn fields in commercial nitrogen rates, 93% did not differ in terms of herbicide application rates, and 67% did not differentiate between corn fields in manure applications.

Farmers were asked to provide nutrient application type and rate information for the representative field. Application of animal manures was included in these calculations when manure was applied to the most productive corn field within 12 months before planting. Estimates of manure nutrients were calculated by having farmers identify the type of manure, size of the manure spreader, number of loads applied to the representative field within 12 months before planting corn, and the size of that field.

Solid manures were credited for inorganic (plant available) nitrogen. This represents approximately 40 % of the total nitrogen in the manure. Manure credits were converted to pounds of available nitrogen per ton regardless of the form in which it was applied (e.g., bushels per acre were converted to tons per acre and in turn converted to SI units). Credits by type of manure were as follows: dairy 3.4 kg/ha of nitrogen, beef manure 4.3 kg/ha nitrogen, swine manure 4.5 kg/ha nitrogen, poultry manure 11.0 kg/ha nitrogen, and sheep manure 14.0 kg/ha nitrogen. For liquid manure (kilograms available per 1,000 gallons) the comparable figures were respectively 9, 13, 13, 39, and 32 kilograms nitrogen per hectare (Madison et al., 1986). No second or third year credits were used in calculating total nitrogen rates.

Nutrient credits for a first-year corn field coming out of a legume rotation were also estimated following University of Wisconsin estimates. In calculating legume credits it was assumed there was a 60% stand at plow down. This results in a nitrogen credit of 146 kg/ha. The 146 kg/ha of nitrogen was based on a recommendation of 45 kg/ha plus 1.7 kg/ha for each percent legume in stand (Wolkowski, 1992; Bundy et al., 1990). Another conservative decision rule was that no nitrogen credits were given for second year corn following alfalfa. Clover was credited at 117 kg/ha nitrogen, soybeans at 39 kg/ha nitrogen, and peas at 20 kg/ha nitrogen.

The underlying goal for UW's crop fertility recommendations has been to supply nutrients to the crop so that economically damaging nutrient stress does not occur at any point during a rotation. This idea is founded in the belief that to avoid stress, a minimum nutrient concentration must be present in the soil or through fertilizer application (Kelling et al., 1981). Recommended nutrient rates were estimated with University of Wisconsin guidelines for corn production after adjusting for specific soil types (Bundy, 1990; and UWEX-WDATCP, 1989). To account for differences in University of Wisconsin-Extension soil test recommendations, soil maps were consulted to find the general soil type for the area surrounding the respondent's farm. A recommended level of 160 kg/ha nitrogen was used for medium textured soils, 112 kg/ha nitrogen for sandy soils, and 157 kg/ha nitrogen for clay textured soils (Bundy, 1990).

The estimated nutrient application rates were calculated to be intentionally conservative in four ways: 1) they do not consider residual soil nitrate other than first-year legume nitrogen credits, 2) they only account for first-year manure credits, with nutrients from manures applied in previous years being ignored, 3) they assume none of the manure was incorporated although this behavior was measured, and 4) only the lowest value was used when a range was presented for manure or legume credits.

Once actual and recommended nutrient application rates were determined, the cost of excess commercial nutrient purchase could be calculated. These costs refer to what farmers paid for commercial nitrogen, phosphate (P_2O_5), and potassium (K_2O) when these nutrients were applied at rates above university recommendations. It is important to note that these costs were calculated only if the farmer purchased commercial nutrients when on-farm nutrient sources were available to meet the recommended crop nutrient need. Costs do not refer to total nutrient values or to the value of on-farm nutrient sources. They only refer to what a farmer could have saved by using on-farm nutrient sources in meeting recommended nutrient levels.

III. Results

A. Overall Nutrient Application Rates

The nutrient application rates used in producing corn are illustrated in Figures 1-3. Nitrogen includes commercial forms, legume credits for those corn fields in the first year out of a legume crop, and manure applications. Phosphorus (expressed as P_2O_5) and potassium (expressed as K_2O) values are based on commercial sources and manure. The extremely high (outlier) values in these graphics were truncated for illustration purposes. Measures of range and variation are provided to provide a sense of the true distribution. Each of these overall nutrient application rates is derived from multiple measures representing the nutrient source. If one or more of these individual measures were missing, as opposed to not being used, the case was deleted from the analysis. The result of this data analysis rule is presented as the number of valid cases still in the calculation.

The average nitrogen (N) application rate in Figure 1 was 242 kg/ha and is based on 1,048 valid cases. This rate varied between 1 kg/ha (a situation where a small amount of manure only was applied) and 1,524 kg/ha (a situation where a field came out of alfalfa, a very large amount of manure was applied, and high rates of commercial nitrogen were applied). This distribution of total nitrogen rates had a standard deviation of 160 kilograms per hectare. The fourth quartile of the distribution is represented by farmers who had applied total nitrogen at rates of at least 309 kg/ha.

The average phosphorus (P_2O_5) application rate was 140 kg/ha as illustrated in Figure 2. Calculations were based on 1,048 valid cases. This rate varied between 1 kg/ha and 1,357 kg/ha with a standard deviation of 125 kg/ha. The fourth quartile of the distribution is represented by farmers who had applied 192 kg/ha or more of phosphorus.

Figure 3 illustrates that potassium (K_2O) was applied at an average rate of 330 kg/ha. This varied between .45 kg/ha and 3,725 kilograms per hectare with a standard deviation of 337 kg/ha. It is based on 1,048 valid cases. Farmers applying potassium at rates of 476 kg/ha or more represented the fourth quartile.

B. Four Popular Beliefs About Manure Management

The remaining analysis is organized around four popular beliefs about manure management. These four beliefs are often used to justify the form and content of remedial policies, technology development, and outreach efforts. The data are analyzed in a fashion to examine validity of each of these beliefs.

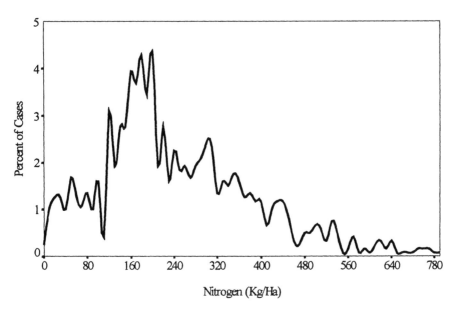

Figure 1. Total nitrogen, all sources.

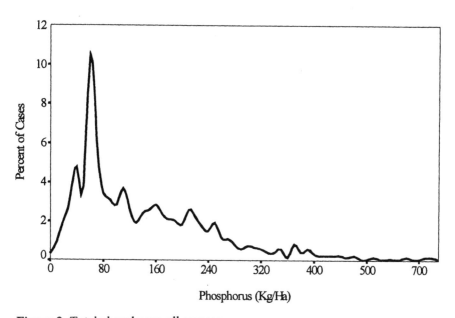

Figure 2. Total phosphorus, all sources.

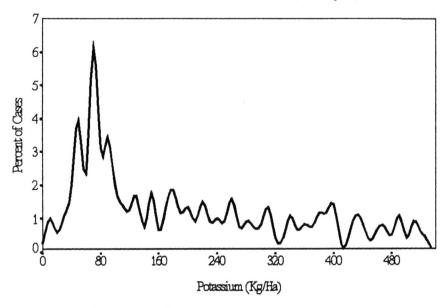

Figure 3. Total potassium, all sources.

1. Farmers Recognize the Value of Manure and Credit Accordingly

Based on informal observations and discussions, most agree that livestock farmers understand that manure has nutrient value, and may increase soil organic matter and enhance soil tilth among other beneficial qualities. Farmers, it is often argued, recognize that manure is "good" for the soil. Yet being able to recognize this "goodness" versus being able to take advantage of this on-farm resource are two separate processes. Figure 4 illustrates both the proportion of farmers crediting manure nitrogen, and the accuracy of that crediting process. Estimates of total manure nitrogen applied to the most productive corn field were determined using the process described earlier. Crediting was measured by determining the amount (kilograms per hectare) that commercial nitrogen was reduced due to available amounts from manure application, i.e., the extent manure was accurately credited.

Of all the farmers spreading animal manures on the most productive corn field, only 29.8 percent made an effort to credit manure nitrogen (left side of Figure 4). Seven out of every ten (70.2%) livestock producers made no effort to credit nitrogen or other nutrients from animal manures spread on their corn fields. Of the 29.8% who do attempt to credit, 66.0% of this group underestimated manure nitrogen by 11% or more while 28.0% of this group overestimated manure nitrogen by 11% or more. Only 6.0% of the 29.8% of farmers who attempted to credit manure were crediting within plus or minus 10% of University of Wisconsin guidelines (right side of Figure 4). In sum, less than 2% of all farmers spreading manure on corn ground are crediting these manures with any degree of accuracy (i.e., ± 10% UW guidelines). While some may argue that livestock farmers

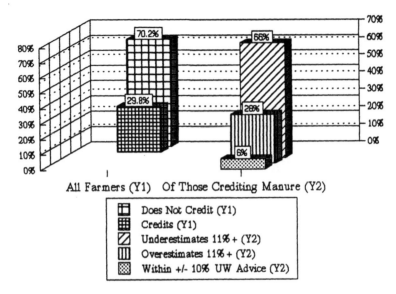

Figure 4. Claims and accuracy in crediting of manure nitrogen.

recognize the inherent value of manure, in fact, few are attempting to take advantage of this on-farm nutrient resource, and fewer still are doing so in an accurate fashion.

2. Manure Crediting Is Uneconomical

There is a cost to distributing manures on cropland. This can include labor, periodic machinery investments, and opportunity costs among others. These costs are found in standard farm budget sheets. There is less evidence, however, on the value of manure other than generalizations on the equivalent worth relative to commercial nutrients. The value of manure in this analysis was calculated as the amount being spent on commercial nutrients when on-farm nutrient sources (manure and legumes) would have provided the recommended nutrient amounts. It is not the total value of the animal manures and legume nutrients. It is the value of the animal manures and legumes up to the amount actually spent on commercial fertilizers required to achieve recommended nutrient levels. The values resulting from this analysis can never exceed the value spent on commercial nutrients on a per acre basis. Farmers in the sample were spending, on average, $15.70 per acre ($38.80 per hectare) on commercial nutrients in the production of corn when on-farm nutrient sources were available (Figure 5). The standard deviation for this value was $15.60 per acre or $38.55 per hectare. The range was between zero and $135.90 per acre ($335.81 per hectare). The fourth quartile of the cost distribution is represented by those farmers spending an average of $22.30 per acre ($55.10 per hectare) on commercial nutrients when on-farm sources were available. Consequently, while there are well-

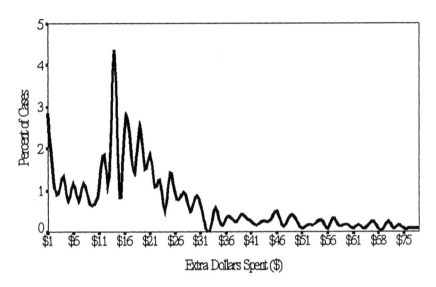

Figure 5. Dollars spent on commercial nutrients when on-farm nutrients available.

nutrients. The value of manure in this analysis was calculated as the amount being spent on commercial nutrients when on-farm nutrient sources (manure and legumes) would have provided the recommended nutrient amounts. It is not the total value of the animal manures and legume nutrients. It is the value of the animal manures and legumes up to the amount actually spent on commercial fertilizers required to achieve recommended nutrient levels. The values resulting from this analysis can never exceed the value spent on commercial nutrients on a per acre basis. Farmers in the sample were spending, on average, $15.70 per acre ($38.80 per hectare) on commercial nutrients in the production of corn when on-farm nutrient sources were available (Figure 5). The standard deviation for this value was $15.60 per acre or $38.55 per hectare. The range was between zero and $135.90 per acre ($335.81 per hectare). The fourth quartile of the cost distribution is represented by those farmers spending an average of $22.30 per acre ($55.10 per hectare) on commercial nutrients when on-farm sources were available. Consequently, while there are well-documented costs to manure management, there are also clear economic benefits that manifest themselves as the potential for a reduction in commercial fertility costs.

3. Storage Structures Improve Manure Management

This is a principal belief currently guiding public manure management programs. There is a significant amount of private and public investment on an annual basis in various types of manure storage structures. Pits, lagoons, tanks, and other types

Table 2. Manure nitrogen crediting/accuracy by manure handling system (%)

Crediting behavior	Daily haul and structure (80)	Structure only (149)	Daily haul only (777)
No credit for manure N	59.3%	61.7%	72.8%
Credits manure N	40.7%	38.3%	27.2%
If credits, underestimates manure N by 11% +	45.4%	52.8%	72.5%
If credits, within + 10% UW recommendations	4.5%	11.1%	4.9%
If credits, overestimates manure by 11% +	50.0%	36.1%	22.5%

Table 3. Nitrogen management (kg/ha) by manure handling system

N sources	Daily haul and structure	Structure only	Daily haul	T-value	2-tail significance
Manure N	99.4	106		-0.45	0.655
Manure N		106	156.7	-3.62	0
Manure N	99.4		156.7	-3.04	0.002
Legume N	119.1	114.7		-1.76	0.083
Legume N		114.7	135.1	0.05	0.963
Legume N	119.1		135.1	-2.33	0.02
Purchase N	102.2	95.5		0.55	0.585
Purchase N		95.5	92.6	0.39	0.698
Purchase N	102.2		92.6	1.02	0.306
Total N	209.2	216.9		-0.44	0.661
Total N		216.9	254	-2.59	0.01
Total N	209.2		254	-2.38	0.018

all collapsed into the "structure" category for this analysis. Three questions were asked to assess whether these types of storage structures lead to better manure management. The results are illustrated in Tables 2 to 4 and Figures 6-7.

The first question attempted to assess whether the manure handling system was related to the manure crediting process (Table 2). Farmers who daily haul manure had the lowest proportion attempting to credit the nitrogen in these manures. A little more than a quarter (27.2%) credited manure nitrogen. This can be contrasted with

Table 4. Phosphorus management (kg/ha) by manure handling system

P_2O_5 sources	Daily haul and structure	Structure only	Daily haul	T-value	2-tail signif- icance
Manure P_2O_5	92.7	98.4		-0.42	0.675
Manure P_2O_5		98.4	144.1	-3.63	0
Manure P_2O_5	92.7		144.1	-3.02	0.003
Purchase P_2O_5	44.6	46.5		-0.61	0.544
Purchase P_2O_5		46.5	49.9	-1.77	0.076
Purchase P_2O_5	44.6		49.9	-2.12	0.023
Total P_2O_5	111.9	118.7		-0.52	0.6
Total P_2O_5		118.7	154.4	-3.16	0.002
Total P_2O_5	111.9		154.4	-2.81	0.005

approximately two-fifths of farmers with structures; 38.3% for those with a structure-only system and 40.7% for those farmers who daily haul and have a storage structure.

The second question assessed the accuracy of those who claimed they were crediting manures (Table 2). The manure nitrogen crediting process was calculated based on the procedures outlined earlier. Accuracy was assessed by comparing the amount that commercial nitrogen was reduced due to crediting versus the amount of first-year manure nitrogen calculated according to the procedures discussed earlier. The general pattern was that a minority of farmers who claim to credit also underestimate this nutrient source. Both daily haul only (72.5%) and structure only (52.8%) farmers were underestimating manure nitrogen by more than 11% of recommended values. Farmers using a structure and a daily haul system who credited had 45.4% of this group underestimating manure nitrogen. An almost equal percentage (50.0%) of these farmers overestimated manure nitrogen by 11% of recommended values. This can be contrasted with 36.1% of structure-only farmers and 22.5% of daily haul-only farmers who overestimated manure nitrogen by this amount. If accuracy is crediting within ± 10% of university guidelines, then structures do not appear to significantly increase the proportion of farmers accurately crediting manure nitrogen. Integrating the percent crediting with the percent within ± 10% of university guidelines results in 1.8% (40.7% X 4.5%) of daily haul and structure farmers, and 4.2% (38.3% X 11.1%) of the structure-only farmers accurately crediting. This can be contrasted with 1.3 percent (27.2% X 4.9%) of the daily haul only farmers. Consequently, investment in structures results in a gain of between 0.5 and 2.9% in the desired behavior of accurately crediting manure.

The third question concerned the overall level of nutrients used in the production of corn. The expectation is that structures would allow farmers to better take advantage of the crop nutrients in manures when compared to their colleagues on a daily haul system. If structures do encourage better manure management, then one

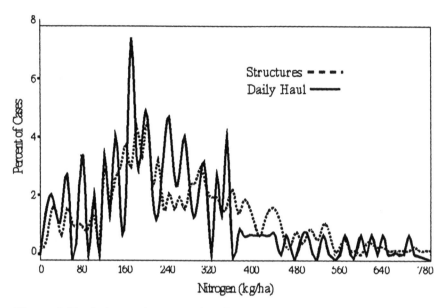

Figure 6. Total nitrogen by manure handling system.

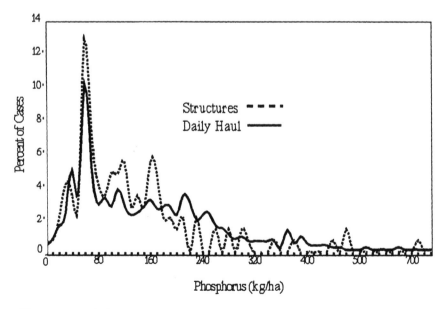

Figure 7. Total phosphorus by manure handling system.

would expect to see those farmers with structures closer to UW and private sector guidelines in crop nutrient rates.

In accord with the procedures discussed earlier, the total nitrogen applied in the production of corn was calculated from manure, legume, and commercial sources. The analysis in Table 3 is laid out to test for statistically significant differences between nitrogen sources and manure management system. Farmers with the hybrid (daily haul and structure) system applied an average of 99 kg/ha of manure nitrogen. Farmers with a structure applied an average of 106 kg/ha, while farmers with a daily haul system applied an average of 156 kg/ha of manure nitrogen. Farmers on a daily haul system had available significantly more manure nitrogen than those farmers with a structure[2] ($\bar{x}_{dh}$ = 156 kg/ha versus $\bar{x}_s$ = 106 kg/ha; t-value = -3.6; 2-tail prob. = .000) or those with a structure and daily haul system ($\bar{x}_{dh}$ = 156 kg/ha versus $\bar{x}_{dh \, s}$ = 99 kg/ha; t-value = -3.0; 2-tail prob. = .002).

First-year corn coming out of a legume had the farmers with the combined structure and daily haul system gaining an average of 119 kg/ha of potential legume nitrogen credit. Farmers with only a structure obtained an average of 115 kg/ha of nitrogen credits from legume sources during first-year corn. Farmers with a daily haul system gained an average of 135 kg/ha of nitrogen credits from legume sources during the first year of corn. Farmers with a daily haul system had significantly more legume nitrogen available in first-year corn than farmers with the combined manure system ($\bar{x}_{dh}$ = 135 kg/ha versus $\bar{x}_{dh+s}$ = 119 kg/ha; t-value = -2.3; 2-tail prob. = .020). There were no other statistically significant differences for this nitrogen source and the manure management systems in Table 3.

Commercial nitrogen varied between an average rate of 102 kg/ha for farmers with both the structure and daily haul system to 92 kg/ha for farmers on a daily haul system. Farmers with a structure purchased an average of 95 kg/ha of nitrogen for the production of corn. None of the combinations between these manure management systems and nitrogen sources represented statistically significant differences.

Combining these three sources produces the average total nitrogen used in the production of corn. Table 3 illustrates that those with a daily haul system applied, on average, significantly more total nitrogen than those with only a structure in the production of corn ($\bar{x}_{dh}$ = 253 kg/ha versus $\bar{x}_s$ = 216 kg/ha; t-value = 2.6; 2-tail prob. = .010). The total nitrogen rates for farmers with the daily haul system were also significantly higher ($\bar{x}_{dh}$ = 253 kg/ha versus $\bar{x}_{dh+s}$ = 208 kg/ha; t-value = -2.4; 2-tail prob. = .018) than those farmers with a combined daily haul and structure system.

The comparable analysis for manure management systems and sources of phosphorus is presented in Table 4. Farmers on a daily haul system had significantly more phosphorus available from manure than those with a structure ($\bar{x}_{dh}$ = 145 kg/ha versus $\bar{x}_s$ = 99 kg/ha; t-value = -3.6, 2-tail prob. = .000) or those with a combination daily haul and structure system ($\bar{x}_{dh}$ = 145 kg/ha versus $\bar{x}_{s+dh}$ = 93 kg/ha; t-value = -3.0, 2-tail prob. = .003). In terms of purchased phosphorus, the only statistically significant difference was between farmers with a daily haul system who purchased more commercial phosphorus than those farmers with a daily

---

[2] The subscript $_{dh}$ refers to daily haul; and the subscript $_s$ refers to a structure.

haul system combined with a structure ($\bar{x}_{dh}$ = 51 kg/ha versus $\bar{x}_{s+dh}$ = 45 kg/ha; t-value = -2.1, 2-tail prob. = .034).

When considering both commercial and manure sources, daily haul farmers are applying more phosphorus than their counterparts with only a structure ($\bar{x}_{dh}$ = 155 kg/ha versus $\bar{x}_s$ = 119 kg/ha; t-value = -3.2; 2-tail prob. = .002), or those with the combination daily haul and structure system ($\bar{x}_{dh}$ = 155 kg/ha versus $\bar{x}_{s+dh}$ = 112 kg/ha; t-value = -2.8, 2-tail prob. = .005).

Farmers with a daily haul system are applying more total nutrients in the production of corn than their counterparts with a structure in the manure management system. There were no statistically significant differences between those farms with structures and those who use both structures and a daily haul system for both nitrogen and phosphorus.

A critical question is whether these statistically significant differences are also meaningful in an economic or environmental sense. This line of analysis asks whether the average difference between mean total nitrogen under a daily haul and a structure system (253.4 - 216.4 = 36.90 kg/ha) makes economic sense when considering the level of public investment in programs promoting structures. Although a detailed analysis is not possible with this data set, the question is partially answered in the next two figures. Here the percent of cases by the total nitrogen (Figure 6) and phosphorus (Figure 7) used in the production of corn on a per hectare basis is plotted for two situations, farms with a structure and those without.

In the nitrogen distribution (Figure 6) the daily haul system has a positive skewness value of 2.0 while the structure-based systems have a positive skewness value of 1.3. Statistically these are different distributions as established by the earlier significance tests. Yet there is enough congruity in the pattern to question the benefits derived from investments in structures. The potential for nitrate-nitrogen leaching — even acknowledging this is a very site-specific process — is roughly the same for both distributions. That is, the proportion of farmers in each category applying excessive (i.e., the right "tail" of the distributions) nitrogen is approximately the same. Measures of dispersion, not central tendency, are the critical indicators when the objective is environmental management. Application rates several standard deviations above the mean probably exceed the capability of the physical setting to buffer, hold, or assimilate the excess nutrient being applied. Figure 6 suggests that investment in structures does not appear to mediate these "tails" in the rate distributions.

Similar conclusions can be reached for the phosphorous distribution in Figure 7. While they are statistically different based on various measures of central tendency, the difference generated by the amount of investment is questionable when considering the potential for non-point pollution processes. Farmers with structures are still applying, on average, approximately two to three times the replacement value of P_2O_5. More important, both types of systems have a significant number of cases that are skewed to the right of the respective mean and median values. As was the case for nitrogen, these farms out on the "right tail" of the distribution represent an even greater potential for water pollution to occur. Generating a statistical difference versus solving an economic or environmental problem are very different

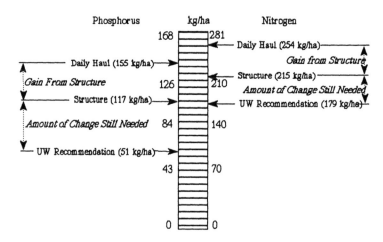

Figure 8. Proportionate gain through investment in manure structures.

outcomes. While structures appear capable of achieving the former, they have not yet accomplished the latter.

Besides the apparent inability of structures to mediate the "extreme" cases in a rate distribution, further evidence against sole reliance on structures can be found with a closer examination in measures of central tendency. This type of analysis is summarized in Figure 8.

Here the average N and P_2O_5 rates by structure-nonstructure manure management systems are graphically scaled against the average recommended rates for these nutrients. That is, the differences in actual nitrogen and phosphorus rates between a structure and daily haul system relative to the average recommended rate are portrayed as scalar functions. While those systems that use structures are lower in a statistical sense than daily haul systems ($\bar{x}_{dh}$ - R) > ($\bar{x}_s$ - R) for both nitrogen and phosphorus, the critical question is the proportionate reduction in the distance between actual and recommended rates. That is, to what extent does investment in structures move the actual rate closer to the recommended rate? While the distance gained on this scalar figure through investment in structures is greater for nitrogen than for phosphorus, both situations fall short of reaching the recommended rate. With phosphorus, the investment in a structure "moved" the farmers 37% of the distance between a daily haul system and the recommended rate. For nitrogen, the investment in a structure is associated with a change of 52% of the distance between a daily haul system and the recommended rate.

There needs to be further policy or economic analysis to assess whether the current level of investment in structures is worth the proportionate gain in manure management. The scaler representation in Figure 8 does not support the hypothesis that investment in structures is the "solution" to manure mismanagement. While they clearly "moved" livestock farmers in the right direction, at issue is the level of investment required for these modest gains in manure management.

4. Farmers with Daily Haul Systems Are More Likely to "Dump" Manure on the Field Closest to the Barn

A common belief is that farmers on a daily haul system will "dump" manure on the field closest to the barn or facility where the manure is generated. This belief was assessed by asking farmers about travel times when spreading manure. They were told to estimate the travel time from the barn or storage facility when the loader (tank, wagon, etc.) was loaded to the edge of the field where the manure was to be spread. These times refer to travel times only and not loading or unloading times. They were asked to provide this travel time for three potentially different situations: the field receiving manure with the shortest travel time, the field receiving manure with the longest travel time, and the travel time to the field that received the most manure. Although this type of analysis cannot account for the spatial relation of the barn or structure relative to the fields, it does provide preliminary estimates of time invested in moving manure away from this barn or facility. The assumption is that longer travel times reduce the likelihood of "dumping" manure on fields close to the barn or storage facility. The results are illustrated in Table 5.

The overall average travel time for all cases to the closest field was 2.9 minutes, the field receiving the most manure 6.6 minutes, and the most distant field 12.5 minutes. The differences between these travel times are disparate in a statistically significant fashion. The closest field was different than the most distant field (t-value = 29.8; 2-tail prob. = .000), the closest field was different than the field receiving the most manure (t-value = 19.9; 2-tail prob. = .000), and the field receiving the most manure was different than the most distant field (t-value = 20.2; 2-tail prob. = .000).

While there were significant differences in travel times between these three field situations, results indicate few statistically significant differences within a travel destination when considering types of manure handling systems. Farmers with structures and daily haul systems spent an average of 2.6 minutes (s.d. = 3.3 mins) traveling to the closest field that received manure. Those with only structures spent 2.4 minutes (s.d. = 2.5 mins), and those on a daily haul system spent an average of 3.1 minutes (s.d. = 3.1 mins) hauling manure to the closest field that received manure. The high standard deviations relative to the means also indicate significant variance within each of these groups. As can be seen in Table 5 there was a significant difference between those with a structure combined with daily haul as well as structure only when compared to those on a daily haul system. Contrary to the popular stereotype, farmers with a daily haul system are transporting manure further to this closest field than those farmers with a structure in the manure handling system.

The field receiving the most manure was just over two times as distant in travel time when compared to the closest field. Those farmers with a daily haul and structure system spend an average of 5.2 minutes (s.d. = 2.8 mins) while those with a structure only spend an average of 5.8 minutes (s.d. = 3.8 mins). Farmers with a daily haul system spend 6.9 minutes (s.d. = 4.9 mins) traveling to the field that received the most manure. The only statistically significant difference was between those with the hybrid system and those on a daily haul. Again, farmers with a daily

Table 5. Average travel times for manure distribution by manure handling system (time in minutes from leaving barn/structure to edge of field)

P_2O_5 sources	Daily haul and structure	Structure only	Daily haul	T-value	2-tail signif-icance
Field closest to barn	2.59	2.4		0.31	0.758
Field getting most manure	5.25	5.82		-0.81	0.419
Most distant field	12.59	11.45		0.7	0.488
Field closest to barn	2.59		3.11	-0.92	0.363
Field getting most manure	5.25		6.91	-2.86	0.006
Most distant field	12.59		13.09	-0.34	0.737
Field closest to barn		2.4	3.11	-2.1	0.038
Field getting most manure		5.82	6.91	-1.86	0.066
Most distant field		11.45	13.09	-1.64	0.104

haul system spent more time traveling to the field that received the most manure when compared to farmers with a hybrid manure handling system.

The field with the longest travel time was approximately twice as far in travel time as the field receiving the most manure. Those with a structure spent 11.4 minutes (s.d. = 7.6 mins) versus 12.6 minutes (s.d. = 8.5 mins) for those with both a structure and daily haul traveling to this most distant field. Farmers relying on a daily haul system spent an average of 13.1 minutes (s.d. = 8.5 mins) traveling to this most distant field that received manure. There were no statistically significant differences between any of these travel times.

There are several limitations of this travel time analysis. First, it does not account for variations in farmstead layout or for variation in field topography and access. Both of these situations could cause significant differences in travel time yet no difference in actual distance traveled. A second limitation is that it does not directly account for the proportion of manure being distributed in these three field situations. While including the field getting the most manure is part of this proportional analysis, it does not account for overall distribution within the farm system. However, it does point out that, in aggregate, the field getting the most manure is more distant in travel time than the field closest to the barn. Moreover, it shows that farmers with a daily haul system are likely to spend more time traveling away from the barn before spreading manure than farmers with structure-based manure handling systems. In sum, this limited analysis finds no support for the belief that farmers with a daily haul system are more likely to "dump" manure on fields that are contiguous to the barn.

IV. Constraints to Proper Manure Management

The second part of this paper is more subjective than the first portion. It will attempt to provide an understanding of why farmers are doing such a poor job in managing animal manures. The central thesis running through this portion of the paper is that farmers cannot be characterized as "villains" based on the poor performance documented earlier unless there is an equal effort to understand why this mismanagement is occurring from the perspective of the farmer. There are a number of reasons for manure mismanagement; some are evident while others are more subtle and complex. These reasons are organized into six categories for presentation purposes. For the livestock farmer, however, these reasons are not neatly packaged or organized into discrete categories. Instead, they may occur simultaneously along temporal (certain seasons or weather patterns), spatial (certain parts of a field or farm), or labor (responsibility and skill levels) dimensions. They are also related to the characteristics of the farm firm and operator as well as the institutional context in which the farm is located. Finally, the relation between these constraints is probably not linear. That is, they can interact and reinforce each other, making proper manure management all the more difficult for the farmer.

A. Institutional

1. The Quality and Quantity of Research on Manure as a Source of Plant Nutrition Is Limited and Biased

Table 6 is derived from the work of Dittrich (1993) and illustrates the emphasis placed on animal manures in all UW corn production or soil fertility bulletins between 1938 and 1991. This graphic is organized into five categories depending on how animal manures were treated in these bulletins. The lowest category is where animal manures were never mentioned in corn production or soil fertility bulletins. This is followed by only a slight mention but no value (i.e., crop nutrient source or good for the soil) was associated with the reference in the text or footnote. Manure was simply mentioned and nothing more. The third category was where manure was mentioned as influencing corn production or soil fertility. Yet there was

P. Nowak, R. Shepard, and F. Madison

Table 6. Role of manure in UW corn/soil fertility bulletins, 1938-1991

Manure value	1938-1943	1944-1949	1950-1955	1956-1961	1962-1967	1968-1973	1974-1979	1980-1985	1986-1991	Total
Nutrient credit	0	0	2	1	0	0	1	1	6	11
Nutrient value	0	0	0	0	0	0	0	1	0	1
Mention	0	4	6	3	0	0	0	0	2	15
Slight mention	0	0	4	2	1	2	1	1	2	13
No mention	1	4	1	1	5	8	7	6	2	35
Total bulletins	1	8	13	7	6	10	9	9	12	75

no information on manure equivalent values or how to credit manures. The fourth category was where manure was discussed as a crop nutrient source. Moreover, values were given on the amount of nutrients per unit measure of different types of animal manures. The final category was where information was provided on how to use manure as a crop nutrition source. This information included equivalent values and how to credit by reducing commercial fertilizers.

Other than three bulletins published in the 1950s, all other corn production and soil fertility bulletins published between 1938 and the mid-1970s failed to provide information on manure as a nutrient source. The lack of reference in the early years was to be expected as manure as a crop nutrient source was the prevailing production model. The critical time period was in the late 1940s and early 1950s when commercial nutrients became available and were widely promoted. From that time up to the early 1970s manure underwent the transition from the primary source of crop nutrition to one where it was largely ignored in crop production. It has only been in recent years that this pattern is being reversed. Since 1974 more bulletins are being developed that give manure nutrient values. Yet it must also be noted that bulletins that ignore or minimize the role of manure continue to be produced during this same period.

A similar pattern emerges when examining bulletins on manure handling and storage between 1938 and 1991 (Table 7). Bulletins developed during this period were classified by whether manure was discussed as a waste, an asset, or whether the characterization was ambiguous — neither waste nor asset (Dittrich, 1993). Manure was largely viewed as a farm asset up until the early 1960s. At that time the theme of manure as a waste, as something to be disposed of, began to emerge, and

Table 7. Emphasis in UW manure management bulletins, 1938-1991

Manure emphasis	1938-1943	1944-1949	1950-1955	1956-1961	1962-1967	1968-1973	1974-1979	1980-1985	1986-1991	Total
Asset	2	0	3	1	1	1	1	2	7	18
Ambig- uous	0	0	0	0	0	1	0	1	0	2
Waste	0	0	0	0	2	1	2	4	0	9
Total bul- letins	2	0	3	1	3	3	3	7	7	29

was consistent with the themes found in the crop production bulletins. From the early 1960s to the mid-1980s manure was largely characterized as a waste. It was not until the mid-1980s when environmental and economic pressures became more intense that manure was again acknowledged as a potential asset.

The above graphs illustrate that approximately the last two cohorts or generations of farmers entering Wisconsin agriculture had little opportunity to learn about the potential nutrient role of manure in corn production or soil fertility. Instead, they were explicitly and implicitly told manure is a waste. While these themes have begun to change in the last few years, the majority of today's farmers have been taught to treat manure as a waste while ignoring it in crop production — a lesson evidently well-learned as evidenced by the data found in the earlier analysis of manure management behaviors.

2. The Transaction Costs for Obtaining Information and Assistance on Manure Management May Be High

Transaction costs refer to the amount of effort and resources required to obtain information and assistance. The previous constraint illustrated that manure management information may be difficult to obtain or even nonexistent. A related issue is conflicting sources of information. That is, there is not a consistent theme or uniform interpretation of data in the information on manure management. The outcome of conflicting or inconsistent information sources in a farm community is often based on the credibility of the source, not necessarily the validity and reliability of the information. In this case it means the majority of farmers will continue to rely on fertilizer dealers for advice on manure management rather than assuming the high transaction costs associated with finding consistent sources of information.

Another dimension of transaction cost is the complexity of remedial information. Manure management is often presented by public sector agencies in complex and very specific models driven by detailed guidelines. The complexity and detailed models are not consistent with the needs of farmers in the early stages of the adoption process, managerial capability, or current views of manure as an on-farm resource. In short, many of these complex models and fact sheets are being developed for fellow technicians and academics, not the typical farmer who needs manure management information. Rather than simple decision aids that will help the majority of farmers begin to manage manures as a valuable resource, we produce complex tools, accurate to the second decimal point, that simply overwhelm most farmers. Rather than recognizing manure management as a process where simple steps must represent the beginning, technical recommendations tend to view it as an "all or nothing" situation where the focus is on the end stage while ignoring the beginning and intermediate steps. As noted earlier, manure management is an ongoing process for farmers. Many technical bulletins, however, treat it as an end objective based on numerous technical details. Farmers have responded to this unneeded technical complexity accordingly by ignoring these high transaction costs.

3. As in Other Areas of Public Outreach and Communication Efforts, There Is a Bias of Information and Assistance Flowing to Larger and More Receptive Farm Audiences

Personnel in public agencies are rewarded (salary, tenure, peer recognition, etc.) for the amount of planned change achieved in their jurisdictional areas. Performance reports are based on criteria that attempt to measure this induced change (i.e., impact indicators). Consequently, these agency representatives have a tendency to work with receptive farmers who are not necessarily those needing assistance the most. The amount of time and effort required to gain the participation of several farmers who are not traditional cooperators — yet who may be the "worst" in terms of manure mismanagement — can be easily offset by continuing to work with a medium or large-sized, cooperating farm. After all, larger farms generate more acres, cows, or pounds of production change for performance reports than do smaller farms. Consequently, and because of this performance evaluation system, professional public change agents are making a correct and rational decision in targeting their efforts to the medium to larger-sized farms that have been traditional cooperators. The private sector selects clients on some form of an "ability to pay" basis. Again, there may be no relation between these clients with investment capabilities and those who have significant manure mismanagement problems. Moreover, sales volume and the potential for higher net returns go with the products and services sold to the larger production units.

4. Prevailing Models of Livestock and Dairy Production Systems Treat It as a Waste at Best, or Ignore it as an Externality at Worst

Animal manure management became synonymous with waste management in the 1960s. Livestock production models during this period became more myopic while focusing on narrow dimensions of economic efficiency and ignoring larger issues associated with the production system. Animal manures were no longer considered part of the general production system equation. That is, a cyclical process where manure is viewed as a herd output used in crop production whose product is used as an input to herd production and so on. The emphasis shifted solely to herd output enhancement instead of balancing this with input reduction goals through using manure. Manure was divorced from the system and reclassified as a system externality or waste. If manure was even mentioned in these production models, it was only considered as detrimental to the herd environment, a waste that should be removed and disposed of. Consequently, students of dairy, beef, swine, or poultry production do not have the opportunity to learn about how manure can be part of an overall production system.

B. Engineering

1. The Box Spreader Is Largely Designed to Get Rid of a Waste, Not Manage On-Farm Nutrient Resources

The conventional box spreader wagon is by far the most common type of manure spreader found in Wisconsin livestock agriculture. Yet this piece of equipment is an engineering anachronism when it comes to manure management. It could be argued that the last major innovation in this equipment occurred when it went from horsepower to tractor power takeoff following the turn of the century. The box spreader is unable to handle variation in manure consistency or provide uniform distribution needed for crediting. Farmers on a daily haul system know that upon reaching the edge of the field, opening the gate on the spreader causes all liquids and fine materials to run off. Clods and large chunks of manure are then spread in a haphazard fashion by the beater bar. Yet crop nutrition would require a fairly uniform distribution of these nutrients. A fundamental principle underlying manure crediting is that there is a certain level of uniform distribution of manure on the cropland field. Lack of uniformity results in a random patchwork of under- and overfertilized portions of the field. This makes consistent crediting of animal manures difficult if not impossible for most farmers.

2. Many Box Spreaders Do Not Have Weight Calibration Needed for Accurate Crediting

Another fundamental principle underlying crediting is the assumption that the farmer has a reasonable accurate estimate of how much (volume or weight) manure is being distributed on a field. Yet box spreaders have no built-in scale that would provide even a crude estimate of the amount of manure in the wagon. Moreover, based on the spreader calibration efforts of the Wisconsin Nutrient and Pest Management Program, there appears to be no relation between the manufacturer's capacity rating and the working loads employed by farmers. Consequently, farmers are put in the position of hauling largely unknown quantities of manures to the field for distribution. Most other industrial and agricultural hauling equipment (e.g., bulk commercial fertilizer spreaders) lists working capacity as a standard manufacturing requirement. This lack of standardization for manure spreaders has forced on farmers the additional task of acquiring a weight calibration for their spreaders if they want to begin crediting manures with any degree of accuracy.

3. The Minimum Rate at Which Liquid Manure Spreaders Can Apply Is Often Above Recommended or Needed Rates

Although there is only preliminary evidence to support this, it appears that many liquid manure spreaders were designed to dispose of a waste rather than credit liquid animal manures. That is, the minimum rate these spreaders inject or eject liquids is often above the maximum rate recommended for crop nutrition. A number of engineering factors (e.g., nozzle and hose size) are responsible for this situation depending on the type of liquid manure spreader. Again, it appears as if this item of machinery is being designed for waste disposal rather than as a tool for managing on-farm nutrient sources. This makes injecting liquid manures at recommended rates very difficult for farmers or for the custom applicators who rely on this type of equipment.

C. Private Sector

1. Fertilizer Dealers Tend to Ignore Manures as a Reliable Source of Plant Nutrition Due to the Uncertainty Associated with Crediting Manures Spread by Customers

This is a sound business decision on the part of the fertilizer dealer. Due to the many factors previously discussed, as well as those that follow, fertilizer dealers simply cannot trust that animal manures will be available on a uniform basis as a source for plant nutrition. A poor crop stand due to a nutrient deficiency will result in a lost customer or may require some form of compensation, a situation that many dealers cannot afford in today's competitive market.

2. Manures Are Viewed and Treated as "Insurance" by Dealers in Event of Ideal Conditions Needed for a "Bumper" Crop

Dealers, like farmers, recognize that manures are "good" for soils and crop nutrition. Yet because of the unreliable nature of manure as a nutrient source, dealers can only view it as insurance. Manure is relegated to a "back-up" nutrient if natural events should cause an excessive loss of the commercial nutrients applied to a cropland field, or if the ideal weather conditions create the opportunity for a "bumper" crop.

3. Fertilizer Salespersons Tend to Follow the "Maximum Output per Acre" Paradigm Rather than Producing Crops to Meet the Needs of Livestock Farms

A difficult concept to communicate is that not all farmers want to achieve maximum production from crop fields. This very idea seems irrational and uneconomical. Yet on a mixed enterprise farm, one where livestock play the central role in generating farm income, maximum yields from crop production often take a secondary role to livestock production. In these cases farmers know the minimum amount of feed and other forages that need to be generated (e.g., "I need enough yield to fill up this bin and half of the other one."). Yet the crop nutrient recommendations are the same for these livestock farms as for cash grain farms participating in commodity programs and the markets. This failure to differentiate by production goals results in manure being underestimated or ignored in many situations where it could play a primary crop nutrition role for the livestock farmer.

4. Custom Manure Spreading Businesses (e.g., Emptying Pits) Use a Volume/Time Price Structure Creating an Incentive to "Dump" as Much Manure as Possible in the Shortest Time in Order to Provide the Best Price on a Per Unit Basis

Farmers with manure storage structures often pay to have a commercial firm come in and empty and distribute the manure from the storage facilities. Yet there is no incentive or legal guidelines for these custom manure spreading firms to distribute manures compatible with crediting and crop nutrition guidelines. Thus, market processes create an incentive to "dump" rather than manage these on-farm nutrient sources.

D. Economics

1. The Cost and Design of "Approved" Manure Storage and Handling Systems Are Biased Toward Larger Farms Due to Investment Capabilities (i.e., Not Scale Neutral)

Investment in a manure storage facility is a major capital decision for most farms. Even with significant government cost-sharing, the farmer's share is often beyond the capability of many farms. Economies of scale can play a role where these facilities have utilities associated with herd health, labor requirements associated with daily haul, and sanitary licensing requirements. Consequently, farms with larger herds often invest in these facilities to capture this utility. These same benefits, however, are not available to the farms with smaller- to medium-sized herds. This process implicitly signals the livestock farmer that the only way to survive is to expand herd size in order to capture the above utilities. This, in turn, exacerbates the existing environmental and economic problems associated with manure mismanagement.

2. Manure Structures Are a Costly and Risky Investment for Farm Families at the Beginning or End of the Farm Family Business Cycle

As noted, manure storage facilities can be a major investment for the farm firm. Younger farm families often face cash flow difficulties due to the start-up debt obligations associated with land, herd facilities, and the herd itself. On the other hand, farm families preparing to retire from farming see little incentive to invest in these type of facilities due to the long payback period, a short planning horizon, and the uncertainty if the realty market will compensate for investment in such a structure. Structures, therefore, may only be appropriate to farms entering the middle portion of farm career cycle.

3. Daily Haul Manure Systems Can Impose Significant Labor Constraints During Certain Critical Periods During the Production Cycle

By definition, daily haul systems require a minimum investment of several hours daily. Those hours invested in manure handling can be critical during the spring planting window, periods when forage is being cut, during fall harvest, and during critical animal production events. The duration of these critical periods may be insignificant when viewing average labor demands on an annual or even seasonal basis. A viable management strategy during one of these critical periods is to minimize time and managerial commitments by disposing of manure in a wasteful yet timely fashion. This management strategy, as we have already seen, is constantly reinforced by the public and private sectors. Consequently, farmers have come to believe that there simply isn't time to treat manure as a valuable on-farm nutrient source on a year-round basis when considering the various constraints they would face.

E. Social-Psychological

1. It Is Difficult to Communicate Problems from Overapplication

Manure is viewed as a "natural" input that has many beneficial qualities: provides nutrients for plant growth, develops soil tilth, increases soil organic matter, encourages microbiological species diversity, and is a renewable resource. Communicating that "too much of a good thing" is a problem can be difficult for public agencies and others promoting sound manure management. The problem of recognizing the limits of excess is a classical dilemma. Public agencies tend to reinforce this situation by focusing education messages on the good qualities of manure. Yet, as noted earlier, the majority of farmers already know this fact. Ignored in this process is the more difficult educational theme of teaching recognition of when there is too much of a good thing. Anecdotal evidence indicates that some farmers informally define this upper limit when tire traction on the tractor deteriorates significantly on fields where thick "blankets" of manure have been spread.

2. Farmers May Face Significant Safety and Weather Concerns Associated with When and Where Manure Can Be Spread

Driving tractors with spreaders up icy or slippery slopes, trying to spread manures under very cold conditions when the manure could freeze in the spreader if it isn't dumped quickly, and spreading manures on muddy fields due to long rainy periods are all real problems for daily haul farmers. All these conditions work against the logical, planned, and timely distribution of manure within a farm setting. These are the "real world" situations farmers face in trying to manage manures, situations that receive little attention in research, outreach, or in the development of rational nutrient management plans.

3. There Are Real Status Issues Associated with Who Spreads Manure

Manure handling is a low status task compared to other concurrent farm tasks such as milking, feeding, and other field operations. Consequently, hired labor, children, or retired parents are often relegated the task of managing the manure. Yet communication and other persuasive efforts are often aimed at the farm manager who only minimally engages in this activity. This results in missing the real target audience and reduces the effectiveness of existing educational programs. The people who are responsible for managing manure on a farm never have the opportunity to learn about alternative methods.

4. The Vocabulary of Manure Management is Biased Toward Mismanagement by Calling These On-Farm Nutrient Sources a "Waste"

It is difficult to get farmers to invest time, money, and effort into managing something constantly referred to by the experts and state agencies as a waste. The etymological basis for the term waste is built around concepts such as a useless or worthless by-product, something that is used in a thoughtless or careless fashion. If we want farmers to continue treating manures literally as a waste, then by all means we should continue calling it that in our technical reports, educational bulletins, legislative rules, and agency regulations. This institutional hypocrisy — calling it one thing while expecting farmers to treat it as something else — is communicating a very mixed message to livestock farmers. While seemingly a minor semantical point, calling animal manures a waste instead of some form of an on-farm resource, the implications are significant to farmers in shaping their behavioral expectations.

F. Environmental

1. Land Constraints May Have Been Created by Past Animal Science Research, Economic Recommendations, and Market Conditions, All of Which Have Accentuated Increasing the Size of Confinement-Type Operations

A large amount of feed may be purchased off-farm with confinement operations. These marketing opportunities mean that a sufficient land base is no longer a requirement for a viable, large operation. This has resulted in operations being larger (number of animals) on proportionately smaller acreage. This increases the probability that the farm will generate more manure than the managed land can safely assimilate. This land constraint hypothesis is often cited as a major reason for existing environmental problems due to manure mismanagement. Future programs and regulations are often debated and designed to address this situation. Yet the fact remains that the extent or distribution of this land constraint situation remains untested. There is no empirical evidence showing either the prevalence or location of manure mismanagement due to land constraints. The continued emphasis on this untested hypothesis is distorting other remedial efforts. That is, while attention and effort is focused on this factor, many other very real constraints and obstacles to proper manure management remain unaddressed.

2. Topography, Soils and Other Surficial Features Can Result in Many Smaller and Spatially Fragmented Fields

This makes distribution of manures under these circumstances managerially complex, time-consuming, and very dependent on seasonal and other weather-related factors. Trying to rationally allocate manure among 20 or 30 small fields, not an unusual situation in the unglaciated area of western Wisconsin, becomes almost a full-time task in which few farmers can afford to invest.

3. Bottom Lands in a Dissected Topography Often Receive the Most Manure While Being in Continuous Corn Due to Fewer Options on Wwhere Corn Can Be Produced

Farmers operating in a dissected topography have learned what fields can be depended upon to reliably produce grain and forage. With the options limited to farming the hill tops and valley bottoms, there is a tendency to produce corn on a continuous basis on the bottom lands. Recent soil conservation and erosion control programs have also supported this selective process. The consequence is that the manure from these operations is concentrated on these bottom lands where the continuous corn is produced. The environmental implication of this action is that these same lands are also those closest and often adjacent to the rivers, streams, and impoundments that follow this bottom land topography.

V. Conclusions

Historically, manure was a primary source for plant nutrition. "Its value for maintaining and improving the productivity of the soil has been recognized from the earliest times. Manure is of value in soil improvement because of its content of fertilizer materials, of humus, and of certain organic constituents" (Salter and Scholenberger, 1938). There is an old German folk saying in Wisconsin that "the manure pit is the farmer's gold pit." A little over a hundred years ago the Reverend Evast preached to his German congregation that, "where there is manure, there is Christ" (Zeitlin, 1977). This remark may appear cryptic today, yet made sense a century ago. Farmers who wanted to practice the biblical notion of stewardship — not just protect or preserve, but enhance the value of their land — relied on manure as a soil amendment to accomplish this spiritual objective. In the last one hundred years we have gone from a situation where manure was the means to agronomic, economic, and spiritual viability to a situation today where it is viewed as a waste. Public research dollars are invested in this topic under the rubric of waste management, public committees and groups debate programs to manage this waste, and farmers largely treat it as a waste.

The premise of this analysis is that designing techniques and programs that will assist farmers to better manage manure is predicated on a good understanding of current manure management behaviors. This premise is derived from the generalization that *altering patterns and consequences of manure management are predicated on the ongoing process of changing human behavior.* This study of current management behaviors has found few farmers taking advantage of the potential benefits associated with manure. The majority of farmers are applying excess nutrients in the production of corn largely due to the unaccounted contribution from animal manures. These excess nutrients, depending on the specific ecological setting of the behaviors, could have significant environmental implications.

This critical analysis was organized around four common and popular beliefs about farmers' manure management. Little support was found for any of these beliefs. That is, assumptions and presumptions about what farmers are doing or why it is being done relative to manure management did not bear up to empirical scrutiny. First, only a very few farmers are crediting manures. Of this minority attempting to credit, few are doing it with any degree of accuracy. For every farmer crediting manures within $\pm$ 10 percent of university recommendations, there are 35 farmers who make no attempt to credit this on-farm nutrient source. While many may believe that farmers recognize the positive aspects of manure, in actuality few farmers are taking advantage of any economic or soil quality benefits to be derived from proper manure management.

The second popular belief is that proper manure management is uneconomical. The limited analysis found that, on average, farmers could save $15.70 per acre ($38.80 per hectare) on commercial nutrients through proper use of on-farm nutrient sources. This finding is limited in that it only examined the short-term, direct benefits of manure and legumes. Long-term benefits such as those associated with improvements in soil quality were not investigated (National Research Council, 1993), nor were opportunity costs associated with managing manures versus other pursuits investigated. The problem is that nutrient replacement values associated with manure may represent only several percent of a livestock operation's total input budget. It is difficult to argue that farmers should allocate significant managerial expertise to recoup such marginal gains. This is especially the case when considering the significant institutional, engineering, and other constraints and complexities a farmer faces when trying to manage manure properly. It makes little economic sense to expend these levels of human capital when the potential returns are marginal at best. Only by making manure crediting simple through reducing or eliminating the constraints listed in the latter portion of the paper would this become a cost-effective activity. It is not a question of teaching farmers how to take advantage of on-farm nutrient sources. Rather the issue is one of making the application of this knowledge convenient and uncomplicated. Consequently, there were mixed results for this second popular belief about the economic value of proper manure management.

Assessing the value of storage structures in manure management, the third popular belief, also had mixed results. Those who used a structure in their manure management system were more likely to credit, and moved in the direction of recommended nutrient rates more so than those relying on a daily haul system. There were statistically significant differences in the overall, average crop nutrient rates between structure and nonstructure operations. Yet it needs to be emphasized that those with a storage structure in their manure handling system still exceeded, on average, recommended nutrient rates to a significant degree. Equally important was the finding that manure structures did nothing to mediate the situation where a very few farmers contribute disproportional to the potential for environmental problems. The positive and skewed "tail" of the nutrient input distributions was still present when structures were in place. All this indicates that structures alone — without concomitant efforts to induce changes in management behaviors — are a poor and costly technological solution to improper manure management. While these devices may be a necessary component of proper manure management, a point yet to be determined, the data demonstrate that by themselves they are not sufficient to achieve this management objective. One is not going to solve problems related to farm economic inefficiency or environmental degradation by simply digging more holes in the ground, pouring more concrete, or installing more tanks. This finding is especially salient in light of market and technological factors that are causing larger operations to be situated on smaller parcels of land with much of the feed being purchased off-farm. The design and finance of structures within this type of concentrated operation have received significant public engineering and financial support. In contrast, public support for behavioral management programs to ensure the proper distribution and crediting of manures from these structures has largely failed to materialize.

The last popular belief examined was that daily haul farmers are more likely to dump manure on the field closest to the barn. There was little support for this belief. In fact, the opposite was the case with daily haul farmers spending, on average, more time traveling away from the barn than their counterparts with a structure.

It needs to be emphasized that the objective of this critical analysis was not to cast blame on the livestock farmer by focusing on levels or extent of manure mismanagement. Rather, it was to document current manure management behaviors in such a way that appropriate solutions could be sought. *Assuming* the behaviors or rationale of livestock farmers is not sufficient for reasoned analysis, nor is it sufficient for the design of remedial technologies or programs. At issue, and an integral part of this critical analysis, is an understanding from the farmer's perspective on why poor manure management is the norm rather than the exception. Twenty-one obstacles or constraints to proper manure management were subjectively described. The strength and distribution of these constraints among livestock farmers are largely unknown due to the lack of research interest, protocols, and financial support. Hence the subjective tone to this latter portion of the critical analysis is appropriate. Even with this subjective limitation, and when trying to view the policy and technical expectations of manure management programs from the perspective of a livestock farmer, it supports the conclusion that these constraints are real and operative.

Examining this list of constraints to proper manure management also generates another conclusion. This is the extent that current policy, research, and debates about future policy are largely ignoring these constraints to proper manure management. It appears that current policy discussions and research can be organized along one or more of three themes: 1) the environmental need for improved manure management based on specifying either various types of damages or technical fixes; 2) the cost and the process of generating and appropriating funds to support research and programs that promote structural "solutions" to manure mismanagement; and 3) the need for various regulatory procedures to force farmers to change manure management due to the apparent failure of the voluntary approach. Yet few of these prevailing policy and research themes attempt to address the previously listed constraints to proper manure management as faced by the typical livestock farmer. Rather than a failure in the voluntary approach, the continued ignoring of these "real world" constraints makes one question whether a voluntary approach has even been tried.

Continued dereliction in trying to understand why farmers are mismanaging manures will subsequently lead to failed policy, production of irrelevant research, and the inefficient use of public dollars. Not only will this current line of program thinking result in inadequate environmental protection, but this well-intended yet misguided program direction could have a significant, negative financial impact on livestock farmers.

The implicit objective of this critical analysis was to generate research, discussion, and analysis of manure management that incorporates the farmer's perspective. The intent was not to provide a solution to manure mismanagement. It was to provide a different perspective on why this problem continues to persist. Solutions to manure mismanagement will not be found solely within engineering, economic, or environmental dimensions. By now it should be clear that insights from the behavioral dimension need to be integrated into these other perspectives. It is in that domain, one where the situation of the livestock farmer is initially and integrally appraised, where realistic and workable solutions will be found.

References

1989. The new Wisconsin nitrogen recommendations. Proceedings, Agricultural Fertilizer, Agchemical and Line Dealers Conference, Madison, Wisconsin.

Bundy, L., K. Kelling, and L. Ward Good. 1990. Using legumes as a nitrogen source. University of Wisconsin Extension Publication (A3517), Madison, Wisconsin.

Dillman, D. 1978. *Mail and Telephone Surveys: The Total Design Method.* John Wiley and Sons, New York, NY.

Dittrich, M. 1993. Corn fertility and manure management: Agricultural and extension bulletin information, 1938-1991. Master of Science thesis, Land Resources Program, Institute for Environmental Studies, Madison, Wisconsin.

Kelling, K., P. Fixen, E. Schulte, E. Liegel, and C. Simson. 1981. Soil test recommendations for field, vegetable and fruit crops. University of Wisconsin Extension Publication (A2809), Madison, Wisconsin.

Madison, F., K. Kelling, J. Peterson, T. Daniel, G. Jackson, and L. Massie. 1986. Managing manure and waste: Guidelines for applying manure to pasture and croplands in Wisconsin. University of Wisconsin Extension Publication (A3394), Madison, Wisconsin.

National Research Council. 1993. *Soil and Water Quality: An Agenda for Agriculture.* National Academy of Science, Board on Agriculture, Washington, D.C.

Salter, R. and C. Scholenberger. 1938. Farm manure. p. 445-461. In: *Soils and Men: 1938 USDA Yearbook of Agriculture.* Government Printing Office, Washington, D.C.

UWEX-WDATCP. 1989. Nutrient and Pesticide Best Management Practice for Wisconsin Farms. University of Wisconsin Extension Publication (A3466) and Wisconsin Department of Agriculture, Trade and Consumer Protection Technical Bulletin ARM-1, Madison, Wisconsin.

Wolkowski, R. 1992. A step-by-step guide to nutrient management. University of Wisconsin Extension Publication (A3578), Madison, Wisconsin.

Zeitlin, R. 1977. *Germans in Wisconsin.* The State Historical Society of Wisconsin, Madison, Wisconsin.

Economic Issues in Animal Waste Management

D.L. Forster

I. Introduction

Economics offers helpful insights regarding livestock waste management. The first section reviews fundamental economic forces at the consumer and producer levels affecting livestock production. The second section reviews the economics of waste management alternatives at the firm level. The third section examines why some waste management alternatives chosen by firms may be economically inefficient from society's viewpoint, and then some public policy mechanisms to improve waste management decisions are summarized. Finally, implications of these economics issues for the research agenda are discussed.

II. Fundamental Economic Forces at the Consumer and Producer Levels

Consumer preferences as reflected in the marketplace are the ultimate driving force for the scale of economic activity in any industry. In the U.S. and other developed countries, consumers have sufficient income to assert a strong preference for red meats. This preference promotes grazing on vast areas of rangeland that otherwise would have little agricultural use. It also encourages grain fed livestock to be concentrated in feedlots that produce enormous quantities of manure with the potential for polluting nearby streams and water bodies.

From 1960 to 1992, total U.S. meat consumption expanded from 150 to 194 pounds per capita of boneless trimmed meat (Stout, 1992). During this period, population grew by 40% and total meat consumption, on a boneless trimmed equivalent basis, grew by over 80%. Consumer preferences shifted toward poultry as per capita consumption increased from 24 pounds to 70 pounds. Consumption of red meat was about 115 pounds per capita both in 1960 and 1992. These changes in consumption have clear implications: more livestock waste to manage and the potential for more point and nonpoint pollution.

Producers' responses to these preferences are shaped by technology, input prices, and institutions, and these responses are the direct cause of pollution and the effectiveness of manure management. Changes in the beef feeding industry offer an interesting example of how waste management is linked to technology, input prices, and institutions.

Beef feeding has concentrated in the Plains states. Five states—Texas, Nebraska, Kansas, Colorado, and Oklahoma—accounted for 50% of U.S. cattle marketings in 1990; their share was 30% in 1960. On the other hand, cow-calf herds are dispersed widely. Cow-calf operations tend to be located on marginal land and provide a method to market forage from pasture land that has little or no alternative use. Once calves reach 450 to 650 pounds, they are shipped to feedlots for fattening with concentrates.

One of the reasons for the increased concentration of cattle feeding in the Plains states was the development of irrigated grain sorghum production in those states to supply needed concentrate feeds. This development was stimulated partly by increased feed grain prices in the early-1970s resulting from increased export opportunities. Until that time, the U.S. was essentially self-sufficient in feed grains. After that, it produced about one-fourth of its feed grains for an export market. Much of the feed grain exports came from the east North Central region, which in essence, increased feed prices for this region's beef feeders. Another factor contributing to cattle feeding concentration in the Plains states was an efficient transportation system that allowed feeder calves to be shipped to feedlots and beef products to be shipped to distant population centers at low cost. Also, fed beef housing and manure disposal costs were substantially less in the Plains states than in the North Central region. Finally, technological advances and economies of scale in Plains states' meat processing facilities helped change industry structure (the size distribution of firms), as well as change the geographic intensity and location of fed beef production and resulting nonpoint pollution.

Gains in beef feeding in the Plains states came, in large part, at the expense of farmers in the eastern portion of the North Central region. The share of U.S. cattle marketings coming from the east North Central region declined from 14.2% in 1960 to 6.4% in 1990. In addition to east North Central states, Iowa, Minnesota, and Missouri lost substantial market share. The market share of these eight states (Ohio, Indiana, Illinois, Michigan, Wisconsin, Iowa, Minnesota, and Missouri) declined from 33% in 1960 to 16% in 1990. So, the location of beef feeding has changed dramatically. Thirty years ago, the eight-state Corn Belt region fed the same amount of beef as the five-state Plains states region; today, it feeds one-third as many. These regional shifts resulted in a more geographically concentrated fed beef industry, and also produced a more concentrated industry structure with fewer, larger firms. The industry's production continues to shift to feedlots with more than 1,000-head capacity, and only 15% of fed beef are produced in feedlots with less than 1,000 capacity.

Of course, institutions, or society's rules of the game, play an important part in either facilitating or constraining producers' responses to their economic environment. Government regulations aimed at lessening pollution are the most obvious examples of institutions affecting waste management and pollution. Other institutions, such as local zoning ordinances or court remedies to odor complaints, reflect communities' tolerance for livestock odors and greatly affect industry location. Some institutions, such as government supported research and development efforts, are proactive in influencing producers' responses.

These exogenous economic forces at the consumer and producer level have a powerful influence on livestock production and point and nonpoint pollution emanating from livestock facilities. Consumer preferences as reflected in the marketplace are the ultimate driving force for the scale of economic activity in any industry, and nonpoint and point source pollution are determined most fundamentally by these preferences. Producers' responses to these preferences are shaped by technology, input prices, and institutions, and these responses are the direct cause of nonpoint and point pollution.

III. Financial Evaluation at the Firm Level

Benefits and costs of alternative waste disposal systems are a primary determinant in selecting a waste management system. For producers, benefits from livestock waste may be derived by resource recovery processes. That is, livestock waste is a potential source of nutrients for plants, a source of energy, or a source of nutrients for livestock feed. Using waste for a nutrient source for plant growth is currently the primary benefit for producers. However, substantial price increases for livestock feed ingredients or fossil fuels would make livestock waste a more attractive source of feed ingredients and energy (Huang, 1979). Under current price relationships, the primary benefit of livestock waste probably will continue to come from its use as a source of nutrients for plants.

Benefits in the form of nitrogen, phosphorus, and potassium differ greatly according to livestock species as well as handling, storage, and application technologies (Table 1). Nitrogen is the nutrient that has the greatest variation in availability among waste disposal systems. Nitrogen may be in the ammonia, nitrate/nitrite, or organic form. The system components used to store, handle, and spread the waste may allow the ammonia portion to be lost to the air or the nitrate/nitrite portion to leach through the soil or the lot surface. Generally, most of the phosphorus and potassium produced by the animal is available to crops when

Table 1. Nutrient availability from livestock waste by mineral species[a]

Species	Nutrients available to crops per animal unit (kg yr^{-1})		
	Nitrogen	Phosphate (P_2O_5)	Potash (K_2O)
Dairy	12-18	15-36	36-72
Beef			
Fed Beef	9-45	10-45	15-52
Cow-calf	36-45	39-45	45-52
Swine			
Fed swine	1-5	4-8	5-8
Swine breeding	3-14	3-6	7-10
Poultry			
Layer	0.22-0.29	0.37-0.39	0.22-0.23
Broiler	0.16-0.19	0.19-0.20	0.24
Turkey	0.46-0.55	0.64-0.68	0.64
Sheep			
Fed lambs	1.8-2.4	0.9-1.2	2.8-3.5
Ewes and lambs	2.9-3.8	1.4-2.0	4.3-5.8

[a]Ranges are attributable to differences in handling, storage and application technologies.
(From White and Forster, 1978.)

spread. Exceptions are open lot and lagoon systems which permit greater loss of potassium and phosphorus than do other systems.

Benefits from these nutrients can be realized if they are effectively utilized. Ineffective use occurs when manure is applied at high application rates to nearby fields in order to reduce transportation distance. Generally, high application rates reduce the economic benefits of manure. As Schnitkey and Miranda (1993) illustrate, for most livestock herd sizes, farmers have economic incentive to apply manure at rates low enough to effectively use available nutrients and incur the transportation costs that are associated with relatively low application rates over large areas.

A producer's waste management system costs include the annual fixed and variable costs attributable to the system. Annual fixed costs include depreciation, interest, repairs, taxes, and insurance charges on capital investments. Capital investments include all waste management system structures and equipment. Variable costs include labor, fuel, electricity, and maintenance charges. Table 2 illustrates a financial budget for a waste system, including returns from nutrients, variable costs, and fixed costs.

Costs of livestock waste systems are sometimes difficult to estimate since the waste disposal system is an integral part of the total livestock enterprise. Where the housing system ends and the waste disposal system begins is often arbitrary. Ideally, analysis of a waste management system would consider its impact on the total business.

Table 2. Waste system budget for dairy: confined free stall, tractor scrape, plank storage, surface spread, 100 cows

Component	Quantity	Capital investment ($)	Annual cost ($/cow yr)	Annual returns ($/cow yr)	Annual net system return ($/cow yr)
Labor	626 hrs		50		-50
Tractor	563 hrs		85		-85
Energy	0 kw		0		0
Tractor scraper/loader	1	5,670	11		-11
Concrete floor	5.1 m²/an	10,611	13		-13
Storage/plank	1.2 x 23 x 45 m	15,444	25		-25
Box spreader/storage	1	7,980	25		-25
Nutrients					
N	65 kg/an yr			29	29
P_2O_5	35 kg/an yr			17	27
K_2O	68 kg/an yr			19	19
Total		39,750	208	65	-143

(From Blauser et al., 1990.)

When analyzing the firm level financial impacts of livestock waste systems, costs of controlling pollution are often underestimated (Johnson and Conner, 1976; Ashraf and Christensen, 1974). Pherson (1974) suggests that the costs of controlling pollution from livestock facilities generally are underestimated due to valuing resources such as labor at a standard wage rate rather than at its opportunity cost. For example, winter storage of manure is one method of reducing runoff. But the time used in the spring in spreading stored waste may come at a high cost since the farmer's alternative use for his time is in field operations that are crucial for a successful crop. Pherson's estimate is that costs used in waste system budgets may underestimate economic costs to the farmer by as much as 40%. Ashraf and Christensen (1974) analyzed dairy farm organizations in the northeast and conclude that cropping systems may need to be reorganized, and as a result, alternative feed sources and feeding systems may be used to accommodate waste management systems that require the intensive use of labor in the spring.

Pricing machinery used in managing waste presents problems since some can be used in other farm enterprises. For example, the tractor used to scrape manure from an open lot may be used primarily in crop enterprises, and charging its full annual fixed cost to the waste management system is clearly inappropriate. Instead, an hourly rate based on a standard leasing arrangement may be the appropriate method of pricing the tractor.

Farmers surveyed by Badger (1977) indicated several practical problems or costs in using manure: compaction of soil due to application equipment, slower rate of application compared to commercial fertilizer application, the necessity to till a field more than once when incorporating manure in the soil, and uneven applications or "skip" problems. These concerns are excluded from most analyses, but they do influence a farmer's choice of waste management systems.

Major changes in a farm's waste management system may burden the farm with substantial costs, but most existing production systems can be modified at a relatively low cost to improve environmental quality (White and Forster, 1978). These modifications include controlling runoff from exposed lot surfaces, controlling runoff from fields where manure is applied, reducing rates of manure application, and using appropriate management practices to lessen odor nuisances. Forster (1975) estimated that beef feedlot producers could comply with relatively severe runoff control regulations and raise the total cost of production by less than 3%.

An important issue in livestock waste management is its effect on industry structure (i.e., distribution of firm size). Smaller producers face higher per unit costs and are at a competitive disadvantage relative to large producers. When faced with added costs accompanying new pollution control regulations, many producers must make the final choice of whether or not to remain in the livestock business. Typically, public policies that force firms to control pollution without compensating them for additional costs result in some firms exiting from the industry and increased concentration of production.

Livestock production firms may have their access to capital affected by lenders' responses to environmental regulations. The financial community has been awakened to the necessity for sound waste management and pollution control practices by a series of recent federal laws and court decisions (Olexa, 1991). Enacted in 1980, the Comprehensive Environmental Response, Compensation, and Liability Act (CERCLA or "Superfund") authorizes the U.S. Environmental Protection Agency to clean up contaminated sites and to recover the cleanup costs from those responsible for the contamination. The act offers lenders some liability protection if the lender does not participate in the management of the facility. However, later court rulings narrowly interpreted this liability protection. Lenders have been held accountable for cleanup costs of environmentally damaged sites where the lender exercised management control over the borrower or where the lender acquired title to real estate by foreclosure and failed to sell the property in a reasonable period.

As a result of legislation and case law, livestock producers' access to capital markets may be affected in three ways (Mazzocco, 1991). Lenders may be liable, as discussed above, when they exercise management control or acquire title. Due to this potential liability, lenders may reduce lending activities to livestock facilities that have the potential to impair water quality or cause odor nuisances. Also, the supply of credit of lenders may be affected by lenders' knowledge that judgments obtained by the government for environmental damage will affect the borrowers' repayment ability. Finally, the cost of the regulatory compliance has an impact on the borrowers' financial strength and ultimately on borrowing capacity. Survey results cited by Mazzocco indicate that 62.5% of surveyed lenders had rejected loan applications because of potential environmental liability, and 45% had discontinued certain types of loans for this same reason.

Net waste management system costs (fixed and variable costs less nutrient benefits) for common waste management systems were identified by White and Forster (1978), and their estimates were updated to 1990 prices by Blauser et al. (1990). Four generalizations can be inferred from their analysis of dairy operations. First, net system costs vary substantially by size of operation. For the typical free stall, open lot system with 50 cows, net system costs average $133 per animal year, using 1990 prices. As size increases to 200 cows, net system costs per animal year decline to $57. Second, net system costs are higher for confined dairy systems than for open lot systems. For the 100 cow herd, net system costs are about $35 per animal year higher for confinement systems. Third, storage does not pay for itself in terms of enhanced benefits from nutrients. More nutrients are available to crops when storage is used, but the value of these nutrients do not approach the cost of storage. Fourth, if storage is present, incorporation of waste by injection or plowdown pays for itself. Net benefits average about $10 per animal year higher for plowdown than for surface spreading.

For swine farms, waste system costs of open lot systems compare favorably with those of confinement systems. While open lot systems are more labor intensive, they require less capital than confinement systems. Another finding of White and Forster (1978) for swine operations is that there are small differences in net system costs for many systems, which implies that a wide range of systems will continue to be used. Third, economies of size are prevalent. For smaller capacity lots, annual net system costs are substantial, but for moderate and large capacity lots, waste management is nearly a breakeven proposition. Fourth, if runoff is to be controlled from open lots, detention and irrigation increase net system costs substantially more than grass infiltration.

Generalizations about beef feedlot waste systems are similar to those of dairy and swine operations. Open lot systems are less costly than confinement, slotted floor systems. Again, economies of size are present with waste management systems, and the larger beef feedlots have a clear cost advantage compared to smaller feedlots. In fact, for moderate and large capacity lots, waste can be profitable. Finally, runoff control using grass infiltration is less costly than using a runoff detention or irrigation system.

IV. Society's Concern: Is Nonpoint Source Pollution Excessive?

Although nonpoint source pollution is costly to downstream water users, government and citizens' groups began exhibiting serious concern over the problem only a few years ago. The environmental consequences of agricultural production began attracting widespread attention in the United States in the 1930s. However, both policy makers and the general public were interested in primarily safeguarding the soil and water resources as a means to safeguard future agricultural production (Rasmussen, 1982).

By the 1970s, a new theme had emerged in agricultural policy debates. Not only was concern raised about the effects of soil and water conservation policies on future food supplies, but also concern was raised about their effects on downstream water users. Runoff from feedlots and erosive agricultural practices were held responsible for fish kills, excessive concentrations of nitrates in both surface waterways and groundwater, eutrophication, and sedimentation of rivers, lakes, and streams. Accordingly, a mandate to reduce point source pollution, such as runoff from larger livestock facilities, was included in the 1972 amendments to the Federal Clean Water Bill (Public Law 92-500).

Nonpoint and point sources of pollution are necessary consequences of economic activity, most importantly food production. It is estimated that agricultural sources are responsible for 66% of total suspended solids, 74% of total phosphorus, 81 percent of BOD, and 95% of pesticide loadings in surface water (Crosswhite and Sandretto, 1991). Recognizing this, a community would opt for total elimination of that form of degradation only in the rarest of circumstances. A more typical problem is to reduce point and nonpoint source pollution, thereby approaching a more socially optimal, yet still positive, level. In order to develop policy that accomplishes this objective, it is essential to understand why individuals choose to "produce" levels of pollution that are, from a social standpoint, excessive.

Among these reasons why economic performance, in general, can be suboptimal is that individual economic agents often fail to either pay the full social costs of their activities or to capture the full social benefits of same (Meade, 1973). In particular, when at least some parts of the social costs of pollution (or the social benefits of resource conservation) are "external," then natural resource management will be suboptimal, from a social standpoint.

Agricultural nonpoint source pollution is an example of suboptimal economic performance (or a classic "externality" problem). If a farmer fails to adopt pollution control measures, then recreators, public utilities, and other downstream water users incur the costs. Similarly, by reducing the rate at which soil or animal wastes run off his land, a farmer benefits other members of society. Given this state of affairs, the agricultural sector will produce excessive nonpoint source pollution. To say the same thing, it will opt for suboptimal level of pollution control.

Using a simple diagram, we can demonstrate the simple inefficiencies that arise when "externalities," like agricultural nonpoint source pollution, are present. Depicted in Figure 1 are three curves, each showing the economic impacts of a slight (or marginal) increase in pollution control. The additional costs that must be incurred to achieve a slight reduction in agricultural nonpoint source pollution are indicated by the marginal cost (MC) curve.

The marginal social benefits (MSB) associated with the control of agricultural pollution are divided into two parts: marginal internal benefits (MIB), which accrue to farmers and ranchers, and marginal external benefits (MEB), which are collected by others. Some benefits of pollution control can be realized by the farmer. For example, the economic value of the plant nutrient content of livestock waste is higher if moderate rates of manure application are used rather than high application rates that also cause nutrient runoff. But some benefits of pollution control are realized by downstream water users.

Having reason to ignore external economic impacts, farmers and ranchers will select a privately optimal level of pollution control (PO in Figure 1), at which MIB equals MC. At that level, their profits (or the internal net benefits of pollution control) are maximized. At the socially optimal level (SO), by contrast, the difference between the full social benefits (external as well as internal) of pollution control and its costs are maximized. The social optimal level (SO) is found where MSB (i.e., MIB plus MEB) equal MC.

This paradigm leads to some general conclusions regarding pollution control. First, from society's viewpoint, downstream demand for water quality (MEB) is as important as the producer's private incentive (MIB) to control pollution, but is generally ignored by producers. Second, downstream demand and the "best" level of pollution control are inherently local or watershed specific. Those areas with demand for higher downstream water quality will require a higher degree of pollution control. If increasing demands are placed on downstream water resources (e.g., if growing communities increase demands for household and recreational water use), then external benefits will comprise a larger share of the full social benefits of nonpoint source pollution control, and increased pressure will be placed on policy makers to intervene in farming operations and do something about the pollution "problem."

Without demand for low levels of nutrients, pesticides, and sediment, the economic incentive for society to reduce pollution is nonexistent. While this fact seems obvious, the demand for water resources is often lost in nonpoint source pollution debates. In principle, the definition of "good" and "bad" water quality changes from one watershed to the next.

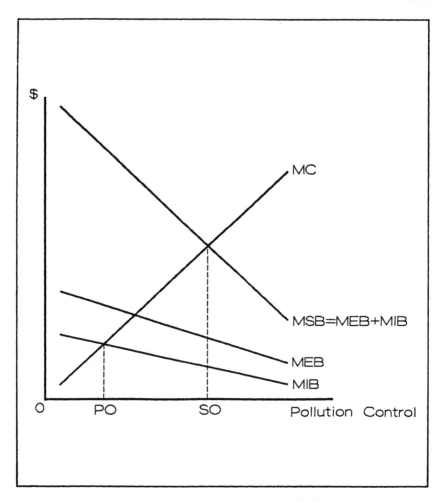

Figure 1. The benefits and costs of nonpoint source pollution control.

Several methods are used to approximate external benefits of water quality improvement (Hitzhusen, 1992). First, pollution may impose direct costs on downstream water users, and <u>market values</u> in related external markets can be estimated. For example, Forster et al. (1987) estimated that soil erosion and nonpoint source pollution from agriculture increased water treatment costs for communities using surface water as the source of their municipal water supplies. Decreasing soil erosion by 10% was estimated to reduce community water treatment costs by 4%.

A second method of measuring externalities is <u>surrogate market measures</u>. For example, willingness to incur travel costs to use or avoid use of a water body might be estimated, as done by Mullen and Mertz (1985) in estimating the economic effect of acidification damage in the Adirondack fishery. Due to acidification damage, the

New York fishermen's "willingness to pay" for travel to Adirondack lakes and for expenditures to fish was reduced by more than $1 million per year (1976 dollars).

Direct surveys of willingness to pay for water quality improvements or willingness to accept payments for a hypothetical deterioration in water quality constitute the third method of measuring externalities. Originally called hypothetical valuation, this method is now called "contingent valuation." For example, Brookshire et al. (1976) used pictures of three alternative levels of visibility to assess the externalities of emissions from a proposed coal-fired power plant in Lake Powell, Arizona. Survey participants estimated willingness to pay for reducing these aesthetic losses.

V. What Can Be Done About Nonpoint Source Pollution?

Pollution from agricultural sources is like any other externality problem. Left to themselves, factory owners would neglect at least a part of the costs society bears from the sewage and air pollution emitted from factories' discharge pipes and smokestacks. Similarly, farmers and ranchers producing nonpoint pollution do not fully take into account the external costs incurred by downstream water users.

Five direct mechanisms that governments can use to affect pollution are: control levels of pollution emission through regulations, levy taxes against polluters, pay subsidies to encourage pollution reduction, establish a market for emission rights, and fund research-education programs. Point source polluters, for example, are typically issued permits that specify, among other things, the rate at which pollutants can be discharged as well as when and by how much that rate should be reduced because of environmental conditions.

Pointing out that regulations are often costly to administer effectively, most economists advocate the use of economic incentives wherever possible (Baumol and Oates, 1975). In addition, regulations fail to give polluters an incentive to reduce emissions lower than what is allowed. Taxes on emissions or subsidies for emission reductions are less costly to administer and provide direct incentives to develop new techniques for emission control. In terms of the economic framework developed in Figure 1, taxes or subsidies would increase the marginal internal benefits (MIB) of pollution control, increase the level of pollution control, and narrow the gap between the social optimum (SO) and private optimum (PO).

A newly developed alternative to affecting emissions is a market-based approach to emission regulation. To emit pollutants, it would be necessary to "own" emission "rights." Emission rights would be granted or sold to producers. These rights could be traded in a market, and their price would reflect the economic value of the right to emit a particular pollutant. The aggregate supply would be controlled by government regulation, and over time this total could be reduced to enhance the environment. If that occurred, rights would be more scarce, their price would rise, and firms would have added incentive to reduce pollution. Under the 1990 Clean Air Act, firms can trade sulfur dioxide emissions rights, a few companies have been actively trading rights, and commodity exchanges have organized legally sanctioned markets (*Wall Street Journal*, 1992).

Affecting nonpoint source pollution through any of these mechanisms presents problems. It would be costly to monitor compliance with runoff control regulations, determine the amount of nonpoint runoff pollution to tax or subsidize, or monitor a firm's pollution relative to its owned rights. These administrative costs would be especially burdensome in agriculture where nonpoint pollution occurs across hundreds of thousands of producers with each having incentive to violate regulations or cheat on payment of taxes, collection of subsidies, or production within certain rights.

In addition to promoting an enhanced environment in an economically efficient manner, the public sector must take into account how policies affect income distribution. For example, a tax that drives diffuse runoff from agricultural land to a socially optimal level at the expense of burdening farmers, many of who might be financially distressed, is apt to be politically unacceptable to policy makers and the general public (Jacobs and Casler, 1979).

Given the administrative costs of these policy alternatives, U.S. soil and water conservation policy has stressed subsidization (or cost-sharing) along with technical assistance of particular farming practices. In the 1984-1990 period, the USDA's Agricultural Conservation Program spent nearly $1 billion on cost sharing agricultural practices to control erosion, conserve water, and improve water quality (USDA, 1991, p. 7). About $40 million of this total went to animal waste management practices.

Few in the United States are entirely satisfied with cost-sharing, however. In part, controversy can be traced to disagreement over what specific natural resource problems should be addressed. In addition, it has been charged that the cost-sharing program has been run more as an income transfer scheme than as a resource conservation initiative (U.S. Comptroller General, 1983).

Much of the public soil and conservation and nonpoint pollution control effort in the United States is based on a research-education approach. Land Grant Universities, Agricultural Experiment Stations, Extension Services, and the U.S. Department of Agriculture work together to develop new management practices and to educate farmers about the effects of these management practices on water quality. The Soil Conservation Service also assists farmers in developing soil and water-conserving farm plans. Other federal government farm programs aimed at enhancing farm incomes and reducing farm production may require these soil and water conservation plans before making payments to participating farmers. Sometimes farm program requirements can conflict with water quality goals. For example, some counties do not allow farmers participating in federal farm programs for selected crops to spread animal waste on cropland-idled Federal Acreage Reduction programs. In some cases, this may prevent timely application rates on cropland.

VI. Other Factors and Government Policies Affecting Nonpoint Source Pollution

Besides these methods of explicitly influencing levels of nonpoint source pollution, other factors have important effects on the extent of nonpoint source pollution. Public policies typically influence these "other factors," and in the end, may have more of an impact on nonpoint source pollution than do explicitly pollution control policies (Henderson and Barrows, 1984).

The actions of producers are shaped, to some extent, by society's changing values. A number of contemporary social movements and groups, including environmentalists, ecologists, and conservationists, stress that nature is to be tended and nurtured rather than dominated; technology must be considered in relationship to the natural environment; and the effect of our production activities on nature must be considered. These values affect nonpoint source pollution through the government policies and farmers' attitudes governing soil conservation, land preservation, chemical use, and animal waste management.

Another influence on nonpoint source pollution is consumer preferences, as mentioned at the beginning of this discussion. Consumers in the U.S. and other developed countries currently have high economic incomes to effectively assert their strong preferences for red meats. They also encourage grain fed livestock to be concentrated in feedlots that produce enormous quantities of manure with the potential for polluting nearby streams and water bodies.

On the production side, technology and input prices define the resources that are used in the production process and the resulting level of pollution. For example, technological advances in transportation, irrigation, and meat processing have dramatically changed industry structure (the size distribution of firms), as well as the geographic intensity, location of fed beef production and resulting nonpoint pollution. Many of these technological changes have been encouraged either directly or indirectly by government policies.

Tenure of farmers and size of farm operations also influence nonpoint source pollution. Farmers having short-term tenure arrangements have less incentive to control nonpoint source pollution than farmers whose future income depends largely on maintaining good relationships with immediate neighbors and downstream water users.

Finally, managerial talent of farmers is an important factor. Increased education, improved entrepreneurial abilities, and improved information systems make farmers more flexible in using resources, less resistant to change, and more willing to reduce nonpoint source pollution when given the proper incentives.

VII. Research Issues

These firm and societal issues have several implications for the research agenda in animal waste management.

1. Foremost is the need for estimating the externalities of livestock waste. That is, what are the effects of livestock waste on water quality (i.e., chemical, biological, and physical characteristics of the water)? These effects are dependent, in part, on characteristics of the waste, which are a function of livestock species, type of livestock production system, and type of waste management system. They also are dependent on the physical environment in which the firm operates. Modeling these relationships would enable the externalities of livestock to be estimated for individual production units, watersheds, or geographic regions.

2. In order to price these externalities, demand relationships for water quality need to be determined for recreational, household, and business users. That is, relationships between users' willingness to pay for water and the water quality characteristics need to be determined. In some densely populated areas and/or areas with high recreational usage for swimming or fishing, the price that water users would be willing to pay for "clean" water (i.e., water having particular chemical, biological, and physical characteristics) might be relatively high. Contrasted to that situation, in sparsely populated areas, the price for clean water would be relatively low.

3. Another need to be addressed by research is an inventory of livestock producers' use of alternative waste management systems. Ideally, this data base would allow use of alternative systems to be categorized by size of operation, by species, and by geographic region. Combining this inventory with the estimated effects of waste management systems on water quality would enable groups of producers or geographic areas to be targeted for efforts to reduce pollution.

4. Policies to encourage control of water pollution from livestock facilities need to be developed. Alternatives to standard regulatory approaches need to be designed which offer economic incentives to producers to control pollution. Regulatory approaches are costly to administer and enforce, politically difficult to implement, and usually have the most adverse consequences for smaller producers.

5. Research at the firm level might concentrate on benefits and costs of alternative waste management systems and their effects on firm profitability. While a substantial amount of this research has been done in the past, these estimates need to be updated. Of course, economic impacts of alternative waste management systems differ between geographic regions, and these impacts would be state and/or region specific.

6. Finally, markets for livestock manure exist in some areas resulting in manure having an explicit economic value to the producer. These markets encourage judicious use of manure since it is treated as a product rather than waste. Another research issue would be to identify factors that make these markets viable and to investigate how they could be developed in areas where they currently do not exist.

References

Ashraf, M. and R.L. Christensen. 1974. Analysis of impact of manure disposal regulations on dairy farms. *Amer. J. Agr. Econ.* 56:331-336.

Badger, D.D. 1977. Economic potential and management considerations in land application of beef feedlot wastes. Proceedings, 1976 Cornell Agricultural Waste Management Conference, Ithaca, NY.

Baumol, W. and W. Oates. 1975. *The Theory of Environmental Policy*. Prentice-Hall, Englewood Cliffs, NJ.

Blauser, R.J., G.D. Schnitkey, and D.L. Forster. 1990. Costs of alternative hog and dairy manure handling systems. ESO# 1777 and 1778. Dept. of Agr. Econ. and Rurl. Soc., The Ohio State University, Columbus, OH.

Brookshire, D.C., B.C. Ives, and W.D. Schultze. 1976. The valuation of aesthetic preferences. *J. Environmental Econ. and Mgmt*. 4:325-346.

Crosswhite, W.M. and C.L. Sandretto. 1991. Trends in resources protection policies in agriculture. Agricultural Resources. U.S.D.A. AR-23:42-49.

Forster, D. Lynn. 1975. Simulated beef feedlot behavior under alternative water pollution control rules. *Amer. J. Agr. Econ*. 57:259-268.

Forster, D.L., C.P. Bardos, and D.D. Southgate. 1987. Soil erosion and water treatment costs. *J. Soil and Water Conservation* 42:349-351.

Henderson, D.R. and R.L. Barrows [eds.]. 1984. Resources, food, and the future. North Central Regional Extension Publication No. 222. Iowa State University, Ames, IA.

Hitzhusen, Fredrick. 1992. Social costs and benefits of recycling coal fired electric power plant FGD by-products. Proceedings, Envirotech Conference. Vienna, Austria.

Huang, Wen-Yuan. 1979. A framework for economic analysis of livestock and crop by-product utilization. *Amer. J. Agr. Econ*. 61:91-96.

Jacobs, J.L. and G.L. Casler. 1979. Internalizing externalities of phosphorus discharges from crop production to surface water: Effluent taxes vs. uniform reductions. *Amer. J. Agr. Econ*. 61:309.

Johnson, J.B. and L.J. Conner. 1976. Economic and regulatory aspects of land application of wastes to agricultural lands. Proceedings, 1976 Cornell Agricultural Waste Management Conference, Ithaca, NY.

Mazzacco, M.A. 1991. Environmental regulation and agricultural lending. *Amer. J. Agr. Econ*. 73:1394-1398.

Meade, J. 1973. *The Theory of Environmental Externalities*. Institut Universitaire de Hautes Etudes, Geneva, Switzerland.

Mullen, John K. and Fredrick Mertz. 1985. The effect of acidification damages on the economic value of the New York fisheries to New York anglers. *Amer. J. Agr. Econ*. 67:112-119.

Olexa, M.T. 1991. Contaminated collateral and lender liability: CERCLA and the new age banker. *Amer. J. Agr. Econ*. 73:1388-1393.

Pherson, C.L. 1974. Beef waste management economics in Minnesota farmer-feeders. Proceedings, 1974 Cornell Agricultural Waste Management Conference. Ithaca, NY.

Rasmussen, Wayne D. 1982. History of soil conservation, institutions, and incentives. In: H.G. Halcrow et al. [eds.], *SoilCconservation Policies, Institutions and Incentives*. Soil Conservation Society of America, Ankeny, IA.

Schnitkey, Gary D. and Mario J. Miranda. 1993. The impact of pollution control on livestock-crop producers. *J. Agr. and Res. Econ.* 18:25-36.

Stout, Thomas T. 1992. U.S. production and consumption of beef, pork, and poultry, 1950-2000. ESO# 1935. Dept. of Agr. Econ. and Rurl. Soc., The Ohio State University, Columbus, OH.

U.S. Comptroller General. 1983. Agriculture's soil conservation programs miss full potential in the fight against soil erosion. Report to Congress, General Accounting Office. GAO/RCED-84-48. Washington, D.C.

U.S. Department of Agriculture (USDA). 1991. Agricultural resources. Economic Research Service. AR-23.

Wall Street Journal. 1992. New rules harness power of free markets to curb air pollution. April 14.

White, R.K. and D.L. Forster. 1978. Evaluation and economic analysis of livestock animal waste systems. U.S. Environmental Protection Agency. EPA-600/2-78-102.

Sources of Manure: Swine

M.C. Brumm

I. Introduction

The production of pork is a major agricultural enterprise in the United States, with a majority of the production occurring in the Midwest (Ohio to Nebraska and Minnesota to Missouri) and North Carolina. Iowa has ranked first in hog inventories since 1880 (Bureau of the Census, 1989). In the most recent USDA survey (USDA, 1994), Iowa had 23.7% of the U.S. inventory of 61,595,000 head followed by North Carolina with 9.3% (Table 1).

The Corn Belt states are expected to remain the primary hog production area although some shifts will occur. Because of historically lower feed grain prices and lower population densities, pork production is expected to expand west of the Mississippi River, especially in the western (Kansas, Colorado and Wyoming) and southwestern (Oklahoma) fringe areas of the Corn Belt (Hurt et al., 1992).

Pork production is accompanied, as might be expected, by the production of animal waste by-products with almost all the waste returned to farm land in some manner. Estimates are that swine manure production accounts for 12-15% of the

Table 1. Major U.S. pork producing states

| | -------------1978[a]------------- | | ---------------1994[b]------------- | |
	Inventory	% of U.S.	Inventory	% of U.S.
Illinois	6,550	10.9	5,600	9.1
Indiana	4,250	7.0	4,200	6.8
Iowa	15,100	25.0	14,600	23.7
Kansas	2,000	3.3	1,350	2.2
Minnesota	4,100	6.8	4,750	7.7
Missouri	4,100	6.8	2,850	4.6
Nebraska	3,650	6.0	4,400	7.1
North Carolina	2,350	3.9	5,700	9.3
Ohio	2,160	3.6	1,580	2.6
South Dakota	1,620	2.7	1,680	2.7
Wisconsin	1,650	2.7	1,170	1.9
11 State total	47,530	78.8	47,880	77.7
U.S. total	60,353		61,595	

[a]December 1 inventory in thousands (Bureau of the Census, 1978).
[b]September 1 inventory in thousands (USDA, 1994).

total livestock waste produced annually in the U.S. (VanDyne and Gilbertson, 1978).

II. Nutrients in Manure

A. Nitrogen

Corn and soybean meal represent more than 80% of the grain and protein supplements fed to swine and poultry in the United States (Cromwell, 1980). Corn, when utilized as the major energy feedstuffs in swine diets, is deficient in meeting the growing pig's need for several essential amino acids. The first and second limiting amino acids in corn are lysine and tryptophan, followed by threonine and isoleucine (Seerley, 1991). Thus, a high quality source of these essential amino acids must be blended with corn to provide a diet containing adequate quantities of these essential amino acids. Soybean meal is the most widely used supplemental protein (i.e., amino acid) source for swine diets in the United States.

When swine diets are formulated with corn and soybean meal, sufficient soybean meal is added to meet the pig's nutritional requirement for lysine. When formulated in this manner, all other essential amino acids are present in the diet in excess of the growing pig's nutritional needs (NRC, 1988). Upon digestion, these excess amino acids are deaminated, with the urea excreted via the kidneys in the urine (Wittemore and Elsley, 1976). Along with spilt and undigested feed, this is a major source of nitrogen (N) in collected manure.

Synthetic amino acids such as L-lysine monohydrochloride are often added to swine diets in an attempt to maintain performance at a reduced cost. The use of synthetic amino acids can significantly reduce the amount of nitrogen excreted. As an example, it is quite common to substitute 45.5 kg of 44% crude protein soybean meal (7.04% N) with 1.4 kg of synthetic lysine (15.3% N) and 44.1 kg of corn (1.28% N) (Goodband et al., 1994) for a savings of 2.4 kg of N per 909 kg (1 ton) of feed compared to conventional corn-soybean meal diets.

As additional synthetic amino acids such as threonine and tryptophan become economic, their use will increase, resulting in a further reduction of nitrogen in swine diets. Swine nutritionists are actively investigating this and other methods to reduce nitrogen excretion by pigs (Dourmand et al., 1992; Gatel and Grosjean, 1992; Kirchgessner et al., 1991). Jongbloed and Lenis (1992) have predicted that by the year 2000, N excretion by pigs in the Netherlands can be reduced 33% through diet and management changes.

B. Phosphorus

The majority of phosphorus (P) in most cereal grains, including corn and soybean meal, is organically bound as phytate (Table 2; NRC, 1988). Phosphorus in this form is poorly available to nonruminants because they lack phytase, the enzyme necessary to cleave the ortho-phosphate groups from the phytate molecule (Cromwell et al., 1993). Thus, only 10-15% of the P in corn and approximately 25% of the P in soybean meal is available to pigs (Table 2; NRC, 1988; Cromwell, 1991; Cromwell, 1992). A majority of the pig's nutritional P needs are met by the addition of inorganic phosphorus sources (generally monosodium or dicalcium phosphate) to swine diets (NRC, 1988; Reese et al., 1992). When diets are formulated in this manner, upwards of 65-75% of the total P in swine diets is excreted in the manure (Table 3; Lei et al., 1993a; Lei et al., 1993b).

An emerging area of research by swine nutritionists is the addition of dietary phytase to swine diets to enhance the utilization of phytate P from the cereal grains in the diet (Table 3) resulting in lower levels of inorganic P additions to the diet. In the Netherlands, it is estimated that P excretion by pigs can be reduced 40% by the year 2000 through such advances in swine nutrition (Jongbloed and Lenis, 1992).

C. Sodium

Sodium chloride additions to swine diets have decreased over the years, partially in response to concerns about the fate of sodium in stored manure. Generally, sodium chloride is supplemented in swine diets at the rate of 025 to 0.5% to prevent deficiency symptoms with .25 to .3% being the most common addition rate (Reese et al., 1992). In anaerobic swine manure storage pits, sodium ranges from 5000 to

Table 2. Phytate phosphorus content of feedstuffs commonly fed to swine

Feedstuff	Total phosphorus %	Phytate phospho- rus %	Phytate as % of total
Corn	0.27	0.19	69
Grain sorghum	0.31	0.21	68
Barley	0.38	0.25	65
Soybean-meal	0.61	0.37	61

(From Cromwell, 1980.)

Table 3. Potential of phytase for reducing phosphorus in excreta of an 80-kg pig

Item[a]	Current practice	Estimated requirement (NRC, 1988)	With added phytase
Dietary phosphorus, %	0.5	0.4	0.3
Phosphorus intake, g/d	15.5	12.5	9.3
Phosphorus retained, g/d[b]	4.0	4.0	4.0
Phosphorus excreted, g/d[c]	11.5	8.4	5.3
Phosphorus excreted, % of uptake	74.2	67.2	57.0
Reduction in phosphorus excretion, %			
From current practice			54
From NRC			37

[a]Assumes a daily gain of 820 g and a daily feed intake of 3,100 g.
[b]Assumes 5 g of phosphorus retained/kg of bodyweight gain.
[c]Includes 1.6 g of endogenous phosphorus.
(From Cromwell, 1991.)

9000 ppm on a dry matter basis for dietary sodium chloride additions of 0.2 to 0.5% (Sutton et al., 1976).

D. Potassium and Trace Minerals

Currently, it is estimated that corn-soy based diets [containing .6% potassium (K) from feed ingredients] supply sufficient K for swine of all sizes such that supplemental additions of K are not normally recommended (NRC, 1988). However, data are beginning to accumulate suggesting a response to K additions in situations that involve changing agronomic practices, changes in feed ingredient processing, and the use of alternative feedstuffs (Mabuduike et al., 1980; Coffey, 1987). Few data exist on the digestibility and/or retention of this increased K in the

Table 4. Effect of dietary copper supplementation on dry matter and volatile solids reduction in anaerobically stored swine waste (wet basis)

| Trial | Dietary copper[a] | % Reduction | |
		Dry matter	Volatile solids
I	0	65	70
	125	55	58
	250	51	57
II	0	41	43
	250	31	33

[a]Addition beyond 2.5 ppm added to prevent deficiency symptoms.
(From Brumm and Sutton, 1979.)

diet. Thus, no predictions can be made as to the impact of supplemental K additions (generally as KCl) on the composition of swine manure.

In addition to calcium, phosphorus, and sodium additions, zinc is added at 50-100 ppm, copper at 5-10 ppm and selenium at 0.3 ppm to prevent deficiency symptoms (NRC, 1988). Unlike the other trace minerals, selenium additions are closely regulated by the U.S. government (Anonymous, 1994). As a percentage of the total mineral content in the diet, excreted swine manure is estimated to contain 86%, 110%, 79%, 40%, 74%, 59%, and 66% of the Cu, Zn, Mn, Ca, Mg, K, and Na, respectively, offered to the pig (Overcash and Humenik, 1976) in traditionally fortified diets.

E. Copper

Copper, as copper sulfate, has been used as a growth promoting additive in swine diets since 1955 when Barber et al. (1955) reported increased gain in pigs fed 250 ppm copper when compared to pigs fed a control diet. Its beneficial effects on animal performance continue to be well documented (Lucas et al., 1962; Castell and Bowland, 1968; Braude and Hosking, 1975; Prince et al., 1979; Cromwell et al., 1989). Copper is routinely recommended in weanling pig diets at 125-250 ppm (Reese et al., 1992; Augenstein et al., 1994; Goodband et al., 1994).

Taiganides (1963) estimated that 80% of the ingested copper was excreted by swine, and biological treatment of manure was limited when 36 ppm of copper was added to the diet. He further theorized that excreted copper from growing-finishing pigs fed 250 ppm copper was the inhibitory substance causing failure of anaerobic waste digestors when present at a concentration of 60 to 80 ppm (wet basis). Ariail et al. (1971) concluded that copper additions to swine diets resulting in copper concentrations in the waste greater than 0.01 mg/liter altered microbial decomposition as measured by the standard BOD_5 test. In contrast, Hobson and Shaw (1973) suggested that excreted copper, if present at less than 50 ppm in the diet, was not inhibitory to the anaerobic digestion process of swine waste.

Copper is a known bactericidal agent at low concentrations of 5 to 10 ppm (Pelczar and Reid, 1972). Kornegay et al. (1976) reported copper concentrations of 312 ppm (wet basis) in moist feces of swine fed 250 ppm copper. Overcash et al. (1975) reported a value of 1,200 ppm copper in raw swine waste (dry solids basis) with the addition of copper to swine diets.

Brumm and Sutton (1979) reported fresh swine waste (feces, urine, spilt feed and water) to contain 61 ppm copper (wet basis) from pigs fed diets containing 250 ppm copper. In addition, these authors reported anaerobic storage pits to contain 46.6 ppm copper (wet basis) when loaded daily with copper-containing fresh manure versus 3.4 ppm copper (wet basis) when loaded daily with fresh manure from pigs offered diets containing <10 ppm copper. These authors concluded that the presence of 250 ppm copper in swine diets (as copper sulfate) caused increased dry matter and volatile solids accumulations in anaerobic pits when compared to waste stored in pits from pigs not fed supplemental copper. With increasing amounts of dietary copper, there was decreased dry matter and volatile solids disappearance in anaerobic manure storage pits (Table 4).

Brumm and Sutton (1979) theorized that the higher levels of copper in excreted waste may directly inhibit activity in the anaerobic manure storage or it may alter the microorganism population in the gut of the pig, indirectly influencing the microbial population in the anaerobic pit which normally biodegrade stored waste organic matter.

F. Zinc

While zinc is routinely added at 50-100 ppm zinc as zinc sulfate to swine diets to prevent deficiency symptoms (NRC, 1988), recent evidence suggests levels of 3,000 ppm zinc as zinc oxide in weanling pig diets may improve performance (Table 5). Nutritionists at Kansas State University recommend the routine addition of zinc oxide to diets for weaned pigs weighing less than 6.8 kg at a level of 0.38% of the diet (Goodband et al., 1994). There are no reports in the literature as to the zinc form or content in manure from pigs offered diets with these higher levels, although it is reasonable as a first approximation to estimate fresh manure zinc values of 700-750 ppm based on copper values of Brumm and Sutton (1979).

G. Chromium

For growing-finishing pigs, additions of 100-800 ppb chromium as chromium picolinate have been reported to enhance carcass lean (percentage of muscling) and reduce carcass fat (Table 6). Similar to zinc, there are no reports in the literature as to the amount of chromium in the manure from pigs fed supplemental chromium. Again, using copper as a reference mineral, fresh manure is predicted to contain 0.02 to 0.2 ppm chromium when pigs are fed diets supplemented with chromium.

Table 5. Performance of weaned pigs fed pharmacological levels of zinc

		Diet	
Item	Basal	3000 ppm Zn (ZnO)	3000 ppm Zn (ZnSO$_4$)
Average daily gain, g/d	445[a]	500[b]	414[a]
Gain:feed, g/g	659	652	651

[a,b]Means with different superscripts differ ($P < 0.05$).
(From Hahn and Baker, 1993.)

Table 6. Effect of chromium picolinate on growth and carcass traits of growing-finishing pigs

	Chromium additions to diet (ppb)					
Item	0	100	200	400	800	SEM
Daily gain, kg/d[a]	0.910	0.899	0.904	0.854	0.827	0.025
10th rib fat, cm[b]	3.15	2.34	2.63	2.20	2.46	0.11
Longissimus muscle area, cm^2 [b]	34.0	40.4	39.9	41.7	40.3	1.2
Percentage of muscling[b]	51.7	56.1	54.7	57.4	56.2	0.7

[a]Linear effect of Cr ($P < 0.05$).
[b]Quadratic effect of Cr ($P < 0.01$).
(From Page et al., 1993.)

III. Manure Volume

The U.S. industry is improving the overall conversion efficiency of the swine herd. Current estimates of manure production and composition (MWPS, 1985) are based on whole herd feed conversion efficiencies of 3.7 to 3.8 kg of feed per kg of gain. However, many producers have made large advances in production efficiency and now report conversions of 3.3 or better (Marsh and Dial, 1992). Advances in reproductive efficiency also mean less waste generated from sows and boars as a percent of the total waste stream. Thus, previous estimates of waste production and composition may prove to be inaccurate estimators and in many cases will overestimate both the total volume of production and the composition of the waste produced.

Feeder and waterer design can have a tremendous impact on the volume and nutrient content of waste to be stored and utilized. Typical feeders utilized by U.S. producers often result in 5-8% of the feed offered not consumed by the pig and entering manure storage devices as uneaten material (Liptrap et al., 1985). Newer feeder designs, based on an understanding of eating behavior, are beginning to appear and promise feed wastage rates less than 1.5% (Baxter, 1989). Recent results in France suggest that incorporating a nipple drinking device in the feeder versus

Table 7. Approximate nitrogen losses from swine manure as affected by handling and storing

Handling and storing methods	Nitrogen loss[a] (%)
Solid systems	
manure pack	20 to 40
open lot	40 to 60
Liquid systems	
anaerobic storage[b]	15 to 30
lagoon[c]	70 to 80

[a]Based on composition of manure applied to the land vs. composition of freshly excreted manure.
[b]Concentrated manure with little water added.
[c]Manure plus dilution water added for biological treatment and odor control.
(From Sutton et al., 1983.)

the customary method of supplying a nipple drinking device separate from the feeder can reduce the amount of slurry up to 30%, primarily due to reduced water waste (Fortune and Lebas, 1991).

As an average value for all phases of production, it is estimated that 41-kg pigs produce 0.18 kg of volatile solids per day with the ratio of volatile solids to total solids equal to 0.81 (Overcash and Humenik, 1976). Generally, growing-finishing pigs from 14 to 91 kg liveweight can be expected to generate 0.39 to 0.45 kg of waste per day on a dry matter basis containing 0.35 kg volatile solids and a pH as collected of 8.17 (Brumm et al., 1980). When corrected for fecal moisture, urine, wasted feed, and spilt drinking water, the daily volume is estimated to be 6.3 liters/d per 64 kg growing-finishing pig and 5.6 liter/d per 159-kg adult breeding animal (Sutton et al., 1983) or 4.1 kg/d for feeder swine and 7.9 kg/d for breeder swine (ASAE, 1990). Mathematical models are being developed to more accurately predict the effects of management on swine manure production (Aarnink et al., 1992).

Depending on the collection and storage system chosen, significant loss of nitrogen often occurs prior to land application (Table 7; Sommer et al., 1993). The final nutrient content at the time of land application has been estimated for swine manure collected and stored by various management systems (Table 8).

IV. Production Systems

Pork production has traditionally occurred with all phases of production located at one site (farrow-finish). Approximately 25% of the U.S. production has changed ownership at 18-23 kg (8-10 weeks of age) and moved to sites that only had finishing facilities (feeder pig finishing) (Brumm et al., 1991).

The production trend is now three-site production with the reproductive herd at one location, newly weaned piglets at a second, isolated site and growing-finishing

Table 8. Approximate dry matter and fertilizer nutrient composition of swine manure at time applied to the land

Manure system	Dry matter	Total N	Ammonium N	P_2O_5	K_2O
			kg/metric ton		
Solid					
Without bedding	18 (15-20)	4.1 (3.7-4.5)	2.9 (2.4-3.7)	3.7 (2.9-5.4)	3.3 (2.4-4.1)
With bedding	18 (17-20)	3.3 (2.9-4.1)	2.4 (2.1-3.3)	2.9 (2.1-4.1)	2.9 (2.4-3.7)
			kg/1000 liters		
Liquid					
Anaerobic storage	4 (2-7)	4.3 (3.4-6.6)	3.1 (2.5-3.7)	3.2 (1.6-3.6)	2.6 (1.4-3.6)
Lagoon	1 (0.3-2)	0.5 (0.4-0.7)	0.5 (0.2-0.6)	0.2 (0.1-0.5)	0.5 (0.2-0.8)

(From Sutton et al., 1983.)

pigs at a third site (Leibbrandt, 1994). The advantage of this system over more conventional systems is healthier pigs (Alexander et al., 1980; Harris, 1988) and pigs with improved growth rates and feed conversion efficiencies (Wiseman et al., 1992).

The impact of this change in production systems on swine waste composition can only be predicted since the practice is too recent for reports to appear in the scientific literature. Whereas previously, manure from weaned piglets offered diets containing elevated amounts of copper or zinc was only a small portion of the waste produced from a production site, with three-site production, all of the manure from the weaned piglet unit can be expected to be elevated for these heavy metals. Similarly, few if any reports are available as to the composition of collected manure at the site housing the adult reproductive herd.

In 1980, 21% of the growing-finishing pigs and 45-50% of the nursing and nursery pigs in the U.S. were housed in confinement facilities (i.e., liquid manure systems; VanArsdall and Nelson, 1984). With the large influx of new confinement construction, especially the construction associated with contract production units in North Carolina and Iowa, it is logical to predict a major increase in the percentage of manure captured and stored as a liquid or semi-liquid.

A major change in the ownership structure of the pork producing industry is also impacting the waste issue. While total pork production continues to increase in the U.S., the number of farms selling hogs/pigs has declined from 1,273,000 in 1959 to under 200,000 in 1990 (Rhodes, 1990). By the turn of the century, the number of farms with pigs is expected to decline to slightly more than 100,000 (Hurt et al, 1992). It was estimated in 1988 that 69% of the commercial hog slaughter (U.S. origin) was from 28,700 operations. It is currently estimated that 25% of U.S. slaughter hog production is controlled by 30 firms (Freese, 1994).

V. Dead Animal Disposal

Along with the usual stream of waste associated with pork production, a new concern is the disposal of dead animal carcasses. In many states, the legal requirements for disposal are incineration, burial, or pickup by a commercial rendering service (Hermel, 1992). Similar to the production sector, there has been a significant decrease in the number of dead animal rendering services in the U.S. (Fats and Proteins Research Foundation, 1992).

Because of decreased access to rendering services, increased charges for rendering services, frozen ground in winter months and high fuel costs associated with incineration, many pork producers are evaluating composting of swine carcasses as a disposal alternative (Morris et al., 1994). Research is limited in support of this practice and it is unclear what the legal aspects of this practice are with regard to current state laws and local health regulations regarding dead animal disposal or what the nutrient composition of the composted material is.

VI. Future Trends

Where do animal welfare and animal rights concerns fit into the equation? In most European countries, an increasing percentage of the breeding herd is being given access to straw bedding during a portion of the gestation and lactation phases of production (MLC, 1994). If the United States follows the European lead at some future time, either through legislation or consumer pressure, an entirely new set of problems is created since little information is available regarding the composition, storage or land application of this high residue waste material.

While the general trend in U.S. swine production is towards increasing confinement production, there is a growing minority of small and not so small producers who are employing intensive outdoor pig production due to the high investment costs associated with confinement production units (Wegehenkel, 1994). In addition to the obvious concern regarding surface runoff from these units, there is the issue of nutrient leaching from the intensive production area, especially if stocking rates result in total removal of all vegetation. While leaching concerns will be specific to each production site, it is of sufficient concern that the United Kingdom has begun to consider this means of nitrogen introduction into agricultural production when it designates "nitrate-vulnerable zones" (Worthington and Danks, 1992).

VII. Summary

As the previous discussion demonstrates, management practices including dietary ingredient selection, feeder selection, manure storage system design, and land application method interact to affect the final fertilizer value of swine manure when land application is used as the utilization method for the collected material.

References

Aarnink, A.J.A., E.N.J. van Ouwerkerk, and M.W.A. Verstegen. 1992. A mathematical model for estimating the amount and composition of slurry from fattening pigs. *Lvstk. Prod. Sci.* 31:133-147.

Alexander, T.J.L., K. Thornton, G. Boon, R.J. Lysons, and A.F. Gush. 1980. Medicated early weaning to obtain pigs free from pathogens endemic in the herd of origin. *Veterinary Record* 106:114-119.

Anonymous. 1994. *Feed Additive Compendium.* Miller Publishing Co., Minnetonka, MN.

Ariail, J.D., F.H. Humenik, and G.J. Kriz. 1971. BOD analysis of swine waste as affected by feed additives. In: *Livestock Waste Management and Pollution Abatement.* Proc. 2nd Int. Symp. Livestock Waste, Columbus, OH, p. 180.

ASAE. 1990. *Manure Production and Characteristics.* ASAE Data:ASAE D384.1. Amer. Soc. Ag. Eng. Standards. St. Joseph, MI.

Augenstein, M.L., L.J. Johnston, G.C. Shurson, J.D. Hawton, and J.E. Pettigrew. 1994. Formulating farm-specific swine diets. Bulletin BU-6496-F, Minnesota Extension Service, University of Minnesota, St. Paul.

Barber, R.S., R. Braude, K.G. Mitchell, and J.C. Cassidy. 1955. High copper mineral mixture for fattening hogs. *Chem. Ind.* 74:601.

Baxter, M.R. 1989. Design of a new feeder for pigs. *Farm Building Progress* 96:19-22.

Braude, R. and Z.D. Hosking. 1975. Feed additives to diets supplemented with copper for growing pigs. *J. Agr. Sci.* 85:263.

Brumm, M.C. and A.L. Sutton. 1979. Effect of copper in swine diets on fresh waste composition and anaerobic decomposition. *J. Anim. Sci.* 49(1):20-25.

Brumm, M.C., A.L. Sutton, and D.D. Jones. 1980. Effect of season and pig size on swine waste production. *Trans. ASAE* 23(1):165-168.

Brumm, M.C., M.Y. Ash, G.W. Jesse, and W.G. Luce. 1991. Starting purchased feeder pigs. Pork Industry Handbook, Fact Sheet PIH-25. Purdue University Coop. Extension, West Lafayette, IN.

Bureau of the Census. 1978. *1978 Census of Agriculture*, U.S. Department of Commerce, Washington, D.C.

Bureau of the Census. 1989. *1987 Census of Agriculture*, U.S. Department of Commerce, Washington, D.C.

Castell, A.G. and J.P. Bowland. 1968. Supplemental copper for swine: growth, digestibility and carcass measurements. *Can. J. Anim. Sci.* 48:403.

Coffey, T. 1987. Potassium, electrolyte balance can impact swine diets. *Feedstuffs*, October 5, p. 14-19.

Cromwell, G.L. 1980. Biological availability of phosphorus for pigs. *Feedstuffs* 52(9):38-41.

Cromwell, G.L. 1991. Phytase appears to reduce phosphorus in feed, manure. *Feedstuffs* 63(41):14-16.

Cromwell, G.L. 1992. The biological availability of phosphorus in feedstuffs for pigs. *Pig News Info.* 13:2:75N.

Cromwell, G.L., T.S. Stahly, and H.J. Monegue. 1989. Effects of source and level of copper on performance and liver copper stores in weanling pigs. *J. Anim. Sci.* 67(11):2996-3002.

Cromwell, G.L., T.S. Stahly, R.D. Coffey, H.J. Monegue, and J.H. Randolph. 1993. Efficacy of phytase in improving the bioavailability of phosphorus in soybean meal and corn-soybean meal diets for pigs. *J. Anim. Sci.* 71(7):1831-1840.

Dourmand, J.Y., D. Guillou, and J. Noblet. 1992. Development of a calculation model for predicting the amount of N excreted by the pig: Effect of feeding, physiological state and performance. *Livestock Prod. Sci.* 31:95-107.

Fats and Proteins Research Foundation. 1992. *Quality Standard for Animal and Plant Fats*. Directors Digest No. 222, Ft. Myers Beach, FL. 22 pp.

Fortune, H. and D. Lebas. 1991. Single-space wet feeder: Effect of adjustment of dosing clapper and number of pigs par device on fattening and slaughter records. *J. Rech. Porcine En France* 23:27-34.

Freese, B. 1994. Pork Powerhouses. *Successful Farming*, October Issue, pp. 20-24.

Gatel, F. and F. Grosjean. 1992. Effect of protein content of the diet on nitrogen excretion by pigs. *Livestock Prod. Sci.* 31:109-120.

Goodband, R.D., M.D. Tokach, and J.L. Nelssen. 1994. Kansas swine nutrition guide. Bulletin C-719 (revised), Kansas State University Cooperative Extension Service, Manhattan.

Hahn, J.D. and D.H. Baker. 1993. Growth and plasma zinc responses of young pigs fed pharmacologic levels of zinc. *J. Anim. Sci.* 71(11):3020-3024.

Harris, D.L. 1988. Alternative approaches to eliminating endemic diseases and improving performance of pigs. *Veterinary Record* 123:422-423.

Hermel, S.R. 1992. Now what? *National Hog Farmer*, March 15, pp. 34-40.

Hobson, P.N. and B.G. Shaw. 1973. The anaerobic digestion of waste from an intensive pig unit. *Water Res.* 7:437.

Hurt, C., K.A. Foster, J.A. Kadlec, and G.F. Patrick. 1992. Industry evolution. *Feedstuffs* 64(35):1, 18-19.

Jongbloed, A.W. and N.P. Lenis. 1992. Alteration of nutrition as a means to reduce environmental pollution by pigs. *Lvstk. Prod. Sci.* 31:75-94.

Kirchgessner, M., F.X. Roth, and M. Kreuzer. 1991. Criteria determining the characteristics of pig manure. 3. Effects of nitrogen intake and dietary amino acid pattern. *Agribiol. Res.* 44:345-356.

Kornegay, E.T., J.D. Hedges, and D.C. Martens. 1976. *Soil and Plant Mineral Levels as Influenced by Four Annual Applications of Manures of Different Copper Contents.* P. 201. Res. Div. Rep. 170, VIP, Blacksburg, VA.

Lei, X.G., P.K. Ku, E.R. Miller, and M.T. Yokoyama. 1993a. Supplementing corn-soybean meal diets with microbial phytase linearly improves phytate phosphorus utilization by weanling pigs. *J. Anim. Sci.* 71(12):3359-3367.

Lei, X.G., P.K. Ku, E.R. Miller, M.T. Yokoyama, and D.E. Ullrey. 1993b. Supplementing corn-soybean meal diets with microbial phytase maximizes phytate phosphorus utilization by weanling pigs. *J. Anim. Sci.* 71(12):3368-3375.

Leibbrandt, V.L. 1994. Collaborative multi-site pork enterprises. In: *Wisconsin Swine Update* (1)2. Wisconsin Coop. Extension Service, University of Wisconsin-Madison.

Liptrap, D.O., G.L. Cromwell, D.E. Reese, H.J. Monegue, T.S. Stahley, and G.R. Parker. 1985. Effect of feeder adjustment on feed wasteage and pig performance. *J. Anim. Sci.* 61(Suppl 1):110.

Lucas, I.A.M., R.M. Livingstone, A.W. Boyne, and I. McDonald. 1962. The early weaning of pigs: VIII. Copper sulfate as a growth stimulant. *J. Agr. Sci.* 58:201.

Mabuduike, F.M., C.C. Calvert, and R.E. Austic. 1980. Lysine-cation interrelationships in the pig. *J. Anim. Sci.* 51(Suppl 1):210.

Marsh, W. and G. Dial. 1992. The PigCHAMP grow-finish herd. *Pigene* 1(3):1-4. University of Minnesota, St. Paul.

MLC. 1994. *Pig Yearbook.* Meat and Livestock Commission, England.

Morris, J.R., T. Acinar, and F. Kains. 1994. An alternative method of dead stock disposal for swine. *Ontario Swine Research Review*, Univ. Guelph O.A.C. Pub. No. 0294, pp. 86-88.

MWPS. 1985. *Livestock Waste Facilities Handbook.* MWPS-18. Midwest Plan Service, Iowa State Univ., Ames.

NRC. 1988. *Nutrient Requirements of Swine* (9th Ed.). National Academy Press, Washington, D.C.

Overcash, M.R. and F.J. Humenik. 1976. State of the art: Swine waste production and pretreatment process. Environmental Protection Technology Series EPA-600/2-76-290, U.S. Environmental Protection Agency.

Overcash, M.R., F.J. Humenik, and L.B. Driggers. 1975. Swine production and waste management: State of the art. In: *Managing Livestock Wastes*. Proc. 3rd Inter. Symp. Livestock Wastes, Amer. Soc. Agri. Eng., St. Joseph, MI, p. 154.

Page, T.G., L.L. Southern, T.L. Ward, and D.L. Thompson. 1993. Effect of chromium picolinate on growth and serum and carcass traits of growing-finishing pigs. *J. Anim. Sci.* 71(3):656-662.

Pelczar, M.J., Jr. and R.D. Reid. 1972. *Microbiology*. McGraw-Hill Book Co., New York.

Prince, T.J., V.W. Hays, and G.L. Cromwell. 1979. Effects of copper sulfate and ferrous sulfide on performance and liver copper and iron stores of pigs. *J. Anim. Sci.* 49(2):507-513.

Reese, D.E., P.S. Miller, A.J. Lewis, M.C. Brumm, and W.T. Ahlschwede. 1992. *University of Nebraska Swine Diet Suggestions*. Nebraska Coop Extension Pub. 92-210-A. University of Nebraska, Lincoln.

Rhodes, V.J. 1990. Structural trends in U.S. hog production. Agricultural Economics Report 1990-5, University of Missouri, Columbia, 62 pp.

Seerley, R.W. 1991. Major feedstuffs used in swine diets. In: *Swine Nutrition* (E.R. Miller, D.E. Ullrey, and A.J. Lewis, ed.). Butterworth-Heineman, Boston.

Sommer, S.G., B.T. Christenson, N.E. Nielsen, and J.K. Schjorring. 1993. Ammonia volatilization during storage of cattle and pig slurry: Effect of surface cover. *J. Agri. Sci.*, Cambridge 1221:63-71.

Sutton, A.L., S.W. Melvin, and D.H. Vanderholm. 1983. Fertilizer value of swine manure. Pork Industry Handbook, Fact Sheet PIH-25, Purdue University Coop. Extension, W. Lafayette, IN.

Sutton, A.L., V.B. Mayrose, J.C. Nye, and D.W. Nelson. 1976. Effect of dietary salt level and liquid handling systems on swine waste composition. *J. Anim. Sci.* 43(6):1129-1134.

Taiganides, E.P. 1963. Characteristics and treatment of wastes from a confinement hog production unit. Unpublished Ph.D. Thesis, Iowa State University, Ames.

USDA. 1994. September 1 USDA Hogs and Pigs Report. National Ag Statistics Service's Ag Statistics Board, USDA, Washington, D.C.

VanArsdall, R.N. and K.E. Nelson. 1984. U.S. Hog Industry. USDA-ERS Ag. Econ. Report No. 511.

VanDyne, D.L. and C.B. Gilbertson. 1978. Estimating U.S. livestock and poultry manure and nutrient production. USDA, ESCS-72.

Wegehenkel, R. 1994. Moorman's Outdoor Swine Technology. Proceedings NPPC Pork Academy, National Pork Producers Council, Des Moines, IA, pp. 81-89.

Wiseman, B., R. Morrison, G. Dial, T. Molitar, B. Frekina, M. Bergeland, and C. Pjoan. 1992. Medicated early weaning: Influence of age on growth, performance and disease. Proceedings Amer. Assoc. Swine Practitioners, p. 469-475.

Wittemore, C.T. and F.W.H. Elsley. 1976. Practical Pig Nutrition. Farming Press Ltd., Warfedale Road, Ipswich, Suffolk, England.

Worthington, T.R. and P.W. Danks. 1992. Nitrate leaching and intensive outdoor pig production. *Soil Use and Management* 8(2):56-60.

Managing Nutrients in Manure: General Principles and Applications to Dairy Manure in New York

D.R. Bouldin and S.D. Klausner

I. Introduction

A. Management of Manure Is Local

The farming system, climate, soil, animal type, rations fed, management of manure from the time it is excreted until the nutrients are either lost or taken up by plants, and other factors influence the effect of manure on crop production and environmental quality. Because of these factors, manure management must be characterized on a local scale.

Notwithstanding the previous statements, there are general concepts that we will discuss, and illustrate how they are used to make specific recommendations for dairy manure in New York. Similar concepts are used in other states/regions of the U.S.; the major differences are the parameters used to implement the concepts.

B. The Management of Manure Is Important to Crop Production, Environmental Quality, and Maintenance of Soil Productivity

Management has been made more difficult by the increase in size of animal production units. Manure is a bulky, low analysis fertilizer, and transportation costs limit the distance it can be economically transported. Most dairy farmers purchase substantial amounts of feed raised far from their farms and hence recycling of most of the nutrients to the land where this feed was grown is impractical under present economic conditions.

Leaching of nitrate into groundwater and phosphorus (P) carried in surface runoff are two common negative environmental effects. Ammonia lost to the atmosphere may have acidifying effects on soils at considerable distance from the source. Odors from production units and from manure spreading create other management problems.

Education programs for farmers in New York are aimed at maximizing the recovery of nutrients in manure by crops with the expectation that this will be environmentally innocuous. However, inevitably there are losses of components of manures to wind and water (e.g., NH_3 and soluble P). At present, descriptions of environmental effects and crop nutrient recovery are qualitative; improvements in management will require quantitative descriptions.

Initially, heavy manure applications may increase crop yields and soil productivity. However, continued applications of manure [which usually supplies P and potassium (K) in excess of crop uptake] may lead to excessive NO_3-N in the groundwater and to such high levels of P in the soil that any surface runoff will be badly contaminated with dissolved P. The latter may lead to excessive algae blooms in receiving waters (Sharpley et al., 1995). Feed additives such as copper may accumulate in the soil to levels high enough to be toxic to crops and thereby lead to decreases in soil productivity.

Beneficial use of nutrients in manure, the effects of applications in excess of crop use, loss of components of manure to air and water, and long-term effects of manure on soil productivity are the threads that will guide the subject matter of our discussion. Another thread is that until we better understand all of the processes, we cannot hope to improve our management to levels expected by our nonfarm neighbors and society in general.

Previous reviews illustrate how the goals of manure management have changed (Salter and Schollenberger, 1939; Bouldin et al., 1984; Blake and Magette, 1992). The major long term trends in the past have been from how to conserve nutrients and recycle them through the cropping system (e.g., Salter and Schollenberger, 1939) to the present when major emphasis has shifted to very large animal production units with emphasis on disposing of the manure without causing major environmental problems. Similar trends in Europe have occurred (Hansen and Henriksen, 1989).

II. Nutrient Management

A. Balance of N, P, and K on Dairy Farms

The nutrients in purchased feed and fertilizers on dairy farms in New York usually exceeds the export in milk and meat by substantial amounts. Mass balances for N, P, and K on dairy farms of different sizes are summarized in Tables 1a, 1b, and 1c. Approximately, inputs minus export as percentage of inputs is relatively constant for different numbers of cows per farm.

Most of the P and K not exported remain on the farm. Implications of this to the environment are discussed in more detail by Sharpley et al. (1995). Nitrogen is subject to losses by several mechanisms and in the following sections some details of N management are discussed.

B. Nitrogen Balance Per Cow

A typical N balance for a New York dairy cow based on records from about 300 dairy farms is shown in Figure 1. The most important conclusion to be drawn from this figure is that about 60% of the N in the feed is purchased. Most dairy farmers will have about 0.4 ha each of corn and alfalfa per cow. These crops will often be in a 6 year rotation of 3 years of alfalfa followed by 3 years of corn for silage. The residual effects from the alfalfa will furnish about 50% of the N needed by the 3 succeeding corn crops and hence the additional N that needs to be supplied by the manure is on the order of 35% of the N excreted by the cow.

C. Evaluation of Ability of Crop to Utilize Manure N

The management of N in manure must be based on an evaluation of the several sources of N available to the planned crop on a specific field in a specific crop sequence. Figure 2 illustrates the several steps in the procedure.

Table 1a. Annual mass N balance for several dairy farms in New York

| | Size of dairy, cows | | | | |
	45	85	320	500	1300
	---------------------- tons/yr-------------------------				
Input					
Purchased feed	3.8	9.7	43.8	78.5	205.0
Purchased fertilizer	1.0	2.2	13.5	26.1	9.8
N fixation by legumes[a]	0.8	0.9	14.6	13.9	16.3
Purchased animals			0.1		
	5.6	12.8	72.0	118.5	231.1
Output					
Milk	2.0	3.8	18.6	26.4	72.8
Meat	0.1	0.4	1.9	1.9	3.2
Crops sold	0.1	0.5			
	2.2	4.7	20.5	28.3	76.0
Remainder					
Tons	3.4	8.1	51.5	90.2	155.1
%	61	63	71	76	67

[a]Assumed as 60% of N content of legume (N fixation in establishment year is $\approx$ 0). (From Klausner, 1993.)

Table 1b. Annual mass P balance for several dairy farms in New York

| | Size of dairy, cows | | | |
	45	85	320	500
	-------------- tons/yr---------------			
Input				
Purchased feed	1.0	1.7	8.4	14.2
Purchased fertilizer	1.2	0.9	2.0	10.0
Purchased animals			0.03	
	2.2	2.6	10.5	24.2
Output				
Milk	0.4	0.68	3.8	5.5
Meat	0.05	0.10	0.5	0.05
Crops sold	0.02	0.06		
	0.47	0.84	4.3	6.0
Remainder				
Tons	1.7	1.8	6.2	18.2
%	79	68	59	75

(From Klausner, 1993.)

Table 1c. Annual mass K balance for several dairy farms in New York

	Size of dairy, cows			
	45	85	320	500
	--------------- *tons/yr*---------------			
Input				
Purchased feed	0.9	2.4	12.3	22.8
Purchased fertilizer	4.7	1.8	7.3	35.1
Purchased animals			0.01	
	5.6	4.2	19.6	57.9
Output				
Milk	0.51	1.0	5.6	8.3
Meat	0.01	0.02	0.1	0.1
Crops sold	0.12	0.4		
	0.64	1.4	5.7	8.4
Remainder				
Tons	5.0	2.8	13.9	49.5
%	89	67	71	85

(From Klausner, 1993.)

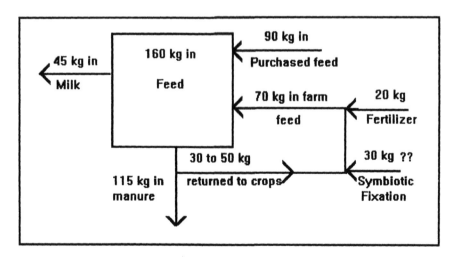

Figure 1. Nitrogen balance for a New York dairy cow based on records for 300 farms (all numbers are in kg N per cow per year).

The estimate of the amount of N that must be accumulated by the crop is based on yield goals; that is, what is the expected yield? Once this is established then the amount of N that must be accumulated by the crop can be estimated from local experimental data. For example, for each Mg/ha of grain, the N content of the aboveground dry matter must contain 20 kg/ha of N or, for example, a grain yield of 7.5 Mg/ha requires that the aboveground OM accumulate about 150 kg N/ha.

The defined N requirement and the several sources of N are shown in Figure 2. In each case, the N requirement and the contribution of the individual sources is expressed as the N content of the aboveground dry matter. This convention is followed because there is not much data on the amount of N mineralized nor the amount of N required for the roots. The experimental data available are sufficient for estimations based on this convention but in general are not sufficient for such things as estimates of losses of N to groundwater or denitrification.

Soil and crop residues are the consequence of past management. They are not subject to current management and yet they are an integral part of the N supply for the current crop and must not be ignored. The more stable forms of organic matter in the soil mineralize at a fairly constant rate over many years. For example, in New York, the soil organic matter may supply on the order of 30 to 60 kg N/ha. Short-term residual effects of previous crops must be added to this amount. In New York we estimate that the short-term effects of the residues from 3 years of alfalfa will supply over 100 kg N/ha to a succeeding corn crop the first year, about 50 to the second corn crop, etc.

The mineralization of the organic N described in the previous paragraph will occur whether or not manure or fertilizer is applied, and manure and fertilizer must be viewed as supplements to these sources. If the sum of the above sources is not adequate for the yield goal, manure may be used to supply the remainder. For example, if our yield goal is 7.5 mg/ha and the sources described in the previous paragraph total 75 kg/ha, then enough manure should be applied to supply an additional 75 kg/ha.

D. Manure as a Source of N

The concept that we believe is most useful is to evaluate the N in manure in terms of its fertilizer equivalents. Fertilizers are of known and invariant composition. Research with fertilizers is easier and in general, the year and site variation are less than with manure. In most regions there is much more contemporary data with fertilizers than with manure. Effects of climate, crop, soil, and other variables on fertilizer N needs can be based on this contemporary data. In effect, the N supplied by manure is equated to a fertilizer standard.

Nitrogen balance studies with livestock and analyses of manures illustrate that 40 to 70% of the N excreted by the animal is in forms converted to NH_4-N in a few days. The remainder is in organic forms that mineralize slowly over several years (Safley et al., 1985; Holmes, 1973; Gale et al., 1991; Schefferle, 1965; Kirchmann and Witter, 1992). These two forms of N behave very differently between the time they are excreted by the animal and incorporation in the soil. For this reason, these

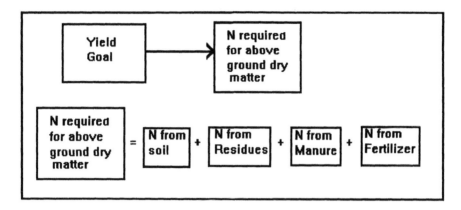

Figure 2. Illustration of procedure for managing manure N for crop production.

components are best considered separately in the integrated chain between animal and plant (Klausner and Bouldin, 1983; Bouldin et al., 1984; Motavalli et al., 1989; Sims, 1987; Bitzer and Sims, 1988).

E. Ammoniacal N

The NH_4-N is rapidly lost by volatilization whenever manure is exposed to air because the partial pressure of NH_3 in the manure almost always exceeds that in air by a factor of 10 or more (Lauer et al., 1976). Substantial losses occur in periods as short as a few hours whether the manure is in barns or feedlots, during spray irrigation, or spread on the soil surfaces (Safley et al., 1985; Holmes, 1973; Gale et al., 1991; Hoff et al., 1981; Beauchamp et al., 1982; Thompson et al., 1987; van der Molen et al., 1990; Safley et al., 1992). van der Molen et al. (1990) developed a comprehensive model of NH_3 volatilization from slurry which incorporates most of the variables influencing loss from surface applications of manure in the field. Hargrove (1988) has summarized the literature on factors influencing NH_3 loss from fertilizer applications.

Given that losses of ammoniacal-N by NH_3 volatilization are dependent on climatic and management factors, how can its value to crops be estimated? One rationale is to a) estimate the losses at each of several stages between excretion by the animal until spread on the soil, b) estimate losses between time of spreading and incorporation into the soil, and c) presume that once the NH_3 is incorporated in the soil, it will behave as an equivalent amount of fertilizer N applied at the same time in the same manner (this will be discussed more fully later).

Safley (1977) estimated losses between excretion until spreading on the soil for 33 management systems for dairy manure. The costs of several systems were summarized in another publication (Safley et al., 1976). One important implication is that the cost of operation of the various systems is very different and that these

Table 2. Comparison of costs and fertilizer N replacement value of two manure systems

Per cow/year	Free stall or stanchion, daily cleaning and spreading, spring incorporation	Free stall, alley scraper, manure pump, pit storage, yearly cleaning, immediate incorporation
Capital costs	$180	$675
Operating costs	$88	$220
Kg N incorporated into soil	36[a]	75[a]
Estimated fertilizer replacement (current + summation of residual)	$180	75
Cost of manure N, $/kg based on operating costs	$2.44[b]	$2.93[b]

[a] Estimated total N in manure as excreted = 90 kg/cow/year.
[b] For comparative purposes, urea fertilizer costs about $.70/kg N.
(Adapted from Safley et al., 1976; Safley, 1977.)

differences in cost of operation usually exceeded any differential in fertilizer costs that might be associated with conservation/loss of N. This is illustrated in Table 2 for two systems.

One way to block most NH_3 loss is to inject or incorporate immediately after spreading. However this is impractical on most farms and incorporation may be as long as several months after spreading. The estimate of NH_3 volatilization loss we use in New York is that 12% of the ammoniacal-N will be lost each day the manure is left on the soil surface (Lauer et al., 1976). Probably this is an overestimate of losses and represents a worst possible case (Brunke et al., 1988; Gordon et al., 1988; Hoff et al., 1981; Beauchamp et al., 1982). The rationale is to protect the farmer from underfertilization in case climatic conditions favor maximum losses for any given application.

As a first approximation, the NH_4-N form can be considered to be equivalent to urea and NH_4-N fertilizers when the fertilizer is applied at the same time and with the same placement. This enables us to tie the nutrient value of the NH_4-N in manure to research results with fertilizers. Note that we specify that the fertilizer N must be as urea or NH_4-N form and it must be applied at the same time (e.g., fall or spring) as the NH_4-N in the manure and it must be applied in the same way (e.g., broadcast or incorporated). In New York we estimate that 80% of fall-applied N will be lost by the following crop year and that 40% of fertilizer N applied in April will be lost by the time corn begins its period of rapid uptake in mid-June (Bouldin and Lathwell, 1968; Bouldin et al., 1971; and Lathwell et al., 1970).

The appropriateness of the approximation described in the previous paragraph depends upon the difference in denitrification between the soil treated with fertilizer and the soil treated with manure; in the latter case, the manure adds energy material which will enhance biological activity and the potential for alternate nitrification-denitrification as the soil goes through normal drying-rewetting cycles (Thompson et al., 1987; Comfort et al., 1990; Motavalli et al., 1985; Rice et al., 1988). The potential for denitrification is further enhanced by irregular distribution of the manure in the soil: the energy material in the manure and the ammoniacal-N will be in the same volumes of soil. Two recent studies illustrate that denitrification with injection is larger than denitrification with surface application of manure (Comfort et al., 1990 and Thompson et al., 1987). The suggested reason is that the injected manure is concentrated in a smaller volume of soil with a corresponding increase in microbial activity and hence increased potential for nitrification-denitrification.

Comfort et al. (1990) found that CO_2 evolution was enhanced for only a few days following application; thus, the effect of the enhanced energy source might be limited to a short time interval following application. If this is true, then the major impact would be on NO_3 already in the soil at time of application since the nitrification of the applied NH_4-N would take several days and would be further delayed by any anaerobic conditions created by the enhanced microbial activity.

F. Organic N

The organic N behaves like partially decomposed plant residues. An important characteristic is that the organic N is mineralized over a period of several years although at a decreasing rate in successive years. The usual way to describe this behavior is with a decay series, which is the rate of mineralization in successive years. In New York we have chosen to describe the mineralization in terms of the amount of side-dressed fertilizer N that a unit of organic N replaces. At present in New York, we estimate the decay series as 0.16, 0.10, 0.03, 0.03, 0.02 for silage yields and 0.21, 0.09, 0.03, 0.03, 0.02 for N uptake (Klausner et al., 1994), where the successive numbers are the amounts of side-dressed fertilizer N replaced by 1 unit of applied organic N in the current, second, third, fourth and fifth crop-years after application, respectively. Figure 3 illustrates the relationship between the amount of N estimated using this set of decay coefficients and experimental values from several location years of data in New York. The results shown in this figure are not as robust as they appear because we used the data illustrated to derive the decay series. The proper conclusion is that this set of decay coefficients is reasonably consistent with our experimental data. The manures used and the soils and climatic conditions are reasonable examples of conditions on New York dairy farms. Thus, they are useful for New York farmers but the reader should understand that they are not necessarily the best estimates for other kinds of manures, soils, or climates.

Their usefulness in other locations needs to be verified by experimentation. This is illustrated in Figure 4 with two examples. In the first example, the procedures used in New York were applied to data from Vermont (Jokela, 1992). In this case the results in Vermont are in reasonable agreement with the results of the procedures used in New York. With similar procedures applied to data from

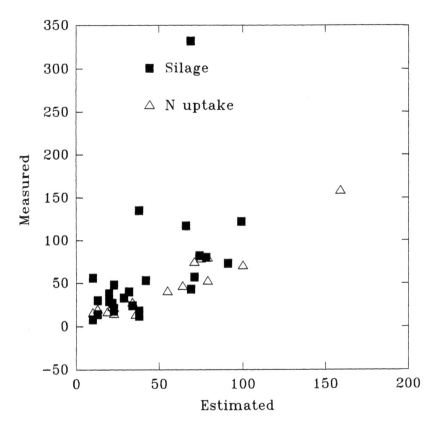

Figure 3. Comparison of the measured and estimated fertilizer equivalents of dairy manure based on decay series for the organic fraction of dairy manure. The values for "estimated" were based on the decay series for silage of 0.16, 0.10, 0.03, 0.03, 0.02 and for N uptake of 0.21, 0.09, 0.03, 0.03, 0.02 for the first through the fifth year after application, respectively. Measured is the value observed in the experiment. (From Klausner et al., 1994.)

Wisconsin (Motavalli et al., 1989), the results were not in agreement (Figure 4). In one case in Wisconsin, the fall-injected manure was not very different from spring applied N and our procedure underestimates the amount of fertilizer N replaced for the fall application but works reasonably well for the spring application in Wisconsin. This illustrates our first statement in the introduction that manure management is local.

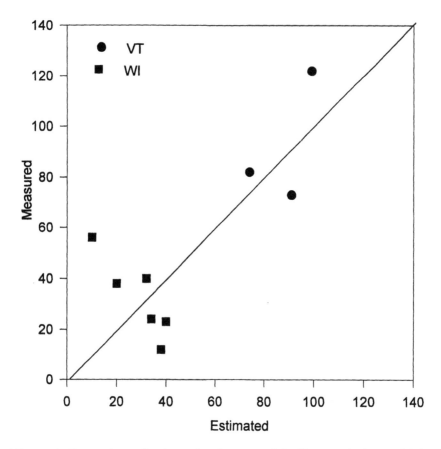

Figure 4. Comparison of estimated and measured fertilizer equivalents of dairy manure. The estimates were derived from the procedure and parameters used in New York for results measured in Vermont (VT) (Jokela, 1992) and Wisconsin (WI). (From Motavalli et al., 1989.)

G. Environmental Consequences: Water Quality

In the foregoing section we have illustrated a conceptual framework for utilization of the N in manure with the objective of supplying enough N for economic optimum corn yields, yet not putting on more than the estimated need. The expectation is that this is a best management practice so far as economic production of corn silage is concerned.

We do not have many direct measurements of NO_3 in recharge from our experimental plots nor P contents of the surface runoff. We do have NO_3 and P contents of stream water and some data on NO_3 in aquifers in New York (Johnson et al., 1976a; Johnson et al., 1976b; Hergert et al., 1981a; Hergert et al., 1981b; Bouldin et al., 1981). These data indicate that the problems are local and that most streams do not have excessive NO_3 or P. About 95% of the aquifers and streams which are sources of domestic water for distribution systems contain less than 2 ppm NO_3-N (Bouldin et al., 1981). A more complete treatment of environmental effects on water quality can be found in Sharpley et al. (1995).

H. Atmospheric Chemistry and Soil Acidification

Unusually high deposition of ammoniacal-N in The Netherlands has enhanced "forest decline" and acidification of parts of the landscape. This experience has led to examination of the adverse consequences of more modest inputs of ammonia in the long run in other areas (Langford et al., 1992; Schlesinger and Hartley, 1992). The most commonly cited adverse effects are acidification of the soil and NO_3-N in the groundwater.

There is ample evidence in the literature that domestic animals are the major source of atmospheric ammoniacal-N, although fertilizers may be significant in some areas (Asman et al., 1988; Buijsman et al., 1987; Langford et al., 1992; Schlesinger and Hartley, 1992). In areas with intensive animal production, the volatilization of ammonia is sufficient to have major impacts on air chemistry and precipitation. One major effect is that the ammonia increases the pH of aerosols and thereby accelerates the conversion of SO_2 to sulfate which results in particulate $(NH_4)_2SO_4$ (Asman and Jaarsveld, 1992; Behra et al., 1989; Kruse-Plass et al., 1993). The fate of the particulate $(NH_4)_2SO_4$ is dependent on weather, and generalizations about where and when it will be returned to the land surface are difficult (Asman and Jaarsveld, 1992; Kruse-Plass et al., 1993).

The major impacts will be seen first on the nonfarmed areas on acid, sandy soils. The worst case scenario occurs when the ammoniacal-N is nitrified and the resulting NO_3 and SO_4 are leached below the rooting depth. These anions will be accompanied by an equivalent amount of cations. The net effect is loss of 4 equivalents of base per mole of S. This will accelerate the loss of Ca and other basic cations with consequent acidification. Another adverse effect is that the NO_3 may find its way into water supplies. The best-case scenario is that the N is used by the vegetation and there is no "enhanced leakage" of NO_3 and SO_4 out of the soil. In the latter case, the enhanced inputs of N may lead to increased growth and/or corresponding decreases in fixation of N by various mechanisms.

To move beyond the extremes stated above, the consequences of ammonia volatilization to acidification of nonfarmed landscapes are difficult to quantify because a) ammonia enhances generation of particulate $(NH_4)_2SO_4$ and as a consequence, dispersal and deposition of both ammoniacal-N and sulfur compounds may be changed, b) likewise the relative concentrations of SO_2 and oxidants such as ozone will influence the dispersal and deposition of ammoniacal-N, and c) if NH_3 is not put into the atmosphere, the inputs of $(NH_4)_2SO_4$ may be replaced by H_2SO_4.

The H_2SO_4 will supply 2 moles of acid per mole of H_2SO_4 while a mole of $(NH_4)_2SO_4$ will supply 4 moles of acid per mole of S if the NH_4 is nitrified. If plants take up NO_3 and their growth is enhanced by the N inputs, some of the acidity will be neutralized by OH from the plants (if they take up anions in excess of cations).

If plant uptake occurs, leaching of both anions and cations will be reduced relative to the situation where uptake does not occur and NO_3 and SO_4 are leached out of the root zone.

At least one criteria for evaluation is the enhanced output of NO_3 from unfarmed areas. In the case of forests, this is referred to as "saturation"; that is, the inputs of N exceed the ability of the biomass to immobilize it. One model predicts that as the inputs of N increase, the time to reach saturation will decrease (Aber et al., 1991). If this model is correct then we may surmise that the time to observe enhanced leakage of NO_3 will be shorter the higher the N input but in any event, it may be several years before any effect on NO_3 outputs in drainage can be measured. This means that decisions based on the potential effects of enhanced ammonia volatilization need to be made before adverse effects are evident. In addition, the effects may be local because there are likely to be differences in susceptibility of different parts of the landscape and deposition, and chemistry of the deposition is likely to vary locally because of source-sink and atmospheric interactions.

There do not appear to be major adverse impacts of "dry deposition" of ammoniacal-N on crops so long as atmospheric concentrations are within the bounds found in most situations (Pearson and Stewart, 1993). There is some perturbation of metabolism, and rates of photosynthesis may be increased slightly by ammonia in the atmosphere. Particulate forms are less important than gaseous forms since the major mode of uptake is gaseous diffusion through open stomates (Hanson and Lindberg, 1991).

We have examined the possibility of enhanced "leakage" of NO_3 on the Teaching and Research Center of the College of Agriculture and Life Sciences near Harford, NY. The farm is located in a landscape typical of southern New York. The valley floor is 1 to 3 km wide composed of well-drained soils developed on gravel outwash. The fairly steep valley sides are mostly medium to somewhat poorly drained soils developed on glacial till. The elevation difference is about 200 meters. The valley floor is intensively farmed and receives most of the manure from dairy and beef operations (on the order of 2 animal units per ha). The valley sides are mostly in pasture while the upper parts are wooded. The NO_3-N has been monitored in streams and wells draining the woods/pasture for about a year and the results are tabulated in Table 3. These data illustrate that the NO_3-N in the stream water from the land that receives no manure is much less than would be expected on the basis of N inputs from precipitation and in any event is comparable to that observed at Hubbard Brook, which is likely as uninfluenced by agricultural operations as any area in the northeastern U.S.

Ammonia concentrations were measured in the air at 2 meters on several days at several locations on the Harford farm. The concentrations have generally been in the range of 5 to 15 micromoles NH_3 per cubic meter except over freshly spread manure and under very stable atmospheric conditions. They were similar to those measured earlier over alfalfa fields at another location near Ithaca (Dabney and Bouldin, 1985 and 1990). The NH_3 from current manure applications is usually quickly dispersed and becomes part of the "background level." The "background level" is probably determined by the cumulative volatilization of NH_3 over a region and the influence of manure from one large farm (e.g., 1000 animal units), does not

Table 3. Comparison of composition of stream water draining wooded areas in New York and New Hampshire and composition of precipitation

Source	mg/liter		
	Cl	NO₃-N	SO₄-S
	Harford, New York		
Streams	0.98	0.15	4.46
Precipitation[a]	0.19	0.45	0.93
Stream/precipitation	5.2	0.3	4.8
	Hubbard Brook, New Hampshire[b]		
Streams	0.54	0.42	2.06
Precipitation	0.51	0.31	0.94
Stream/precipitation	1.0	1.4	2.2

[a] Knapp et al., 1988.
[b] Likens et al., 1977.

extend much beyond the farm boundaries (perhaps a few hundred meters) under most atmospheric conditions. Thus, any controls on NH_3 would appear to require action over a whole region (several states?) and not on a local basis. So far there is no evidence indicating leakage of NO_3-N out of wooded areas.

III. Future Research Needs

The parameters we have derived from our experiments are described. While these parameters are useful for New York conditions, they may not be reliable when conditions are somewhat different. Considering the literature, our knowledge of NH_3 volatilization, denitrification, and leaching are such that prediction of losses of manure N are determined by weather as well as soil and manure properties. At time of application, the farmer does not know what the weather will be following application and therefore he cannot adjust application to account for post-application weather factors. However, conceptually at least, post-application calculations could be useful in estimating what has been lost and presumably supplemental N could be added at a later date to compensate for more than expected losses.

A very substantial amount of experimental work is required to derive the empirical parameters we use in New York. Is the flame worth the candle? The funding required for the empirical experiments in New York was substantial; is society willing to pay for continuing of even this research in the foreseeable future? If we were to improve our understanding of the many processes involved (e.g., denitrification, riparian denitrification), then even more funding would be needed. How can this be paid for? Perhaps equally important is the question: what do we need to know? In the following pages, the need for information is analyzed for funding required for the empirical experiments in New York was substantial; is society willing to pay for continuing of even this research in the foreseeable future? If we were to improve our understanding of the many processes involved (e.g., denitrification, riparian denitrification), then even more funding would be needed. How can this be paid for? Perhaps equally important is the question: what do we need to know? In the following pages, the need for information is analyzed for several situations. The importance of detailed knowledge of each component of management increases with each successive situation.

A. No Major Constraints Imposed by Environmental Quality

In this situation we suppose the environmental consequences of excessive N and P applications are not important. The economics of replacing fertilizer N, P, K, etc. with manure is the major factor. In general, at present with most livestock operations, fertilizer nutrients are less expensive than conservation of nutrients with elaborate manure management systems. Table 2 illustrates the cost of fertilizer N replacement with manure N. Not included is the value of the P and K; in general they are far in excess of needs with all manure management systems (Sharpley et al., 1995). Most soils on livestock farms test high or very high in available P and K and hence these nutrients are usually of no consequence in terms of economic benefits for different management systems.

This situation will exist a) where the climate and water management preclude leaching of NO_3 (and as a consequence there is no surface runoff), and b) where any excess NO_3 is denitrified before it can contaminate an aquifer.

Present information is sufficient for this situation.

B. Modest Environmental Constraints

In this situation the environmental consequences are not of overriding importance but the farmer is constrained to use manure to the extent that it replaces the fertilizer N needed to provide economic optimum yields. As long as manure applications are not in excess of agronomic crop needs, he will not be subjected to regulation. This is the situation we have used in most of our examples.

The following are circumstances that make this situation common in New York. First, most of the intensively cropped land is well to very well drained so that most of the manure is applied to soils where surface runoff is not large. This means that the buildup of P is of little consequence to water quality. Second, the landscape is a mosaic of cropped land, pasture, abandoned farmland, and wooded areas. Corn generally will not occupy more than perhaps half of cropped land and it is the major crop which receives large amounts of N. Crops such as alfalfa and grass ordinarily receive relatively small amounts of N and are not likely to accumulate much NO_3 in the soil profile. The net result is that the corn land is the major source of NO_3 but dilution by water from surrounding fields mitigates its effects on any but very small and local aquifers. Third, precipitation in excess of evaporation is on the order of

0.5 m/yr. This furnishes large dilution factors (on the average 50 kg of NO_3 dissolved in 0.5 meter of recharge will result in 10 ppm NO_3-N).

In this situation, the soil test for available N (PSNT) at sidedressing time is likely to be very useful (Klausner et al., 1993). Briefly, a soil sample is taken when the corn is about 0.15 to 0.30 m tall; if the amount of extractable NO_3 is less than a threshold value, then additional sidedressed N is needed. By using this procedure for several years and fields, farmers can develop management appropriate for their particular situation.

The research needs for this situation are a continuation of the present research programs in most states. The empirical approach seems adequate and additional emphasis on using our present information seems warranted.

C. Stringent Environmental Constraints

In this situation the farmer must be sure that the water leaving his farm/fields is always "clean." We suppose that this means less than 10 ppm NO_3-N, very low levels of dissolved and total-P, low levels of organic matter, etc. In this case all of the parameters described above must be applied with accuracy to each of his fields for all loads of manure. This is difficult. Two examples follow.

First, the estimation of NH_3 volatilization from time of excretion to incorporation is not easily quantified. Second, the estimates of denitrification and mineralization are very difficult to make and they are subject to much spatial and temporal variation. Most experimental estimates of these parameters have been done in such a way that the results cannot be extrapolated. The results shown in Figure 3 are an example; they are summaries of experimental observations useful to New York dairy farmers but they cannot be extrapolated to other situations without further experimental verification.

The research required for the detailed management described above is very extensive. There is no reason that it cannot be done, but it will require financial support for experimental work far beyond present levels. It is a task that needs to be done since it will be very useful for all phases of N management and not just manure management.

D. Odor Control Is Essential

In this situation odor control is very important and storage under closed conditions followed by injection into the soil are likely to be the best management options. However, this means most of the NH_3 will be conserved and injected into the soil. The following are some possible consequences if this were to be used in New York with daily spreading and spring incorporation on a typical dairy farm. Based on Table 2, we suppose that about 75 kg N per cow per year would be applied to 0.4 ha of corn. Based on our experience, manure applications would need to supply about 10, 80, and 120 kg N/ha for the first, second, and third corn crops, respectively, following 3 years of alfalfa. There will be about 0.14 ha of each per cow, or a total of about 30 kg per 0.4 ha (or per cow) would need to be supplied by the manure. Thus, the daily spread shown in Table 2 would supply slightly more than this. But the other system, where most of the N was conserved, would supply about twice the amount needed. In New York we have about 1 m of rain and about 1/2 of this is lost by evapotranspiration. If the excess N were dissolved in the recharge water of about 0.5 m, the concentration in the groundwater would be 7 ppm higher than the daily spread management. This calculation assumes that denitrification is zero and that the excess N is leached into the recharge water. Any denitrification would reduce the input into groundwater and so the assumption of no denitrification is a worst-case scenario. The important conclusion to be drawn from this exercise is that injection will reduce odors and at the same time increase N inputs; the increase in N inputs may have important consequences if NO_3 loading is important. The impact on the groundwater can also be mitigated by having more land under non-legume crops per unit of livestock.

The major research needs for improvement of this management option are better tools for describing denitrification and leaching.

E. Off-Site Effects of NH₃ Are Important

In this situation the off-site effects of ammonia volatilization are important. In several parts of the preceding discussion, we illustrated that a) many animal production units have manure-N far in excess of their crop needs, b) conservation and incorporation of all of this N will very likely lead to excessive NO_3-N in the groundwater, and c) NH_3 volatilization is a convenient way to "dispose" of excess N. Restrictions on NH_3 volatilization would have many of the consequences described for odor control.

IV. Possible Changes in the Future

Some possible future developments are discussed in the following several paragraphs.

A. Continuing of Present Trends

The present trends will continue into the near future. The concentration of livestock production will likely continue. We cannot foresee the economic consequences of many technologies now under development. However, based on the emphasis on environmental quality, the future farm will of necessity have to control odors, composition of water leaving the farm, and general appearance of the operations.

B. Reduce Size of Operations and Produce Most Feed Locally

Locally means that the transportation and other management costs of returning the manure to the crop land from which it was derived will not be excessive. These costs place a ceiling on the size of operations. A further restriction is that if the NH$_3$ is conserved, some method of storage for a year is required since fall applications of manures high in ammoniacal N does not conserve the N for crop use. Storage is expensive and the massive amounts of manure which must be spread in the spring create logistical and labor problems. A useful tool would be a nitrification inhibitor. If such a tool were available, then fall application would become feasible since the NH$_4$-N would not be converted to NO$_3$.

The economic consequences of reduction/restrictions on size of operations are not apparent.

C. Ration Modification

Feed additives commonly enrich the feed so that the ratio of elements in the feed are unbalanced with respect to crop uptake. If the manure is added in amounts sufficient to supply the N needs of nonlegume crops such as corn, the soil will be enriched with other elements in the long run. Two examples are described below.

The accumulation of P in excess of crop uptake is a common characteristic of most management systems (Sharpley et al., 1995). Not only is the feed imported but the feed is usually enriched by addition of inorganic forms of P such as dicalcium phosphate. This is further exacerbated by loss of N by volatilization. For example, the P to N ratio of corn grain is about 0.2 (based on 0.3% P and 10% protein). The P to total N ratio of poultry litter is on the order of 0.47 while the P to organic N ratio is 0.63 (Malone, 1992). Corresponding ratios for dairy manure are 0.25 to 0.45 (Sutton et al., 1986). As the P accumulates in soil, the soil test P levels increase with a corresponding increase in potential for increased concentration of P in surface runoff (Daniel et al., 1992). An additional potential problem is the effect of the P on the chemistry of micronutrients.

Pig rations are commonly supplemented with copper and the manures contain between 600 and 2400 mg Cu/kg dry weight (Payne et al., 1988; Sutton et al., 1983). To put this in perspective, the Cu/N ratio in pig manures will be on the order of 0.01 to 0.03 while the corresponding ratios for corn grain are on the order of 0.00015. The corn grain will generally contain less than 0.1 kg Cu/ha while an amount of manure needed to supply the N for this crop will contain 25 to 100 times this much Cu. Finally, the median concentration of Cu in soil in the U.S. is 18 mg/kg or about 50 kg Cu in the surface 30 cm of a ha of soil (Holmgren et al., 1993). If pig manure is applied to supply the N needed for a corn crop over a 10 year period, the total Cu in the soil could be increased by a factor of 2 to 10. As might be expected, the question has been raised about how long this can be continued without decreasing soil productivity. Several long-term studies have shown no decrease in crop yields even with cumulative additions of Cu totaling 320 kg/ha (Sutton et. al., 1983; King et al., 1990; Anderson et al., 1991; Payne et al.,

1988). Surely there is some upper limit but so far it has not been defined and hence the restrictions this might place on swine operations is not clear.

One solution to problems such as these is modification of rations. Some alternatives for broiler production have been discussed by Malone (1992). Characteristics of swine manure are influenced by environmental and management factors (Clanton et al., 1991). Perhaps there are practical alternatives to additions of Cu.

D. Develop Methane Generators and Concurrently Recover the Ammoniacal N

This would concentrate the N so that it could be transported and used in the same way as fertilizer N. This is mostly a hypothetical alternative; the technical problems which must be solved to make this feasible are formidable.

E. Abandon the 10 ppm NO_3 Standard for Drinking Water

Technically, this standard applies only to major distribution systems, but it has become a widely accepted standard for all potable water. While this standard is accepted almost worldwide, the consequences of increasing it by a factor of 2 or 5 or 10 should be examined. This would change the nature of many "Best Management Practices" substantially.

V. Summary

1) Most manure management is based on using the N for crop production without causing unacceptable adverse effects on the environment. The environmental effects are qualitative since they have not been measured often enough nor well enough to be quantitative. Very often the amount of N available is far in excess of beneficial recycling.

2) The most uncertain parts of management of the N in manure are a) loss of ammoniacal-N by NH_3 volatilization between excretion and incorporation in the soil, b) fate of the NH_4-N once it is incorporated in the soil (e.g., what fractions are taken up by plants, denitrified, and leached), and c) rate of mineralization of organic N.

3) Between 40 to 60% of the excreted manure is in forms that are converted to ammoniacal-N within a very few days; the fraction of this N that is spread on the soil surface varies from almost none to about 80% in various management systems.

4) The additional losses of ammoniacal-N between the time it is spread on the soil surface and incorporated into the soil are variable and depend primarily on weather factors which prevail for several days after spreading unless the manure is incorporated within a few hours after spreading or injected into the soil. The most

important factors which promote volatilization of NH_3 are wind speed and drying conditions.

5) The fate of the ammoniacal-N once incorporated in the soil is also uncertain and determined in part by weather factors for periods of up to about a year.

6) The best management is to apply the proper amount of N just previous to the period of most rapid crop uptake. The adverse environmental effects derive from excess amounts of N applied too far in advance of the period of most rapid crop uptake. The seriousness of inappropriate applications is mitigated as follows. The leaching potential increases as the excess of precipitation/irrigation over evapotranspiration increases but at the same time, it also dilutes the NO_3. In the case of very small excess, there may be very low potential for leaching, but salt buildup and other problems may occur. Leaching can be described with any of several models but for most situations, they are of uncertain value because denitrification cannot be predicted with any degree of precision. Measurement of denitrification is difficult, and how to extrapolate any such measurements over time and space is not evident.

References

Aber, J.D., J.M. Melillo, K.J. Nadelhoffer, J. Pastor, and R.D. Boone. 1991. Factors controlling nitrogen cycling and nitrogen saturation in northern temperate forest ecosystems. *Ecol. Appl.* 1:303-315.

Anderson, M.A., J.R. McKenna, D.C. Martens, S.J. Donohue, E.T. Kornegay, and M.D. Lindemann. 1991. Long-term effects of copper rich swine manure application on continuous corn production. *Commun. Soil Sci. Plant Anal.* 22:993-1002.

Asman, W.A.H. and H.A. Jaarsveld. 1992. A variable resolution transport model applied for NHx in Europe. *Atmos. Environ.* 26A:445-464.

Asman, W.A.H., B. Drukker, and A.J. Janssen. 1988. Modelled historical concentrations and depositions of ammonia and ammonium in Europe. *Atmos. Environ.* 22(4):725-736.

Beauchamp, E.G., G.E. Kidd, and G. Thurtell. 1982. Ammonia volatilization from liquid dairy cattle manure in the field. *Can. J. Soil Sci.* 62:11-19.

Behra, P., L. Sigg, and W. Stumm. 1989. Dominating influence of ammonia on the oxidation of aqueous sulfur dioxide. The coupling of ammonia and sulfur dioxide in atmospheric water. *Atmos. Environ.* 23(12):2691-2708.

Bitzer, C.C. and J.T. Sims. 1988. Estimating the availability of nitrogen in poultry manure through laboratory and field studies. *J. Environ. Qual.* 17:47-54.

Blake, J., J. Donald, and W. Magette (eds). 1992. *National Livestock, Poultry and Aquaculture Waste Management.* ASAE Publication 03-92. American Society of Agricultural Engineers. St. Joseph, MI.

Bouldin, D.R. and D.J. Lathwell. 1968. Timing of application is key to crop nitrogen use. *New York's Food and Life Sciences* 1:9-12.

Bouldin, D.R., S.D. Klausner, and W.S. Reid. 1984. Use of nitrogen from manure. In: R.D. Hauck (ed.), *Nitrogen in Crop Production*. Am. Soc. of Agron., Madison, WI, pp. 221-245.

Bouldin, D.R., K. Porter, and G. Casler. 1981. *Nitrate in Water in Upstate NY*. Agronomy Mimeo no.81-26. New York State College of Agriculture and Life Sciences. Cornell University, Ithaca, NY 14850.

Bouldin, D.R., W.S. Reid, and D.J. Lathwell. 1971. Fertilizer practices which minimize nutrient loss. In: Proceedings of Cornell University Conference on Agricultural Waste Management. New York State College of Agriculture and Life Sciences, Ithaca, NY.

Brunke, R., P. Alvo, P. Schuepp, and R. Gordon. 1988. Effect of meteorological parameters on ammonia loss from manure in the field. *J. Environ. Qual.* 17:431-436.

Buijsman, E., H.F.M. Maas, and W.A.H. Asman. 1987. Anthropogenic ammonia emissions in Europe. *Atmos. Environ.* 21(5):1009-1022.

Clanton, C.J., D.A. Nichols, R.L. Moser, and D.R. Ames. 1991. Swine manure characterization as affected by environmental temperature, dietary level intake, and dietary fat addition. *Transactions of the ASAE*. 34:2164-2170.

Comfort, S.D, K.A. Kelling, D.R. Keeney, and J.C. Converse. 1990. Nitrous oxide production from injected liquid dairy manure. *Soil Sci. Soc. Am. J.* 54:421-427.

Dabney, S.M. and D.R. Bouldin. 1985. Fluxes of ammonia over an alfalfa field. *Agron. J.* 77:572-578.

Dabney, S.M. and D.R. Bouldin. 1990. Apparent deposition velocity and compensation point of ammonia inferred from gradient measurements above and through alfalfa. *Atmos. Environ.* 24A:2655-2666.

Daniel, T.C., A.N. Sharpley, and T.J. Logan. 1992. Effect of soil test phosphorus on the quality of the runoff water: Research needs. In: J. Blake (ed.), *National Livestock, Poultry and Aquaculture Waste Management*. Am. Soc. of Agric. Engin., St. Joseph, MI.

Gale, P.M., J.M. Phillips, M.L. May, and D.C. Wolf. 1991. Effect of drying on the plant nutrient content of hen manure. *J. Prod. Agric.* 4:246-250.

Gordon, R., M. Leclerc, P. Schuepp, and R. Brunke. 1988. Field estimates of ammonia volatilization from swine manure by a simple micrometeorological technique. *Can. J. Soil Sci.* 68:369-380.

Hansen, J.A. and K. Henriksen (eds.). 1989. *Nitrogen in Organic Wastes Applied to Soils*. Academic Press, New York.

Hanson, P.J. and S.E. Lindberg. 1991. Dry deposition of reactive nitrogen compounds: A review of leaf, canopy and non-foliar measurements. *Atmos. Environ.* 25A:1615-1634.

Hargrove, W.L. 1988. Evaluation of ammonia volatilization in the field. *J. Prod. Agric.* 1:104-111.

Hergert, G.W., S.D. Klausner, D.R. Bouldin, and P.J. Zwerman. 1981a. Effects of dairy manure on phosphorus concentrations and losses in tile effluent. *J. of Environ. Qual.* 10:345-349.

Hergert, G.W., D.R. Bouldin, S.D. Klausner, and P.J. Zwerman. 1981b. Phosphorus concentration-water flow interactions in tile effluent from manured land. *J. of Environ. Qual.* 10:338-344.

Hoff, J.D., D.W. Nelson, and A.L. Sutton. 1981. Ammonia volatilization from liquid swine manure applied to cropland. *J. Environ. Qual.* 10:90-95.

Holmes, B.J. 1973. *Effect of Drying on the Losses of Nitrogen and Total Solids from Poultry Manure.* M.S. Thesis. Cornell University Library, Ithaca, NY, 97 pp.

Holmgren, G.G.S., M.W. Meyer, R.L. Cheney, and R.B. Daniels. 1993. Cadmium, lead, zinc, copper and nickel in agricultural soils of the United States of America. *J. Environ. Qual.* 22:335-348.

Johnson, A.H., D.R. Bouldin, E.A. Goyette, and A.M. Hedges. 1976a. Nitrate dynamics in Fall Creek, N.Y. *J. Environ. Qual.* 5:386-391.

Johnson, A.H., D.R. Bouldin, E.A. Goyette, and A.M. Hedges. 1976b. Phosphorus loss by stream transport from a rural watershed: Quantities, processes and sources. *J. Environ. Qual.* 5:148-157.

Jokela, W.E. 1992. Nitrogen fertilizer and dairy manure effects on corn yield and soil nitrate. *Soil Sci. Soc. Amer. J.* 56:148-154.

King, L.D., J.C. Burns, and P.W. Westerman. 1990. Long-term swine lagoon effluent applications on 'coastal' Bermuda grass: II. Effect on nutrient accumulation in soil. *J. Environ. Qual.* 19:756-760.

Kirchmann, H. and E. Witter. 1992. Composition of fresh, aerobic and anaerobic farm animal dungs. *Bioresource Technology* 40:137-142.

Klausner, S.D. 1993. Managing nutrients responsibly. In: 1993 Cornell Dairy Nutrition Conference. Dept. Animal Sci., Cornell Univ., Ithaca, NY 14853.

Klausner, S.D. and D.R. Bouldin. 1983. Managing animal manure as a resource. Part I: Basic principals. Agronomy Fact Sheet Series, p. 100, Dept. Soil, Crop and Atmospheric Sci., Cornell Univ., Ithaca, NY 14853.

Klausner, S.D., V.R. Kanneganti, and D.R. Bouldin. 1994. An approach for estimating a decay series for organic N in animal manure. *Agron. J.* 86:897-903.

Klausner, S.D., W.S. Reid, and D.R. Bouldin. 1993. Relationship between late spring soil nitrate concentrations and corn yields in New York. *J. Prod. Agric.* 6:350-354.

Knapp, W.W., B.I. Chevone, J.A. Lynch, V.C .Bowersox, S.V. Krupa, and W.W. McFee. 1988. *Precipitation Chemistry in the United States, Part 1: Summary of Ion Concentration Variability, 1979-1984.* Water Resources Institute. Center for Environmental Research, Cornell University, Ithaca, NY 14853.

Kruse-Plass, M., H.M. ApSimon, and B. Barker. 1993. A modelling study of the effect of ammonia on in-cloud oxidation and deposition of sulphur. *Atmos. Environ.* 27A:223-234.

Langford, A.O., F.C. Fehsenfeld, J. Zachariasssen, and D.S. Schimel. 1992. Gaseous ammonia fluxes and background concentrations in terrestrial ecosystems of the United States. Global Biogeochem. Cycles 6:459-483.

Lathwell, D.J., D.R. Bouldin, and W.S. Reid. 1970. Effects of nitrogen fertilizer applications in agriculture in relationship of agriculture to soil and water pollution. Cornell University Conference on Agricultural Waste Management. College of Agriculture and Life Sciences, Ithaca, NY.

Lauer, D.A., D.R. Bouldin, and S.D. Klausner. 1976. Ammonia volatilization from dairy manure spread on the soil surface. *J. Environ. Qual.* 5:134-141.

Likens, G.E., F.H. Bormannn, R.S. Pierce, J.S. Eaton, and N.M. Johnson. 1977. *Biogeochemistry of a Forested Ecosystem.* Springer-Verlag, New York, NY, p. 47.

Malone, G.W. 1992. Nutrient enrichment in integrated broiler production systems. *Poultry Science* 71:1117-1122.

Motavalli, P.P., K.A. Kelling, and J.C. Converse. 1989. First year nutrient availability from injected dairy manure. *J. Environ. Qual.* 18:180-185.

Motavalli, P.P., S.D. Comfort, K.A. Kelling, and J.C. Converse. 1985. Changes in soil profile N, P, K from injected liquid dairy manure. In: American Society of Agricultural Engineers Publication 13-85. Agricultural Waste Utilization and Management, pp. 200-210.

Payne, G.G., D.C. Martens, E.T. Kornegay, and M.D. Lindeman. 1988. Availability and form of copper in three soils following eight annual applications of copper-enriched swine manure. *J. Environ. Qual.* 17:740-746.

Pearson, J. and G.R. Stewart. 1993. The deposition of atmospheric ammonia and its effects on plants. *New Phytologist* 125(2):283-305.

Rice, C.W., P.E. Sierzega, J.M. Tiedje, and L.W. Jacobs. 1988. Simulated denitrification in the microenvironment of a biodegradable organic waste injected into soil. *Soil Sci. Soc. Amer. J.* 52:102-108.

Safley, L.M. 1977. *System Selection and Optimization Models for Dairy Manure Handling Systems.* Ph.D. Thesis. Cornell Univ., Ithaca, NY 14853.

Safley, L.M., J.C. Barker, and P.W. Westerman. 1992. Loss of nitrogen during sprinkler irrigation of swine lagoon liquid. *Bioresource Tech.* 40:7-15.

Safley, L.M., D.R. Price, and D.C. Ludengton. 1976. Network analysis for dairy waste management alternatives. *Trans. ASAE.* 19:920-924.

Safley, L.M., P.W. Westerman, and J.C. Barker. 1985. Fresh dairy manure characteristics and barnlot nutrient losses. In: American Society of Agricultural Engineers Publication 13-85. Agricultural Waste Utilization and Management, pp. 191-199.

Salter, R.M. and C.J. Schollenberger. 1939. *Farm Manure.* Ohio Agric. Exp. Sta. Bull. no. 605.

Schefferle, H.E. 1965. The decomposition of uric acid in built up poultry litter. *J. Appl. Bact.* 28:412-420.

Schlesinger, W.H. and A.E. Hartley. 1992. A global budget for atmospheric ammonia. *Biogeochemistry* 15(3):191-211.

Sharpley, A., J.J. Meisinger, A. Breeuwsma, J.T. Sims, T.C. Daniel, and J.S. Schepers. 1995. Impacts of animal manure management on ground and surface water quality. In: J. Hatfield (ed.), Effective management of animal waste as a soil resource, *Advances in Soil Science* (this volume).

Sims, J.T. 1987. Agronomic evaluation of poultry manure as a nitrogen source for conventional and no-tillage corn. *Agron J.* 79:563-570.

Sutton, A.L., D.W. Nelson, V.B. Melrose, and D.T. Kelly. 1983. Effect of copper levels in swine manure on corn and soil. *J. Environ. Qual.* 12:198-203.

Sutton, A.L., D.W. Nelson, D.T. Kelly, and D.L. Hill. 1986. Comparison of solid vs. liquid dairy manure applications on corn yield and soil composition. *J. Environ. Qual.* 15(4):370-375.

Thompson, R.B., J.C. Ryden, and D.R. Lockyer. 1987. Fate of nitrogen in cattle slurry following surface application of injection to grassland. *J. Soil Sci.* 38:689-700.

van derMolen J., H.G. van Faassen, M.Y. Leclerc, R. Vriesma, and M.J. Chardon. 1990. Ammonia volatilization from arable land after application of cattle slurry. 1. Field estimates. *Neth. J. Agr. Sci.* 38:145-158.

Best Management Practices for Poultry Manure Utilization That Enhance Agricultural Productivity and Reduce Pollution

P.A. Moore, Jr.

I. Introduction

A. Poultry Manure Production in the U.S.

Recently there has been an increase in demand for low-cholesterol meat products, which has resulted in significant increases in poultry production. One of the agricultural benefits of this tremendous production is the simultaneous production of poultry manure, which is an excellent organic fertilizer. At present, over 13 billion kg of poultry manure and/or litter are produced each year in the U.S. (Table 1). Over half of this production is in five states: Arkansas, North Carolina, Georgia, Alabama, and California. Since the industry is geographically concentrated, there are certain areas that have a tremendous amount of manure production, which has resulted in water quality problems in some cases.

Table 1. Number of birds and manure generated (dry basis) on U.S. farms in 1990, ranked according to total amounts generated

State	Broilers		Layers[a]		Turkeys		Total	
	Number produced[b]	Manure generated[c]	Number produced[b]	Manure generated[d]	Number produced[b]	Manure generated[e]	Number produced	Manure generated
	Millions	Mg x 10³	Millions	Mg x 10³	Millions	Mg x 10³	Millions	Mg x 10³
Arkansas	951	1427	15.3	52.8	22.0	239.8	989	1719
North Carolina	540	810	12.5	53.4	58.0	632.2	611	1496
Georgia	855	1282	18.0	55.6	2.0	21.9	875	1359
Alabama	847	1270	9.5	34.1	_f	_f	856	1304
California	231	347	29.0	136.9	32	348.8	292	832
Mississippi	413	620	6.1	24.4	_f	_f	419	644
Virginia	297	445	3.4	12.1	17.0	185.3	317	643
Minnesota	41	62	10.2	41.7	46.3	504.7	98	608
Texas	338	507	14.0	50.9	_f	_f	352	558
Maryland	265	398	3.3	8.6	0.1	1.2	269	408
Missouri	88	132	6.6	26.0	18.0	196.2	113	354
Delaware	232	348	0.6	1.5	_f	_f	232	349
Pennsylvania	116	173	18.7	54.3	8.4	91.9	143	320
Oklahoma	142	213	3.7	14.8	_f	_f	146	228
Florida	120	179	11.2	45.1	_f	_f	131	224
South Carolina	84	125	5.7	20.7	5.5	60.0	95	206
Ohio	21	31	17.7	74.1	4.8	51.8	43	157
Tennessee	99	149	1.1	3.6	_f	_f	100	152
Iowa	9	14	8.6	33.3	8.8	95.94	27	143
West Virginia	41	62	0.7	2.0	3.9	42.0	46	105

Table 1. continued

Oregon	24	36	2.6	11.8	2.3	25.1	29	72
Washington	33	50	5.0	21.5	[f]	[f]	38	71
Michigan	1	1	5.4	18.5	4.3	46.9	10	67
Nebraska	3	4	5.1	21.8	2.1	22.9	10	49
Wisconsin	14	21	3.4	18.3	[f]	[f]	17	39
New York	2	4	3.7	12.7	0.5	5.2	7	22
Kentucky	2	2	1.7	6.0	[f]	[f]	3	8
Hawaii	2	3	0.9	4.9	[f]	[f]	3	8
Other states	156	233	47.9	182.9	47.1	513.4	251	930
Total	5966	8948	272	1044	283	3085	6520	13078

[a]Includes laying hens and pullets of laying age; pullets of laying age represent 56% of the total number produced.

[b]Adapted from USDA, 1991.

[c]Broiler litter; based on 1.5 kg litter bird^{-1} yr^{-1}.

[d]Based on 7.00 kg manure bird^{-1} yr^{-1} for laying hens and 1.4 kg manure bird^{-1} yr^{-1} for pullets of laying age (Sims et al., 1989).

[e]Based on 10.9 kg manure bird^{-1} yr^{-1} (Sims et al., 1989).

[f]Included in totals for "Other states."

(From Moore et al., 1995b.)

The objectives of this paper were: (1) to provide an overview of current poultry manure management practices, (2) to discuss potential problems associated with land application of poultry litter, and (3) to delineate some possible solutions to these problems.

Numerous publications have been written on poultry litter. For more information on the various aspects of litter, the reader is directed to the following: agronomic value of poultry manure (Bosch and Napit, 1992; Miller et al., 1991; Salter and Schollenberger, 1939; Smith and Wheeler, 1979; Stephenson et al., 1990; Wilkinson, 1979), the value of poultry litter as an animal feed (Bhattacharya and Taylor, 1975; Fontenot and Webb, 1975; Fontenot and Jurubescu, 1980), environmental aspects of poultry manure use (Edwards and Daniel, 1992b; Moore et al., 1995c; Sims and Wolf, 1994), manure handling (Malone et al., 1992; Moore et al., 1995c; Sims and Wolf, 1994), manure production (Malone et al., 1992) and manure composition (Malone et al., 1992; Sims and Wolf, 1994; Stephenson et al., 1990).

B. Current Poultry Manure Management Practices

Before discussing the current practices used in poultry manure management, it is important to define the relevant terms. *Poultry manure* is a mixture of poultry feces and urine. *Poultry litter* is a mixture of manure, bedding material, feathers, wasted feed, and soil (which is usually inadvertently included during the cleanout operation). *Bedding materials* are used to absorb the liquid fraction of the excreta. Materials typically utilized for bedding include wood shavings, sawdust, rice hulls, peanut hulls, and oat straw (Carpenter, 1992). Poultry litter is synonymous with *broiler litter* and is usually produced in broiler houses. On the other hand, caged laying hen operations are usually constructed over a large pit, which receives undiluted manure from the hens. Due to these differences in management, poultry litter is usually much drier than laying hen manure. Litter associated with broiler production, manure generated from laying operations (hens and pullets), and dead birds are the three wastes of primary concern in poultry production (Edwards and Daniel, 1992b). The majority of manure (68%) produced in the U.S. is in the form of poultry litter (Table 1).

In most states, the litter in broiler houses is totally removed once a year, normally in April or May. At this time the house is disinfected and the waterers and feed troughs are cleaned. Fresh bedding (5 to 10 cm) is then spread on the floor of the houses. One day old broilers are then placed in the house at a density of approximately 13.5 birds m^{-2}. After each flock of birds are harvested, the top layer of hardened manure, which is referred to as cake, is removed using a "de-caker" which is pulled behind a tractor. Malone et al. (1992) reported an average of 0.32 kg (wet) of cake manure was produced per bird in birds grown to 51 days of age. During that same time frame, they reported 1.08 kg of litter would be produced per bird. Normally five or six flocks of chickens are grown over a one year period and the cycle is repeated. However, in the Delmarva area (Delaware, Maryland, and Virginia) a total cleanout of the litter may only occur once every three to five years.

When the litter is removed from the houses it is either spread on the fields, stacked in deep piles (deep stacking), or composted. If weather conditions permit, direct application of manure and/or litter is preferable to deep stacking or composting, since both of these processes result in considerable loss of nitrogen via volatilization. Although composting of poultry litter has received a tremendous amount of attention in recent years, it is a waste of time, money, and nutrients, unless the litter is being used as a feed supplement. Composting of dead birds, on the other hand, provides a fairly economical solution to a major waste product.

Except for small amounts of poultry manure used in animal feed and other uses, the major portion (>90%) is applied to agricultural land (Carpenter, 1992), within a few miles from where it is produced (Moore et al., 1995b). Land application offers the best solution to management of the enormous amounts of manures generated on U.S. poultry farms each year. In most cases, the land base available for application of manure is limited. This limitation mainly arises from restrictions imposed by the high cost of transporting manure long distances.

The University of Arkansas Cooperative Extension Service recommends an annual maximum application rate of 11.2 Mg of poultry litter ha^{-1} (5 tons acre^{-1}), with no more than 5.6 Mg ha^{-1} (2.5 tons acre^{-1}) in a single application. This recommendation is based on meeting the N requirements of forage, a common approach in animal manure application programs (Wallingford et al., 1975). Recommendations based on forage N requirements do not consider other possible limiting factors in land application of litter, such as P or heavy metal content.

II. Production Problems Associated with Poultry Litter

Potential problems associated with land application of poultry litter can be divided into two categories: production problems and environmental problems. The production problems associated with poultry litter are: (1) excess salinity (or nitrogen) resulting in crop damage, (2) an imbalance of calcium (Ca), magnesium (Mg), and potassium (K) in soils which results in grass tetany in cattle, (3) copper (Cu) toxicity in animals being fed poultry litter, and (4) ammonia volatilization, which causes high atmospheric ammonia levels in poultry rearing facilities.

A. Salinity Damage to Crops

Poultry litter is generally considered the most valuable animal manure for use as a fertilizer, due to its low water content and relatively high content of macro, secondary, and trace elements. However, under certain conditions, N and K salts can build up from excessive poultry litter applications, causing salinity damage to crops (Liebhardt, 1976; Weil et al., 1979). Reported damage to crops includes reduced germination, leaf burn, stunted root growth, and decreased production (Hileman, 1971; Weil et al., 1979). Normally, this damage has only been observed when excessive rates of litter have been utilized; therefore, applications of litter based on the nutrient requirements of plants will normally eliminate this problem. In fact, it should be noted that poultry litter has been used for reclamation of salt-affected soils in areas of south Arkansas where brine spills, associated with oil exploration, have occurred (Hileman, 1973). The response to litter on brine-impacted soils is probably due to a decrease in the sodium saturation of the cation exchange capacity (CEC), increased available calcium, and an improvement in soil physical properties, such as hydraulic conductivity and water holding capacity.

B. Grass Tetany in Cattle

Mineral imbalances in forages due to excessive litter applications are typical in areas of the U.S. where concentrated poultry production and cattle production are linked, such as northwest Arkansas. Grass tetany in ruminants, which is related to the K/(Ca + Mg) ratio in forages, is much more likely on soils that have received excessive rates of poultry litter, which increases soil K levels in litter to a greater extent than divalent cations (Stuedemann et al., 1975; Wilkinson et al., 1971). Shreve et al. (1995) reported K concentrations as high as 12% in tall fescue leaf tissue that had been heavily fertilized with litter in the past. This is four times higher than the National Academy of Sciences (1980) toxicity threshold for K in cattle feed (3% K). Stuedemann et al. (1975) suggested that litter application rates should be limited to 9 Mg ha^{-1} or less for use on fescue to avoid this problem. Mineral supplements (particularly Mg) should be provided to cattle grazing pastures fertilized with poultry litter.

C. Copper Toxicity in Sheep

Poultry litter, when mixed with feed grains, has been found to be a successful feed for cattle. Approximately 4% of the poultry litter produced in the U.S. is fed to cattle (Carpenter, 1992). Although disease problems have not been reported from feeding manures to animals under acceptable conditions, Cu toxicity has been reported to be a problem in sheep (Fontenot et al., 1971). Most poultry producers feed an excess of copper sulfate. Although this results in an increase in weight gain in broilers, the gains are not due to a change in diet *per se*, but rather to a change in litter composition (Johnson et al., 1985). There are two possible explanations for this phenomenon: (1) high Cu levels reduce populations of pathogenic microorganisms in the litter, and (2) other processes, such as NH_3 volatilization, are affected.

D. Ammonia Volatilization

Ammonia volatilization from poultry litter causes both production and environmental problems. The production problems are twofold: (1) NH_3 volatilization causes high levels of atmospheric NH_3 in poultry houses, which is detrimental to both farm workers and birds, and (2) NH_3 results in N loss from the litter, which reduces the fertilizer value of the litter.

1. Factors Affecting Ammonia Volatilization

The dominant form of inorganic N in manure is ammonium (NH_4). Ammonium is converted to ammonia (NH_3) as pH increases. Ammonia ($NH_3°$) is in equilibrium with $NH_3(g)$ (ammonia gas), which diffuses from the litter into the atmosphere. This process is referred to as ammonia volatilization. The amount of N that is lost from poultry manure via NH_3 volatilization during land application varies with the application method, manure and/or soil pH, soil CEC, moisture content, type of manure, weather conditions, and ammonia gradient between soil/manure and the atmosphere (Reddy et al., 1979; Moore et al., 1995a).

Sims and Wolf (1994) stated that in excess of 50% of the total N in poultry manure may be lost via NH_3 volatilization. In a laboratory study, Moore et al. (1995a) found that 33% of the total N volatilized from broiler litter in six weeks. Schilke-Gartley and Sims (1993) found up to 31% of the total N was lost from broiler litter in 12 days, if the litter was surface applied. These losses were greatly decreased when the litter was incorporated into the soil.

2. Impact of Ammonia Volatilization on Poultry Production

The majority of NH_3 loss from broiler litter and laying hen manure probably occurs when the litter or manure is still in the houses, since uric acid conversion to NH_3 is a quick process. In the U.S., poultry houses are normally only cleaned out once a year, as stated earlier. This reuse of poultry litter for several flocks results in tremendous amounts of NH_3 volatilization, which can cause very high concentrations of NH_3 in the atmosphere of poultry houses. Researchers have known for over 30 years that NH_3 levels build up in poultry rearing facilities and this buildup adversely affects poultry. Valentine (1964) observed NH_3 levels in the 60 to 70 μL L^{-1} (ppm) range in the atmosphere of poultry houses, whereas Anderson et al. (1964b) found that NH_3 levels were often as high as 100 uL L^{-1} in the atmosphere of commercial poultry houses.

In Europe, COSHH (Control of Substances Hazardous to Health) has set the limit of human exposure to NH_3 at 25 uL L^{-1} for an eight hour day and 35 uL L^{-1} for a 10 minute exposure (Williams, 1992). Reece et al. (1979) and Anderson et al. (1964b) indicated that high NH_3 concentrations in poultry houses are more common in the winter, since the curtains on the houses are closed and high ventilation rates increase energy costs.

Research on the effects of high NH_3 levels on poultry has shown it causes decreased growth rates (Reece et al., 1980; Charles and Payne, 1966a; Quarles and Kling, 1974), decreased egg production (Deaton et al., 1984; Charles and Payne, 1966b), reduced feed efficiency (Caveny and Quarles, 1978; Caveny et al., 1981), damage to the respiratory tract (Anderson et al., 1964a; Nagaraja et al., 1983), increased susceptibility to respiratory diseases, such as Newcastle disease (Anderson et al., 1964a), increased levels of Mycoplasma gallisepticum (Sato et al., 1973), increased incidence of airsaculitis (Kling and Quarles, 1974; Quarles and Kling, 1974; Oyetunde et al., 1978), impaired immunosuppression (Nagaraja et al., 1984), and increased incidence of blindness due to keratoconjunctivitis (Bullis et al., 1950; Faddoul and Ringrose, 1950).

3. Effects of Litter Amendments on Ammonia Volatilization

a. Ammonia Volatilization from Chemically-Treated Litter

Carlile (1984) indicated that 25 μL L^{-1} NH$_3$ should not be exceeded in poultry houses. Attempts to inhibit NH$_3$ volatilization from poultry litter were first reported in the 1950s (Cotterill and Winter, 1953). Since then, many different chemicals have been tested for their effectiveness to inhibit NH$_3$ release from litter. Carlile (1984) indicated that these chemicals fall into two categories, those that act by inhibiting microbial growth (which would slow uric acid decomposition) and those that combine with the released NH$_3$ and neutralize it. These chemicals include calcium chloride (Witter and Kirchmann, 1989), paraformaldehyde (Seltzer et al., 1969), zeolites like clinoptilolite (Nakaue et al., 1981), superphosphate (Cotterill and Winter, 1953; Reece et al., 1979), phosphoric acid (Reece et al., 1979), ferrous sulfate (Huff et al., 1984), hydrated lime (Cotterill and Winter, 1953), limestone (Cotterill and Winter, 1953), gypsum (Cotterill and Winter, 1953), magnesium salts (Witter and Kirchmann, 1989), yucca saponin (Johnston et al., 1981), acetic acid (Parkhurst et al., 1974), propionic acid (Parkhurst et al., 1974), and antibiotics (Kitai and Arakawa, 1979).

Moore et al. (1995a and 1996b) tested the efficacy of a variety of chemical amendments to reduce NH$_3$ volatilization from litter in laboratory studies, including Ca(OH)$_2$ (calcium hydroxide); Al$_2$(SO$_4$)$_3$·18H$_2$O (alum); alum + CaCO$_3$, FeSO$_4$·7H$_2$O (ferrous sulfate); H$_3$PO$_4$ (phosphoric acid); a proprietory product, composed of water, ethylene glycol and sodium silicate; NaHSO$_4$ (sodium bisulfate); Ca-Fe silicate; and yucca plant extract. An average of 14.8 and 14.4 g N kg^{-1} litter was lost from the controls (unamended litter) during the 42-day incubation period in the first and second study, respectively (Figures 1 and 2). This corresponds to an NH$_3$ volatilization rate of approximately 350 mg N kg^{-1} day^{-1}. Ammonia volatilization from litter treated with ethylene glycol/sodium silicate, sodium bisulfate, and Ca-Fe silicate was not significantly different from the controls, even though the commercial products were applied at recommended rates.

Moore et al. (1995a and 1996a) found that compounds that decreased volatilization included alum, ferrous sulfate, and phosphoric acid. This is due to a decrease in litter pH caused by these compounds, which shifts the NH$_4$/NH$_3$ equilibria towards NH$_4$. The rate of NH$_3$ volatilization is highly dependent on pH. As pH increases, the NH$_3$/NH$_4$ ratio increases, causing volatilization to increase. Reece et al. (1979) indicated that NH$_3$ volatilization in poultry litter increases dramatically at pH levels above 7.

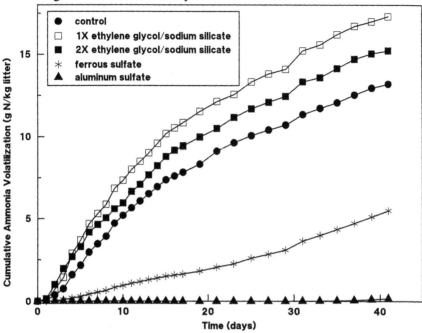

Figure 1. Cumulative ammonia volatilization from poultry litter with and without chemical amendments as a function of time. (From Moore et al., 1995a.)

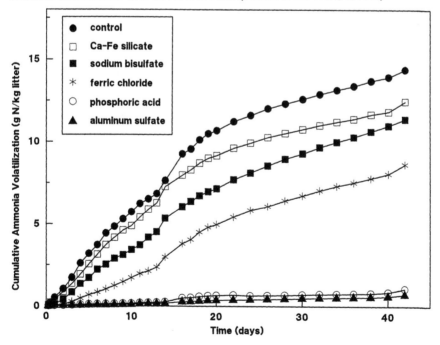

Figure 2. Cumulative ammonia volatilization from poultry litter with and without chemical amendments as a function of time. (From Moore et al., 1996a.)

Although ferrous sulfate does control NH_3 emissions, there is a major problem with its utilization as a litter amendment. Research by Moore et al. (1996b) on the effects of litter amendments showed that after four weeks of growth, the cumulative mortality rates of broilers were significantly lower in alum and ferric chloride treated litter compared to that treated with ferrous sulfate (3.6, 4.4, 8.3, and 10.2% for the alum, liquid ferric chloride, control, and ferrous sulfate amendments, respectively). Low mortality rates in the alum and ferric chloride treatment compared to the controls were probably due to decreased levels of atmospheric NH_3, whereas high mortality in the ferrous sulfate treatment was probably due to Fe toxicity (the rate of Fe applied from the liquid ferric chloride was much lower than for ferrous sulfate). These results support the findings of Wallner-Pendleton et al. (1986) and indicate that ferrous sulfate is not a suitable amendment for poultry litter. Wallner-Pendleton et al. (1986) reported a large die-off of one day old chicks in a commercial flock after the litter had been treated with ferrous sulfate. Apparently, chickens normally consume a certain amount of the litter and when it has been treated with ferrous sulfate, it can result in increased mortality of the birds.

Alum and phosphoric acid are the most effective compounds in reducing NH_3 volatilization. Nitrogen losses at 42 days with alum were 99 and 95% lower than the controls for the first and second study, respectively (200 and 130 g alum kg^{-1}, respectively). Phosphoric acid is very effective at reducing NH_3 losses, with decreases of 93% at treatment rates of only 4 g kg^{-1} litter. However, phosphoric acid increases P solubility in litter by an order of magnitude, aggravating P runoff problems. Therefore, it should not be used in areas containing P sensitive watersheds.

b. Effects of Chemical Amendments on Litter Nitrogen Contents and Crop Yields

There are other advantages of reducing NH_3 volatilization with chemical amendments like alum, besides improving the in-house environment for birds and decreasing acid rain. One major agronomic advantage is the increase in N content of the litter. Moore et al. (1995a) showed that treatments resulting in lower NH_4 volatilization had higher N contents, as would be expected (Figure 3). Most of the additional N in the treatments with low volatilization was NH_4-N (Table 2). The total N concentration of the 200 g alum kg^{-1} treatment was 41.5 g N kg^{-1}. This was somewhat higher than the original N concentration (38.5 g N kg^{-1}) and significantly higher than any of the other treatments. If the weight of alum present had been taken into account, the N concentrations of the litter itself would have been in excess of 50 g N kg^{-1}. Moore et al. (1995a) attributed higher total N concentrations at the end of the study as the result of losses of C via CO_2 evolution from microbial decomposition. The controls contained 26.1 g total N kg^{-1} at the conclusion of the study. Therefore, the addition of alum at the higher rate resulted in a doubling of the N concentrations in the litter, which would greatly increase the value of poultry litter as a fertilizer source.

Increased available N and decreased soluble P make treated litter more valuable as a fertilizer, which may help make it economically feasible to transport longer

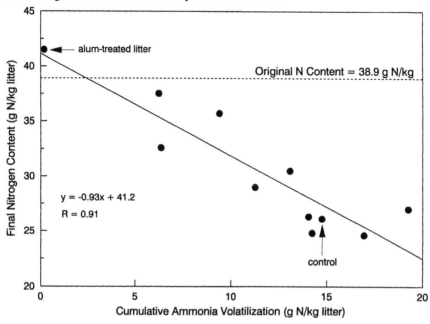

Figure 3. Total N content of the litter after 42 days of incubation as a function of the cumulative ammonia volatilization from the various treatments. (From Moore et al., 1995a.)

Table 2. Effect of litter amendments on selected litter characteristics after 42 days

Treatment	pH	TKN	NH₄ᵃ	TP	SOC	DRP	TDP
		--------g kg⁻¹--------			-----mg kg⁻¹-----		
Control	8.89	26.1	3.72	24.8	27.4	2022	2621
25 g Ca(OH)₂ kg⁻¹	9.09	24.8	3.62	23.3	26.7	1305	1798
50 g Ca(OH)₂ kg⁻¹	9.03	26.3	5.80	21.9	30.3	989	1324
100 g Al₂(SO₄)₃·18H₂O kg⁻¹	8.37	35.7	13.6	22.5	20.7	467	734
200 g Al₂(SO₄)₃·18H₂O kg⁻¹	7.07	41.5	17.6	21.6	22.0	111	261
100 g Al₂(SO₄)₃·18H₂O + 50 g CaCO₃ kg⁻¹	8.07	29.0	10.9	22.5	14.8	431	605
200 g Al₂(SO₄)₃·18H₂O + 50 g CaCO₃ kg⁻¹	7.88	32.6	16.1	21.4	11.1	194	286
100 g FeSO₄·7H₂O kg⁻¹	8.37	30.5	12.1	22.3	19.9	748	978
200 g FeSO₄·7H₂O kg⁻¹	8.09	37.5	19.9	21.1	14.6	529	727
10 g MLT kg⁻¹	9.11	27.0	5.21	24.4	24.8	1827	2361
20 g MLT kg⁻¹	9.09	24.6	3.54	23.5	26.4	1788	2245
LSD₀.₀₅	0.36	2.7	3.16	ns	3.5	211	295

ᵃSum of water soluble and exchangeable NH₄-N.
(From Moore et al., 1995a.)

Table 3. Mean yields of fescue forage as affected by chemically-amended poultry litter

Treatment	Forage yield		
	First harvest	Second harvest	Total
	--------------------(kg ha^{-1})--------------------		
Litter + alum	802	1557	2358
Litter + ferrous sulfate	681	1292	1974
Litter alone	735	1112	1847
Control	279	454	733
LSD (0.05)	84	298	290

(From Shreve et al., 1995.)

distances. By changing soluble litter P to less-soluble forms, land application of litter can be made based on meeting crop N requirements, reducing the need for commercial N fertilizers. It should be recognized that with the increase in N content of litter treated with alum, some reduction in litter application rate would be expected while still supplying adequate N to crops.

Shreve et al. (1995) found tall fescue yields were significantly higher when fertilized with alum-treated broiler litter, compared to normal litter or litter treated with ferrous sulfate (Table 3). They also found the fescue had higher N contents than fescue fertilized with normal litter and attributed the growth differences to differences in N uptake.

c. Effects of Aluminum Sulfate on Broiler Production

Recently, a large commercial-scale evaluation of alum treatment of poultry litter has been completed (Moore et al., 1995c). This study was conducted on two commercial farms in northwest Arkansas; one of the farms had six houses, the other four. Alum was applied at a rate of two tons per house to half of the houses on each farm, following each growout, for a full cycle (one year). Results from this study show that atmospheric NH_3 levels were significantly lower in the alum-treated houses than in the controls. This decrease in NH_3 was due to a decrease in litter pH (Moore et al., 1995c). Use of alum resulted in increases in weight gains in the broilers (Figure 4; Moore et al., 1995c). These differences in growth rates allowed the complex manager at company A to harvest the birds in the alum-treated houses one day earlier than the controls. Since 20,000 birds eat approximately 7,000 pounds of feed in one day when they are six weeks of age, this early harvest can potentially save as much as $700 on feed (feed cost $0.10 per pound). Moore et al. (1995c) also found feed conversion was better for alum-treated houses than houses with normal litter (1.83 and 1.89 kg/kg, respectively). Although the propane utilization by the control and alum-treated houses in this study was not known at the time of

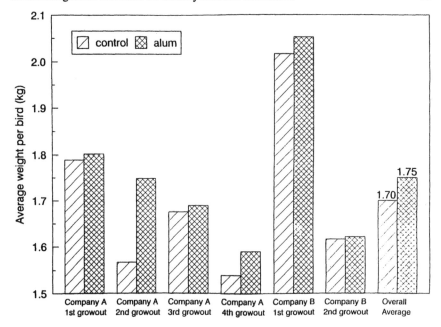

Figure 4. Weight gains of broilers grown on alum-treated and normal poultry litter. (From Moore et al., 1995c.)

this writing, we suspect that the alum-treated houses will have much lower energy use, since these houses required much less ventilation for NH_3 control in the winter months.

Some poultry producers argue that increased ventilation will solve most of the health problems associated with high NH_3 levels in poultry houses, and that chemical additions to litter are not needed. This is somewhat true; however, ventilization in the winter is an expensive solution, due to energy costs. Carr and Nicholson (1980) studied three ventilation rates in poultry houses and found that weight gains increased as the ventilation rate increased. They calculated that the highest rate of ventilation (that needed to keep NH_3 levels below about 40 µL L^{-1}) resulted in an increase in fuel consumption of 172% compared to the medium rate (which had ventilation rates 50% lower). Attar and Brake (1988) developed a computer program that modeled the economic benefits of controlling NH_3 in poultry houses. They calculated that if the outside temperature is 7°C, the cost of producing broilers increases by $0.11 kg^{-1} when NH_3 concentrations increase from 25 to 80 µL L^{-1}. In a poultry house with 19,000 birds weighing 1.82 kg each, this cost would be roughly equivalent to $3800 per flock. The cost of treating the litter with aluminum sulfate would be almost an order of magnitude lower than this. It should also be noted that while increased ventilation may improve the atmospheric NH_3 levels in the houses, it would contribute more NH_3 to the external atmosphere, and thus, cause more acid rain.

III. Environmental Problems Associated with Poultry Litter

Potential environmental problems associated with poultry manure can be divided into three categories: (1) leaching of substances (particularly nitrate) into groundwater, (2) surface runoff of pollutants (bacteria, carbon, metals, pesticides, and/or phosphorus), and (3) NH_3 volatilization, which is important in acid deposition.

A. Leaching of Substances into Groundwater

1. Nitrate Leaching

Nitrate leaching into the groundwater is a potential threat to human health from land application of poultry litter. Infants less than three months old drinking water contaminated with high levels of NO_3-N are susceptible to "blue-baby syndrome." This condition, also known as methemoglobinemia, is characterized by a bluish skin coloration which arises from a lack of oxygen in the blood. Poor oxygen transport is due to the oxidation of Fe^{2+} to Fe^{3+} by NO_2-N, which causes the conversion of hemoglobin to methemoglobin, which cannot provide oxygen transport. In order for this to occur, NO_3-N must first be converted to NO_2-N (nitrite) by bacteria in the digestive tract. This normally only occurs in infants, since the bacteria that reduce nitrate to nitrite are very pH dependent and cannot function in older humans, due to the higher levels of acidity produced in the stomach. The EPA limits nitrate concentrations in drinking water supplies to 10 mg NO_3-N L^{-1} (U.S. EPA, 1985). Since animals are not as susceptible to methemoglobinemia as humans, the health advisory level for them is 40 mg NO_3-N L^{-1} (Sims and Wolf, 1994).

Overapplication of poultry litter has been shown to cause elevated levels of nitrate in soil solutions and groundwater (Adams et al., 1994; Kingery et al., 1993; Ritter and Chirnside, 1982; Weil et al., 1979). Ritter and Chirnside (1982) indicated that 32% of 200 water wells sampled in Sussex County, Delaware had high nitrate levels (>10 mg N L^{-1}) due to improper poultry litter applications. Kingery et al. (1993) found that high loading rates of poultry litter resulted in buildup of nitrate in the soil to 3 meters depth or to bedrock (Figure 5).

Adams et al. (1994) evaluated NO_3 leaching in soils fertilized with both poultry litter and hen manure at 0, 10, and 20 Mg ha^{-1}. They found that the amount of NO_3 leaching into the groundwater was a function of litter application rate, with higher rates resulting in increased nitrate concentrations in soil solutions. They also showed that applications of litter or manure at or below the recommended rate in Arkansas (11.2 Mg ha^{-1}) resulted in nitrate concentrations in soil solutions that were usually below the 10 mg NO_3-N L^{-1} standard (Figure 6; Adams et al., 1994).

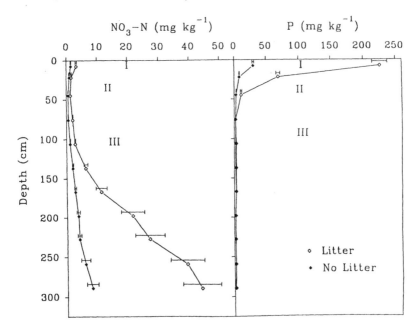

Figure 5. Nitrate-N and soil test P concentrations as a function of depth in soils that have received excessive poultry litter applications and in soils fertilized with chemical fertilizers. (From Kingery et al., 1994.)

2. Phosphorus Leaching

Historically, P has been considered immobile in soils and not subject to leaching. However, when most of the P sorption sites in a soil become saturated with P, then P leaching into groundwater can occur. Breeuwsma and Reijerink (1992) stated that leaching of P from soil has the major features needed to be described as a "chemical time bomb" (i.e., there is a significant time delay between high application rates of P and adverse environmental impacts). The adverse impact is eutrophication of surface water bodies, since they would have high P loading from both surface runoff and groundwater recharge. Breeuwsma et al. (1989), as cited by Breeuwsma and Reijerink (1992), found in one catchment area the P loads to surface water were sufficient to achieve a mean annual P concentration of 1 mg P/L, 87% of which they attributed to leaching and groundwater recharge.

B. Surface Runoff of Pollutants

Nonpoint source runoff from agricultural lands is now believed to be responsible for the water quality problems in over 70% of the lakes and rivers in the U.S. (U.S. EPA, 1994). Potential contaminants in runoff water from fields fertilized with poultry litter include bacteria, C compounds, metals, pesticides, and phosphorus.

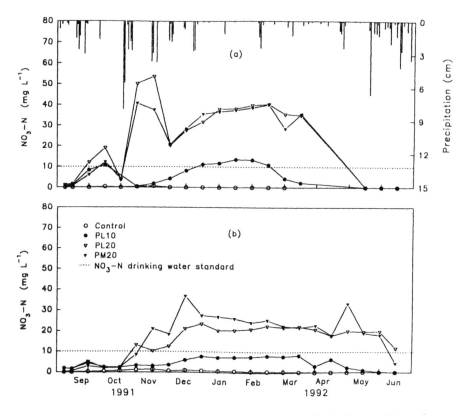

Figure 6. Nitrate concentrations in the vadose zone of soils fertilized with poultry litter. (From Adams et al., 1994.)

1. Bacteria and Virus Runoff

Poultry manure contains many potential pathogens. In Arkansas, the nation's leading poultry producing state, 90% of the surface water bodies (statewide) sampled by the Arkansas Department of Pollution Control and Ecology contained fecal coliform counts in excess of the primary contact standards (200/100 ml). However, fecal coliform counts prior to the rise in poultry in this state are not available. Therefore, it is unknown whether these levels are indigenous or, in fact, due to runoff from animal manures.

Giddens and Barnett (1980) indicated that moderate amounts of surface-applied poultry litter should not cause a water quality problem (with respect to bacterial contamination) unless excessive amounts of rainfall occur. However, several researchers have pointed out that poultry manure contains many pathogens which are responsible for human diseases (Bhattacharya and Taylor, 1975; Fontenot and Webb, 1975; McCaskey and Anthony, 1979). Runoff studies, using fecal coliform as an indicator organism, have shown that runoff water from pastures fertilized with poultry manure may have from 10^4 to 10^6 fecal coliform/100 ml (Giddens and Barnett, 1980; Quisenberry et al., 1981). In a literature review of bacterial runoff, Baxter-Potter and Gilliland (1988) concluded that bacterial densities in runoff water from agricultural lands often exceed water quality standards, but were lower if the sources of fecal pollution were far enough from the receiving stream to minimize overland flow. Therefore, vegetative buffer strips would be expected to decrease fecal coliform levels from lands receiving manure, as was found by Doyle et al. (1975). However, Chaubey et al. (1994) found vegetative buffer strips had no effect on fecal coliform numbers in runoff. Addition of aluminum sulfate to poultry litter (10% w/w) has been shown to decrease fecal coliform numbers in poultry litter in commercial broiler houses (Scantling et al., 1995). In a recent runoff study, fecal coliform counts in runoff water from fescue plots, either not fertilized, fertilized with ammonium nitrate, alum-treated poultry litter, and normal poultry litter were 1,500, 1,700, 12,000 and 63,000 CFU/100 ml, respectively, indicating alum-treated litter may help reduce fecal coliform runoff from pastures fertilized with poultry litter (P.A. Moore, Jr., unpublished data).

Viruses have also been reported in poultry litter and may represent a greater problem than bacteria (Sims and Wolf, 1994). These include viruses responsible for New Castle disease and Chlamydia. At present, little information on virus runoff from fields receiving litter is available.

2. Carbon Runoff

Edwards and Daniel (1992b) indicated that C runoff from poultry litter could negatively impact aquatic life by decreasing dissolved oxygen levels in waterways. Moore and Miller (1994) and Moore et al. (1995a) showed soluble organic C (SOC) levels decreased dramatically with addition of various Al and Fe amendments to poultry litter. Shreve et al. (1995) found SOC concentrations were significantly lower in runoff water from fescue plots receiving alum-treated litter than litter alone, whereas litter treated with ferrous sulfate did not display this decrease. This should help improve the runoff water quality from fields receiving poultry litter by decreasing the biochemical oxygen demand (BOD) of the runoff.

3. Metal Runoff

At present, the poultry industry adds heavy metals, such as arsenic (As), cobalt (Co), copper (Cu), iron (Fe), manganese (Mn), selenium (Se), and zinc (Zn) to poultry feed (Tufft and Nockels, 1991). Kingery et al. (1993) found elevated levels of Cu and Zn in soils heavily fertilized with poultry litter. Elevated levels of heavy metals in the soil will result in increased uptake by plants, which will be consumed by animals or man. However, normally concentrations do not reach toxic levels. For example, Wilkinson and Stuedemann (1990) found that application of up to 68 kg Cu ha^{-1} from broiler litter resulted in only small increases in Cu contents of Bermuda grass and fescue.

Sims and Wolf (1994) indicated that chelation of metals in poultry litter is probably an important process, since poultry litter contains extremely high concentrations of soluble organics. When metals are complexed, their solubilities increase, resulting in higher metal availability. Moore et al. (1995a) showed that water-soluble Cu in poultry litter was highly correlated with soluble organic C, and found that alum-treated litter had the lowest levels of soluble Cu. This was confirmed in a recent runoff study comparing alum-treated to normal poultry litter, which showed that treating with alum resulted in lower concentrations of As, Cu, Fe, and Zn in the runoff water (P.A. Moore, Jr., unpublished data). Decreases in metal runoff with alum-treated litter is most likely due to a decrease in C runoff, since most of the metals appear to be associated with organics.

4. Pesticide Runoff

Pesticide contamination of surface and groundwater is not normally associated with poultry production. However, there are a few pesticides used by the poultry industry, mainly to kill flies and litter beetles. Flies are a major problem when manure has a high water content, such as laying hen or breeder operations. The main pesticide used to control flies under these circumstances is cyromazine, which is a larvicide that is mixed with the poultry rations. Pote et al. (1994) found cyromazine in the runoff water from small fescue plots fertilized with caged layer manure; however, it is not known whether the levels of cyromazine in the runoff represent a major environmental threat or not.

5. Phosphorus Runoff

Phosphorus, unlike NO_3, is not toxic to humans. Likewise, P normally does not have a direct negative impact on land to which it is applied if in excess, although it can adversely impact surface waters if it is moved off-site by runoff or erosion (Sharpley and Menzel, 1987). Phosphorus is considered to be the primary element of concern with respect to eutrophication of freshwater systems (Schindler, 1977, 1978). Eutrophication is derived from the Greek word meaning well-nourished; it describes a condition of lakes or reservoirs involving excess algal growth, which may eventually lead to severe deterioration of the body of water.

The relationship between the total P concentration of lake water and eutrophication has been well documented (Effler et al., 1985; Vollenweider, 1975). In Arkansas, the Department of Pollution Control and Ecology (DPC&E) has set guidelines on P in surface waters of 0.05 mg L^{-1} for streams and 0.10 mg L^{-1} for lakes and reservoirs (ADPC&E, 1988).

The importance of manure as a potential source of P to lake water was demonstrated by Biggar and Corey (1969), who attributed 30% of the P entering Lake Mendota to farm manures, and by Duda and Finan (1983), who report up to 50 fold increases in total P (TP) runoff from watersheds with high livestock populations compared to mostly forested watersheds. Recent studies have shown extremely high P concentrations in the runoff water from pastures receiving low to moderate levels of poultry litter (Edwards and Daniel, 1992a; Edwards and Daniel, 1992b; Edwards and Daniel, 1993; Shreve et al., 1995). The majority (80-90%) of the P in the runoff water is water soluble, which is referred to as dissolved reactive P (Edwards and Daniel, 1993), the form that is most readily available for algal uptake (Sonzogni et al., 1982).

In most states, manure application rates are based primarily on the management of N to minimize nitrate losses by leaching. This has usually led to an increase in soil P levels in excess of crop requirements, due to the generally lower ratio of N:P added in poultry litter than in crops. For example, poultry litter has an N:P ratio of 2 or 3, while the N:P requirement of major grain and hay crops is 8 (Moore et al., 1995b). Since the plants can utilize more N than P, the soil test P level in these soils builds up and after many years far exceeds that required for 100% sufficiency of many crops (Sims, 1992; Sims, 1993; Wood, 1992). This low N:P ratio is further aggravated by N loss mechanisms, such as nitrate leaching, ammonia volatilization, and denitrification. The net result of this process is a large surplus of P in the soil. Breeuwsma and Reijerink (1992) stated that in the intensive livestock areas of the Netherlands, excess P_2O_5 applied to corn ranged from 150 to 450 kg/ha/yr; excess P_2O_5 applied to pasture ranged from 50 to 150 kg/ha/yr.

Although P applications in intensive livestock areas of the U.S. are not as high as that in Holland, P accumulation is still occurring. Kingery et al. (1993) observed soil test P levels as high as 225 mg P/kg soil in the soils in the Sand Mountain area of Alabama. In a similar study of continual long-term poultry litter application to 12 Oklahoma soils, Sharpley et al. (1993) found that P accumulated in the surface meter of treated soil, to a greater extent than N. This reflects the differential mobility, sorption, and plant uptake of N and P in soil.

Phosphorus concentrations in runoff water from fields that have been recently fertilized with poultry litter/manure are often very high (Edwards and Daniel, 1992a,b, 1993; Shreve et al., 1995). However, after the second or third rainfall that results in runoff, P concentrations drop to background levels (Shreve et al., 1995). After this time, the level of P in the runoff water is directly correlated to the level of P in the soil (Pote et al., 1995).

a. Best Management Practices to Reduce P Runoff

Phosphorus runoff from land application of poultry litter may be prevented by implementation of effective BMPs. Examples of BMPs include: (1) proper nutrient management, such as correct timing and placement of manure, and applying when there is a low likelihood of rainfall in the near future, (2) basing manure application rates on P, rather than N, (3) implementing a cut-off level for soil test P, above which manure is not applied, (4) utilizing buffer strips or grass hedges between areas receiving manure and waterways, (5) adding phytase enzymes to feed to reduce inorganic P requirements of birds, (6) adding chemical amendments that precipitate P to the soil, and (7) adding chemical amendments that precipitate to the litter.

b. Nutrient Management Plans

One of the primary factors affecting P runoff from fields fertilized with poultry litter is the rate of application, with higher rates resulting in higher concentrations of P in the runoff water (Edwards and Daniel, 1992a, 1993). Therefore, it is important that excessive rates of application be avoided. Growers should calculate the nutrient requirements needed for maximum (realistic) crop yields, taking into account the P present in the soil (using soil test P values), and apply the litter based on the amount of nutrients present in the manure. In most states, the Natural Resources Conservation Service (formerly Soil Conservation Service) or the Cooperative Extension Service will help growers develop a nutrient accounting system for their farm, referred to as a *nutrient management plan*. Since the amount of litter produced in broiler operations is relatively constant (one kg/bird), and the P content of litter is relatively constant (1.5 to 2.0% P), the amount of P generated in a typical broiler production farm can easily be calculated. A farm-wide nutrient budget such as this will help the grower increase yields, by applying litter on the fields that can utilize the nutrients most efficiently and protect surface water resources by reducing P runoff.

Another very important parameter affecting P runoff from poultry manure applications is timing. Manure should be applied when crops can utilize the nutrients most efficiently. For most crops, this is in the spring or early summer. Manure should not be applied in the winter on snow or frozen ground, since this does not allow P infiltration into the soil, but facilitates P runoff. Similarly, manure should not be applied shortly before a large storm, since P runoff increases with increasing storm intensities (Edwards and Daniel, 1992a; Edwards et al., 1994) and the majority of P runoff occurs during the first few runoff events (Shreve et al., 1995). Edwards et al. (1992) developed a model to determine the best time for application of poultry litter.

Another variable influencing P mobility is the method of application. Surface broadcast applications of manure on pastures or on row crops with minimum tillage, can lead to a buildup of P near the surface, enhancing P runoff. This method of application also enhances ammonia volatilization. Therefore, many European countries have begun injecting manures, when they are in the form of slurries.

c. Phosphorus-Based Manure Management Strategies

Historically, strategies for land application of animal manures have been based on meeting the N needs of the crop being produced. Perhaps this approach can be justified on the basis of groundwater protection but little can be gleaned on the basis of surface water protection. Therefore, the question as to whether poultry litter applications should be based on P loading, rather than N loading, has arisen.

In areas such as northwest Arkansas, where producers have a limited land base for application of poultry manure, this can become an important consideration. For example, Simpson (1991) calculated that 18.2 ha of pasture land is required to dispose of manure produced annually in a 20,000 bird house if litter application is limited by N, whereas if P limits on litter application are considered, 91.1 ha of pasture are required. If application rates are limited by P, producers would need to buy supplemental commercial N fertilizer to ensure adequate crop nutrition.

Moving poultry litter to areas where soil N and P levels are low would not only improve crop production, but would decrease the likelihood of environmental problems associated with excess litter. However, the cost of transportation would prohibit this practice, unless the government or the poultry industry provides the growers subsidies for such a program as indicated by Bosch and Napit (1992).

d. Upper Limit for Soil Test Phosphorus

The University of Arkansas Extension Service is recommending growers not apply poultry litter on soils which test 150 kg P ha^{-1} (Mehlich III extractable) or higher due to the concern over eutrophication. Many fields in northwest Arkansas already test above 500 kg P ha^{-1} and the numbers continue to grow. This could eventually force growers to transport litter to areas which test low in soil P. Although this will probably help the environment, it will, in effect, lower the value of the litter (due to high transportation costs). The growers will also have to buy commercial N fertilizers to meet the N requirements of their pastures.

The time required to lower soil test P levels from excessive to that level needed to maintain maximum yields depends upon how much P is present. McCollum (1991) found that maximum yields of corn and soybeans were maintained as long as soil test P levels (Mehlich I) were above 22 g P m^{-3}. If soils tested 50 to 60 g P m^{-3}, it would take 8 to 10 years of cropping to reduce to 22 g P m^{-3}, whereas if the initial level were 100 to 120 g P m^{-3}, it would take 14 years to reach this level. McCollum indicated that P reversion to unextractable forms was a larger factor than crop removal.

As mentioned earlier, P concentrations in runoff water from agricultural fields have been shown to be correlated to soil test P (Sims, 1993; Pote et al., 1995). However, Breeuwsma and Silva (1992) indicated that the most important parameter with respect to P runoff was the percent P saturation of the soil, rather than the total P content or the levels of extractable P. Research conducted by Sharpley (1995) supports this theory (Figures 7 and 8). He showed that P runoff from soils fertilized with poultry litter was correlated to soil test P levels (Mehlich III extractable P), with higher soil test P levels resulting in higher P concentrations in runoff water. However, the slopes and intercepts for the correlations varied with soil type (Figure 7). This was not the case for the percent P saturation, which showed a single linear relationship with P runoff (Figure 8). Sharpley (1995) indicated that the amount of P in the soil or the P adsorption capacity was not as important in determining if P runoff would occur, but rather the amount of adsorption sites already occupied by P (percent P saturation).

Basing manure applications on P, rather than N soil contents and crop requirements may mitigate the excessive build-up of soil P and at the same time lower the risk for NO$_3$ leaching to groundwater. However, a soil test P based strategy would eliminate much of the land area with a history of continual manure applications, as

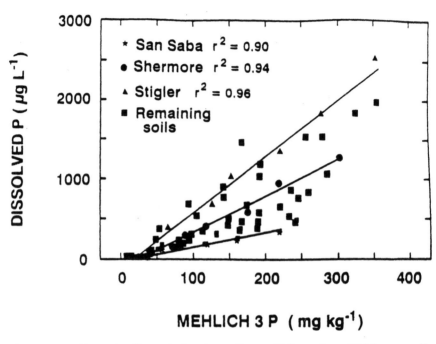

Figure 7. Effect of soil test P levels on P runoff from three Oklahoma soils fertilized with various rates of poultry litter. (From Sharpley, 1995.)

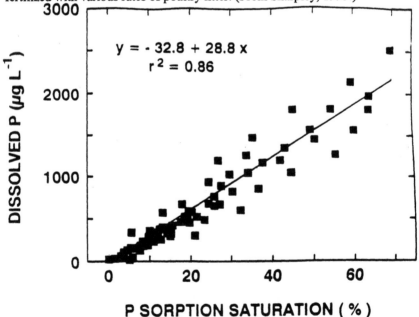

Figure 8. Effect of percent P saturation on P runoff from three Oklahoma soils fertilized with various rates of poultry litter. (From Sharpley, 1995.)

many years are required to lower soil P levels, once they reach excessive levels (Wood, 1992). In addition, farmers relying on poultry manure to supply most of their crop N requirements will have to purchase commercial fertilizer N, instead of using their own manure N. Using a soil test P based strategy may resolve potential environmental issues, but at the same time may be placing unacceptable economic burdens on farmers.

e. Vegetative Filter Strips

Vegetative filter strips are a low-cost management strategy for reducing P runoff from land receiving poultry litter. They function by providing an area for infiltration and adsorption of soluble pollutants, as well as deposition of sediment and sediment-bound pollutants (Chaubey et al., 1994). Examples of vegetative filter strips include terraces, grass waterways, grass hedges, grass buffer strips, sediment catch basins, and so forth. Chaubey et al. (1993) found that vegetative filter strips reduced both soluble and total P concentrations dramatically in runoff water from pastures fertilized with poultry litter (Figure 9).

Although vegetative filter strips are one of the simplest, most effective, and economical means of reducing nonpoint source pollution, there are three drawbacks to this BMP. The primary drawback is that they only work properly if sheet flow of water is occurring. When runoff is channelized into gully flow, their effectiveness is greatly diminished. Another drawback is that these areas of land will have lower productivity than areas receiving manure. The third drawback is the reduction of land the grower will have to apply manure on, which would affect those growers who do not have a large enough land base to receive the manure produced on their farm.

f. Phytase Addition to Feed

There are two reasons why poultry litter has a high N:P ratio: (1) NH_3 volatilization causes low N concentrations, and (2) addition of large quantities of inorganic P (usually dicalcium phosphate) to feed. This addition of inorganic P is necessary, since poultry lack the phytase enzyme needed to break down phytate P compounds (Nelson, 1967). Normally, about 60% of the P in corn and soybeans, the main components of poultry feed, is in the phytate form. Research on the addition of the phytase enzyme to poultry feed was first reported by Nelson et al. (1968, 1971), who showed that phytase additions would successfully liberate phytate P in the digestive system of poultry. Efforts are still continuing today in the U.S. and the E.C. to find an economical method of delivering phytase enzyme to the digestive tract of poultry, with most of the research focusing on enzymes produced by fungi. However, this process is still cost prohibitive, at least in the U.S. More cost-effective methods of phytase delivery would be to: (1) give one day old chicks an oral challenge of bacteria or fungi, that would live in their digestive tract, that have high phytase production, (2) genetically engineer chickens that have the phytate

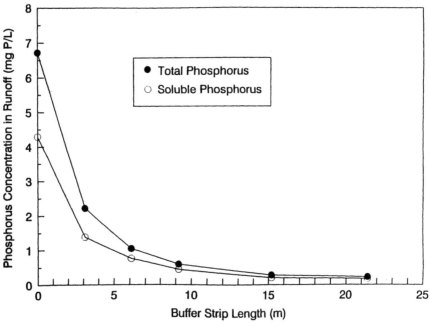

Figure 9. Effect of vegetative buffer strip length on total and soluble P concentrations in runoff water from fields fertilized with poultry litter. (From Chaubey et al., 1993.)

enzyme in their digestive system, or (3) develop corn and soybeans through genetic engineering that has less phytate P. Pioneer seed company has reportedly developed a low-phytate corn hybrid through genetic engineering, but has not tested it as a poultry feed at present.

g. Soil Amendments

Another method of reducing P runoff from fields receiving poultry litter is the use of soil amendments to reduce P solubility. Addition of alum and ferrous sulfate have been shown to decrease P solubility in soils from 10 mg P kg^{-1} to less than 0.01 mg P kg^{-1}, particularly when applied with calcium hydroxide to buffer soil pH (Dave Miller, personal communication). Since there is not normally an economic return associated with this practice, waste products which have low or no costs are more attractive for this use than refined products, such as alum or ferrous sulfate. One possibility is spent alum from water treatment plants, which is currently being tested on Oklahoma soils testing high in P (Nick Basta, personal communication).

h. Litter Amendments

To minimize runoff losses of P while maintaining adequate N fertility, the available nutrient content of the litter should more nearly match plant uptake. One possible means of accomplishing this is to increase N availability or content relative to that of P in the litter, or conversely to decrease the availability of P relative to N. Most P in runoff from pastures receiving poultry litter is soluble; thus, converting litter P to less-soluble or less-available forms would effectively reduce the potential for P loss in runoff.

Alum, calcium hydroxide, and ferrous sulfate have all been shown to be effective in decreasing P solubility in poultry litter (Moore and Miller, 1994). Similar decreases in P solubility have been reported for sewage sludge treated with Al, Ca, and Fe compounds, with the largest fraction of P being in Al-bound forms (McCoy et al., 1986; Soon and Bates, 1982).

It is unclear whether these decreases in solubility are due to mineral precipitation, adsorption, or a combination of both. Possible precipitation reactions for alum, ferric sulfate, and calcium hydroxide are as follows:

$$Al_2(SO_4)_3.14H_2O + 2PO_4^{3-} \longrightarrow 2AlPO_4 + 3SO_4^{2-} + 14H_2O$$
alum

$$Fe_2(SO_4)_3.14H_2O + 2PO_4^{3-} \longrightarrow 2FePO_4 + 3SO_4^{2-} + 14H_2O$$
ferric sulfate

$$5Ca(OH)_2 + 3H_2PO_4^- + 3H^+ \longrightarrow Ca_5(PO_4)_3OH + 9H_2O$$
slaked lime

Since the major form of P in runoff from fields fertilized with poultry litter is water soluble P (Edwards and Daniel, 1993), P runoff from fields receiving litter amended with these compounds should be lower than that from fields fertilized with normal litter. This was confirmed in a field study using rainfall simulators by Shreve et al. (1995) who showed that the P concentrations in runoff from small plots receiving alum-treated poultry litter were 87% lower than plots receiving the same rates of normal litter (Figure 10). Amending poultry litter with alum resulted in an 87% reduction in the soluble reactive phosphorus (SRP) concentrations compared to litter alone for the first runoff event and a 63% reduction for the second runoff event. Poultry litter amended with ferrous sulfate also caused a decrease in SRP concentrations (77% and 48% for the first and second runoff events, respectively). Mean total P concentrations were reduced by the addition of chemical amendments to litter in the same manner.

The likelihood of this practice causing environmental problems appears to be remote (Moore et al., 1995c). If alum is added to poultry litter after each flock of chickens is grown, then the NH_3 levels in the houses will be lowered for the first few weeks of the next flock. Normally, one day old chicks, which are very susceptible to high NH_3, are placed in the houses. If the litter has been treated with alum, then the pH will start out low (5 to 6) and increase with time as the acidity

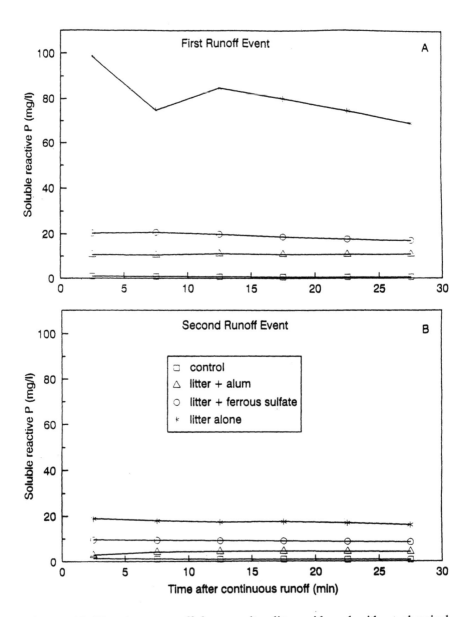

Figure 10. Phosphorus runoff from poultry litter with and without chemical amendments. (From Shreve et al., 1995.)

from the alum reacts with bases, such as NH_3. Ultimately all of the acidity will be consumed, causing the pH to increase to over 7 after about five weeks. At such high pH levels, concentrations of Al^{3+} are too low to cause problems. Rather, the Al is in the $Al(OH)_3$ form, which is insoluble. In fact, runoff studies on alum-treated and normal litter have indicated that Al runoff is significantly different from these two sources.

C. Ammonia Volatilization

Another negative environmental impact associated with poultry litter is NH_3 volatilization, which enhances atmospheric acid deposition and helps contribute to eutrophication.

1. Atmospheric Acid Deposition from Ammonia

Ap Simon et al. (1987) indicated that atmospheric NH_3 pollution plays a very important role in acid rain in Europe. They indicated that the dominant source of NH_3 in Europe was livestock wastes, with long-term trends showing a 50% increase in NH_3 emissions from 1950 to 1980. Anthropogenic NH_3 emissions in the U.S. were approximately 840,000 tons/year in 1980, with 64% (540,000 tons) associated with livestock waste management (U.S. EPA, 1987).

The impact of NH_3 on acid rain has often been overlooked, particularly in the U.S., since NH_3 initially raises the pH of rainwater. However, when the pH of rainwater is increased, it increases the amount of SO_2 that will dissolve in it. Ammonium sulfate then forms and when it reaches the soil, the NH_4-N is oxidized to NO_3-N by microorganisms, releasing nitric and sulfuric acid (van Breemen et al., 1982). This process produces two to five times the acid input to soils previously described for atmospheric acid deposition, resulting in extremely low pH values (2.8 to 3.5) and high levels of dissolved aluminum (Al) in noncalcareous soils (van Breemen et al., 1982; van Breemen et al., 1989).

2. Impact of Ammonia Volatilization on Eutrophication

a. Atmospheric Nitrogen Loading to Lakes and Rivers

Ammonia volatilization can also contribute to eutrophication. Nitrogen deposited via wet fallout tripled in Denmark from 1955 to 1980 and corresponded to N losses from agriculture during this period (Schroder, 1985). The rising levels of N in the fallout were also shown to be highly correlated to the NO_3-N content in Danish streams (Schroder, 1985). Ammonia loss also causes low N/P ratios in litter, which increases the likelihood of excessive P runoff into adjacent water bodies, thus increasing eutrophication.

b. Effects of Low Nitrogen/Phosphorus Ratios in Litter on Phosphorus Runoff

Moore et al. (1995b) indicated that poultry litter had N:P ratios of 2:1, whereas most crops needed N:P ratios of 8:1. These low N:P ratios are due mainly to NH_3 volatilization, as stated earlier. When fields are fertilized with litter having a low N:P ratio, the plants can utilize most of the plant available N, but not the P. This results in a buildup in soil P, as mentioned earlier, which causes P runoff and increases the chance for soils to become saturated with P, which results in P leaching into the groundwater.

As a result of acid rain from NH_3 and P saturated soil conditions, the Dutch government has implemented a plan to reduce NH_3 emissions from manure by 90% relative to the levels released in 1980. Although the law imposes heavy taxation for growers exceeding accepted emission levels in the future, at present there are few BMPs being recommended by the government for the growers to utilize.

IV. Summary

Over 13 billion kg of poultry litter and/or manure are produced each year in the U.S. More than half of this manure is concentrated in five states (AR, NC, GA, AL, and CA). As a result, there are certain areas of the country where poultry manure has been overapplied, resulting in high nitrate levels in groundwater, heavy metal accumulation in soils, high soil test P levels, and elevated concentrations of P in runoff water. The objectives of this paper were to provide an overview of current poultry manure management practices, to discuss potential problems associated with land application of poultry litter, and to delineate some possible solutions to these problems. Potential environmental problems associated with poultry manure can be divided into three categories: (1) leaching of substances, such as NO_3, into groundwater, (2) surface water runoff with high concentrations of P, C, metals, and/or bacteria, and (3) NH_3 volatilization, which is important in acid deposition. Due to the low N/P ratio in litter, P runoff can be a problem even when litter is applied at recommended rates and at the proper time. Best management practices (BMPs) which reduce P runoff from poultry litter include vegetative buffer strips, establishment of an upper level for soil test P, feed additives to reduce dietary P requirements of poultry, and manure additives which precipitate P in the manure. At present, one of the most promising BMPs to decrease P runoff appears to be the addition of alum to poultry litter. Alum has also been shown to reduce NH_3 volatilization from litter (which should decrease atmospheric acid deposition), as well as decrease fecal coliform, C, and metal runoff. Decreasing NH_3 volatilization lowers NH_3 levels in poultry houses, resulting in better feed conversion and weight gains in broilers. Reducing NH_3 loss also increases the N content of litter, which results in higher crop yields.

References

Adams, P.L., T.C. Daniel, D.R. Edwards, D.J. Nichols, D.H. Pote, and H.D. Scott. 1994. Poultry litter and manure contributions to nitrate leaching through the vadose zone. *Soil Sci. Soc. Am. J.* 58:1206-1211.

Anderson, D.P., C.W. Beard, and R.P. Hanson. 1964a. The adverse effects of ammonia on chickens including resistance to infection with Newcastle Disease virus. *Avian Dis.* 8:369-379.

Anderson, D.P., F.L. Cherms, and R.P. Hanson. 1964b. Studies on measuring the environment of turkeys raised in confinement. *Poult. Sci.* 43:305-318.

Ap Simon, H.M., M. Kruse, and J.N.B. Bell. 1987. Ammonia emissions and their role in acid deposition. *Atmospheric Environment* 21:1939-1946.

Arkansas Department of Pollution Control and Ecology. 1988. Regulation No. 2, as amended. Regulation establishing water quality standards for surface waters for the state of Arkansas.

Attar, A.J. and J.T. Brake. 1988. Ammonia control: Benefits and trade-offs. *Poultry Digest*, August 1988, pp. 362-365.

Baxter-Potter, W.R. and M.W. Gilliland. 1988. Bacterial pollution in runoff from agricultural lands. *J. Envir. Qual.* 17:27-34.

Bhattacharya, A.N. and J.C. Taylor. 1975. Recycling animal waste as a feedstuff: A review. *J. Anim. Sci.* 41:1438-1457.

Biggar, J.W. and R.B. Corey. 1969. *Eutrophication: Causes, Consequences, Correctives.* National Academy of Sciences, Washington, D.C. p. 404-445.

Bosch, D.J. and K.B. Napit. 1992. Economics of transporting poultry litter to achieve more effective use as fertilizer. *J. Soil and Water Cons.* 47:342-346.

Breeuwsma, A. and J.G.A. Reijerink. 1992. Phosphate-saturated soils: A "new" environmental issue. In: G.R.B. ter Meulen, W.M. Strigliani, W. Salomons, E.M. Bridges, and A.C. Imerson (eds.), *Chemical Time Bombs*, Proc. European State-of-the-Art Conference, Foundation for Ecodevelopment, Sept. 1992, Hoofddorp, The Netherlands.

Breeuwsma, A. and S. Silva. 1992. *Phosphorus Fertilization and Environmental Effects in The Netherlands and the Po Region, Italy.* Rep. No. 57. DLO, The Winand Staring Centre for Integrated Land, Soil and Water Research. Wageningen, The Netherlands.

Breeuwsma, A., J.G.A. Reijerink, O.F. Schoumans, D.J. Brus, and H. van het Loo. 1989. Fosfaatbelasting van bodem, grond- en oppervlakterwater in het stroomgebied van de Schuitenbeek. Wageningen, Staring Centrum. Rapport 10.

Bullis, K.L., G.H. Snoeyenbos, and H. Van Roekel. 1950. A keratoconjunctivitis in chickens. *Poult. Sci.* 29:386-399.

Carlile, F.S. 1984. Ammonia in poultry houses: A literature review. *World's Poultry Science Journal* 40:99-113.

Carpenter, G.H. 1992. Current litter practices and future needs. p. 268-273. In: *Proc. National Poultry Waste Management Symposium*, Birmingham, AL. Auburn University, AL.

Carr, L.E. and J.L. Nicholson. 1980. Broiler response to three ventilation rates. *Transactions Am. Soc. Agric. Eng.* 23:414-418.

Caveny, D.D. and C.L. Quarles. 1978. The effect of atmospheric ammonia stress on broiler performance and carcass quality. *Poultry Sci.* 57:1124-1125.

Caveny, D.D., C.L. Quarles, and G.A. Greathouse. 1981. Atmospheric ammonia and broiler cockerel performance. *Poultry Sci.* 60:513-516.

Charles, D.R. and C.G. Payne. 1966a. The influence of graded levels of atmospheric ammonia on chickens. I. Effects on respiration and on the performance of broilers and replacement growing stock. *Br. Poult. Sci.* 7:177-187.

Charles, D.R. and C.G. Payne. 1966b. The influence of graded levels of atmospheric ammonia on chickens. II. Effects on the performance of laying hens. *Br. Poult. Sci.* 7:189-198.

Chaubey, I., D.R. Edwards, T.C. Daniel, and D.J. Nichols. 1993. Effectiveness of vegetative filter strips in controlling losses of surface-applied poultry litter constituents. Paper No. 932011, ASAE Meeting, June 1993, Spokane, WA.

Chaubey, I., D.R. Edwards, T.C. Daniel, P.A. Moore, Jr., D.J. Nichols. 1994. Effectiveness of vegetative filter strips in retaining surface-applied swine manure constituents. *Trans. ASAE* 37:845-850.

Cotterill, O.J. and A.R. Winter. 1953. Some nitrogen studies on built-up litter. *Poultry Sci.* 32:365-366.

Deaton, J.W., F.N. Reece, and B.D. Lott. 1984. Effect of atmospheric ammonia on pullets at point of lay. *Poultry Sci.* 63:384-385.

Doyle, R.C., D.C. Wolf, and D.F. Bezdicek. 1975. Effectiveness of forest buffer strips in improving the water quality of manure polluted runoff. *Managing Livestock Wastes*, Proc. 3rd Int. Symp., ASAE, St. Joseph, MI. pp. 292-302.

Duda, A.M. and D.S. Finan. 1983. Influence of livestock on nonpoint source nutrient levels of streams. *Trans. ASAE* 26:1710-1716.

Edwards, D.R. and T.C. Daniel. 1992a. Potential runoff quality effects of poultry manure slurry applied to fescue plots. *Am. Soc. Agric. Eng.* 35:1827-1832.

Edwards, D.R. and T.C. Daniel. 1992b. Environmental impacts of on-farm poultry waste disposal - A review. *Bioresource Technology* 41:9-33.

Edwards, D.R. and T.C. Daniel. 1993. Effects of poultry litter application rate and rainfall intensity on quality of runoff from fescuegrass plots. *J. Environ. Qual.* 22:361-365.

Edwards, D.R., T.C. Daniel, and O. Marbun. 1992. Determination of best timing for poultry waste disposal: A modeling approach. *Water Res. Bull.* 28:487-494.

Edwards, D.R., T.C. Daniel, P.A. Moore, Jr., and A.N. Sharpley. 1994. Solids transport and erodibility of poultry litter surface-applied to fescue. *Trans. ASAE* 37:771-776.

Effler, S.W., C.T. Driscoll, M.C. Wodka, R. Honstein, S.P. Devan, P. Jaran, and T. Edwards. 1985. Phosphorus cycling in ionically polluted Onondaga Lake, New York. *Water Air Soil Pollut.* 24:121-130.

Faddoul, G.P. and R.C. Ringrose. 1950. Avian keratoconjunctivitis. *Vet. Med.* 45:492-493.

Fontenot, J.P. and V. Jurubescu. 1980. Processing of animal waste by feeding to ruminants. *Digestive Physiology and Metabolism in Ruminants.* (Y. Ruckebusch and P. Thivend, eds.), AVI Publ. Co., Inc., Westport, CT.

Fontenot, J.P. and K.E. Webb, Jr. 1975. Health aspects of recycling animal wastes by feeding. *J. Anim. Sci.* 40:1267-1277.

Fontenot, J.P., K.E. Webb, Jr., K.G. Libke and R.J. Buehler. 1971. Performance and health of ewes fed broiler litter. *J. Anim. Sci.* 33:283.

Giddens, J. and A.P. Barnett. 1980. Soil loss and microbiological quality of runoff from land treated with poultry litter. *J. Environ. Qual.* 9:518-529.

Hileman, L.H. 1971. Effect of rate of poultry manure application on selected soil chemical properties. *Livestock Waste Management Pollution Abatement*, Proc. Int. Symp., 1971, pp. 247-248.

Hileman, L.H. 1973. Reclaiming land polluted by brine from oil fields. *Ark. Farm Res.* 22:12.

Huff, W.E., G.W. Malone, and G.W. Chaloupka. 1984. Effect of litter treatment on broiler performance and certain litter quality parameters. *Poultry Sci.* 63:2167-2171.

Johnson, E.L., J.L. Nicholson, and J.A. Doerr. 1985. Effect of dietary copper on litter microbial population and broiler performance. *British Poultry Sci.* 26:171-177.

Johnston, N.L., C.L. Quarles, D.J. Faberberg, and D.D. Caveny. 1981. Evaluation of yucca saponin on performance and ammonia suppression. *Poultry Sci.* 60:2289.

Kingery, W.L., C.W. Wood, D.P. Delaney, J.C. Williams, and G.L. Mullins. 1994. Impact of long-term application of broiler litter on environmentally related soil properties. *J. Environ. Qual.* 23:139-147.

Kingery, W.L., C.W. Wood, D.P. Delaney, J.C. Williams, G.L. Mullins, and E. van Santen. 1993. Implications of long-term land application of poultry litter on tall fescue pastures. *J. Prod. Agric.* 6:390-395.

Kitai, K. and A. Arakawa. 1979. Effect of antibiotics and caprylohydrozamic acid on ammonia gas from chicken excreta. *Bri. Poultry Sci.* 20:55.

Kling, H.F. and C.L. Quarles. 1974. Effect of atmospheric ammonia and the stress of infectious bronchitis vaccination on Leghorn males. *Poultry Science* 53:1161-1167.

Liebhardt, W.C. 1976. Soil characteristics and corn yield as affected by previous applications of poultry manure. *J. Environ. Qual.* 5:459-462.

Malone, G.W., J.T. Sims, and N. Gedamu. 1992. *Quantity and Quality of Poultry Manure Produced under Current Management Programs.* Final report to Delaware Dept. Natural Resources & Environ. Control. p. 96.

McCaskey, T.A. and W.B. Anthony. 1979. Human and animal health aspects of feeding livestock excreta. *J. Anim. Sci.* 48:163-177.

McCollum, R.E. 1991. Buildup and decline in soil phosphorus: 30-year trends on a Typic Umprabuult. *Agron. J.* 83:77-85.

McCoy, J.L., L.J. Sikora, and R.R. Weil. 1986. Plant availability of phosphorus in sewage sludge compost. *J. Environ. Qual.* 15:403-409.

Miller, D.M., B.R. Wells, R.J. Norman, and T. Alvisyahrin. 1991. Response of lowland rice to inorganic and organic amendments on soils disturbed by grading in eastern Arkansas. pp. 57-61. In: T.C. Keisling (ed.), *Proc. of the 1991 Southern Conservation Tillage Conference*. Ark. Agri. Exp. Sta., Fayetteville, AR.

Moore, P.A., Jr. and D.M. Miller. 1994. Reducing phosphorus solubility in poultry litter with calcium, iron and aluminum amendments. *J. Environ. Qual.* 23:325-330.

Moore, P.A., Jr., T.C. Daniel, D.R. Edwards, and D.M. Miller. 1995a. Effect of chemical amendments on ammonia volatilization from poultry litter. *J. Environ. Qual.* 24:293-300.

Moore, P.A., Jr., T.C. Daniel, C.W. Woods, and A.N. Sharpley. 1995b. Poultry manure management. *J. Soil and Water* 50:321-327.

Moore, P.A., Jr., T.C. Daniel, A. Waldroup, and D.R. Edwards. 1995c. Evaluation of aluminum sulfate on broiler litter characteristics and broiler production in commercial houses. *Poultry Sci.* 74 (supplement 1):135.

Moore, P.A., Jr., T.C. Daniel, D.R. Edwards, and D.M. Miller. 1996a. Evaluation of chemical amendments to inhibit ammonia volatilization from poultry litter. *Poultry Sci.* (in press).

Moore, P.A., Jr., J.M. Balog, G.R. Bayyari, N.C. Rath, W.E. Huff, and N.B. Anthony. 1996b. Effect of litter amendments on ammonia volatilization and broiler production. *Poultry Science* (in prep.).

Nagaraja, K.V., D.A. Emery, K.A. Jordan, V. Sivanandan, J.A. Newman, and B.S. Pomeroy. 1983. Scanning electron microscopic studies of adverse effects of ammonia on tracheal tissues of turkeys. *Am. J. Vet. Res.* 44:1530-1536.

Nagaraja, K.V., D.A. Emery, K.A. Jordan, V. Sivanandan, J.A. Newman, and B.S. Pomeroy. 1984. Effect of ammonia on the quantitative clearance of *Escherichia coli* from lungs, air sacs, and livers of turkeys aerosol vaccinated against *Escherichia coli*. *Am. J. Vet. Res.* 45:392-395.

Nakaue, H.S., J.K. Koelliker, and M.L. Pierson. 1981. Studies with clinoptilolite in poultry. 2. Effect of feeding broilers and the direct application of clinoptilolite (zeolite) on clean and re-used broiler litter on broiler performance and house environment. *Poultry Sci.* 60:1221.

National Academy of Sciences. 1980. *Mineral Tolerance of Domestic Animals*. National Academy Press, Washington, D.C.

Nelson, T.S. 1967. The utilization of phytate phosphorus by poultry - a review. *Poultry Sci.* 46:862-871.

Nelson, T.S., T.R. Shieh, R.J. Wodzinski, and J.H. Ware. 1968. The availability of phytate phosphorus in soybean meal before and after treatment with a mold phytase. *Poultry Sci.* 47:1842-1848.

Nelson, T.S., T.R. Shieh, R.J. Wodzinski, and J.H. Ware. 1971. Effect of supplemental phytase on the utilization of phytate phosphorus by chicks. *J. Nutr.* 101:1289-1294.

Oyetunde, O.O.F., R.G. Thomson, and H.C. Carlson. 1978. Aerosol exposure of ammonia, dust and *Escherichia coli* in broiler chickens. *Can. Vet. J.* 19:187-193.

Parkhurst, C.R., P.B. Hamilton, and G.R. Baughman. 1974. The use of volatile fatty acids for the control of micro-organisms in pine sawdust litter. *Poultry Sci.* 53:801.

Pote, D.H., T.C. Daniel, D.R. Edwards, J.D. Mattice, and D.B. Wickliff. 1994. Effect of land-applied caged-layer manure on cyromazine loss. *J. Environ. Qual.* 23:101-104.

Pote, D.H., T.C. Daniel, A.N. Sharpley, P.A. Moore, Jr., D.R. Edwards, and D.J. Nichols. 1995. Relating extractable soil phosphorus to phosphorus losses in runoff. Submitted to *Soil Sci. Soc. Am. J.*

Quarles, C.L. and H.F. Kling. 1974. Evaluation of ammonia and infectious bronchitis vaccination stress on broiler performance and carcass quality. *Poultry Sci.* 53:1592-1596.

Quisenberry, V.L., R.O. Hegg, L.E. Reese, J.S. Rice, and A.K. Torrence. 1981. Management aspects of applying poultry or dairy manures to grasslands in the Piedmont region. In: *Livestock Waste: A Renewable Resource.* Proc. 4th Int. Symp. on Livestock Wastes. ASAE, St. Joseph, MI, pp. 170-177.

Reddy, K.R., R. Khaleel, M.R. Overcash, and P.W. Westerman. 1979. A nonpoint source model for land areas receiving animal wastes: II. ammonia volatilization. *Trans. ASAE* 22:1398-1405.

Reece, F.N., B.J. Bates, and B.D. Lott. 1979. Ammonia control in broiler houses. *Poultry Sci.* 58:754.

Reece, F.N., B.D. Lott, and J.W. Deaton. 1980. Ammonia in the atmosphere during brooding affects performance of broiler chickens. *Poultry Sci.* 59:486-488.

Ritter, W.E. and A.E.M. Chirnside. 1982. Groundwater quality in selected areas of Ken and Sussex counties, Delaware. Delaware Agric. Exp. Stn. Project Completion Report.

Salter, R.M. and C.J. Schollenberger. 1939. *Farm Manure.* Ohio Agri. Exp. Sta. Bull. 605. pp. 69.

Sato, S., S. Shaya, and H. Kobayashi. 1973. Effect of ammonia on Mycoplasma gallisepticum infection in chickens. *Natn. Inst. Anim. Hlth. Qt., Tokyo* 13:45-53.

Scantling, M., A. Waldroup, J. Marcy, and P.A. Moore, Jr. 1995. Microbiological effects of treating poultry litter with aluminum sulfate. *Poultry Sci.* (in press).

Schilke-Gartley, K.L. and J.T. Sims. 1993. Ammonia volatilization from poultry manure-amended soils. *Biol. Fertil. Soils* 16:5-10.

Schindler, D.W. 1977. The evolution of phosphorus limitation in lakes. *Science* 195:260-262.

Schindler, D.W. 1978. Factors regulating phytoplankton production and standing crop in the world's freshwaters. *Limnol. Ocean.* 23:478-486.

Schroder, H. 1985. Nitrogen losses from Danish agriculture - Trends and consequences. *Agri. Ecosystems and Environment* 14:279-289.

Seltzer, W., S.G. Moum, and T.M. Goldhaft. 1969. A method for the treatment of animal wastes to control ammonia and other odors. *Poultry Science* 48:1912-1918.

Sharpley, A.N. 1995. Dependence of runoff phosphorus on soil phosphorus content. *J. Environ. Qual.* (submitted).

Sharpley, A.N. and R.G. Menzel. 1987. The impact of soil and fertilizer phosphorus on the environment. *Adv. Agron.* 41:297-234.

Sharpley, A.N., S.J. Smith, and W.R. Bain. 1993. Effect of poultry litter application on the nitrogen and phosphorus content of Oklahoma soils. *Soil Sci. Soc. Am. J.* (in press).

Shreve, B.R., P.A. Moore, Jr., T.C. Daniel, D.R. Edwards, and D.M. Miller. 1995. Reduction of phosphorus in runoff from field-applied poultry litter using chemical amendments. *J. Environ. Qual.* 24:106-111.

Simpson, T.W. 1991. Agronomic use of poultry industry waste. *Poult. Sci.* 70:1126-1131.

Sims, J.T. 1992. Environmental management of phosphorus in agriculture and municipal wastes. p. 59-64. In: F.J. Sikora (ed.), *Future Directions for Agricultural Phosphorus Research.* Nat. Fert. Environ. Res Cent., TVA, Muscle Shoals, AL.

Sims, J.T. 1993. Environmental soil testing for phosphorus. *J. Prod. Agric.* 6:501-507.

Sims, J.T. and D.C. Wolf. 1994. Poultry manure management: Agricultural and environmental issues. *Advances Agron.* 52:1-83.

Sims, J.T., D. Palmer, J. Scarborough, and R. Graeber. 1989. Poultry manure management. Coop. Bull. No. 24, Delaware Coop. Extn. Ser., University of Delaware, Newark, DE. p. 16.

Smith, L.W. and W.E. Wheeler. 1979. Nutritional and economic value of animal excreta. *J. Anim. Sci.* 48:144-156.

Sonzogni, W.C., S.C. Chapra, D.E. Armstrong, and T.J. Logan. 1982. Bioavailability of phosphorus inputs to lakes. *J. Environ. Qual.* 11:555-563.

Soon, Y.K. and T.E. Bates. 1982. Extractability and solubility of phosphate in soils amended with chemically treated sewage sludges. *Soil Sci.* 134:89-96.

Stephenson, A.H., T.A. McCaskey, and B.G. Ruffin. 1990. A survey of broiler litter composition and potential value as a nutrient resource. *Biol. Wastes* 34:1-9.

Stuedemann, J.A., S.R. Wilkinson, D.J. Williams, H. Giodia, J.V. Ernst, W.A. Jackson, and J.B. Jones, Jr. 1975. Longterm broiler litter fertilization of tall fescue pastures and health and performance of beef cows. In: *Managing Livestock Wastes*, 264-268. Proc. 3rd. Int. Symp. Livest. Wastes. ASAE Pub. Proc., 275. ASAE, St. Joseph, MI.

Tufft, L.S. and C.F. Nockels. 1991. The effects of stress, *Escherichia coli*, dietary EDTA, and their interaction on tissue trace elements in chicks. *Poultry Sci.* 70:2439-2449.

USDA. 1991. *Agricultural Statistics 1991.* U.S. Government Printing Office, Washington, D.C. 524 pp.

U.S. Environmental Protection Agency (EPA). 1985. National primary drinking water regulations: Synthetic organic chemicals, inorganic chemicals, and microorganisms: Proposed Rule. *Fed. Regist.* 50:46935-47022.

U.S. Environmental Protection Agency (EPA). 1987. *Ammonia Emission Factors for the NAPAP Emission Inventory.* Research Triangle Park, North Carolina, Office of Research and Development (Report No. PB87-152336).

U.S. Environmental Protection Agency (EPA). 1994. *National Water Quality Inventory. 1992 Report to Congress.* US EPA 841-R-94-001. March, 1994. U.S. Govt. Printing Office, Washington, D.C.

Valentine, H. 1964. A study of the effect of different ventilation rates on the ammonia concentrations in the atmosphere of broiler houses. *Br. Poultry Sci.* 5:149-159.

van Breemen, N., P.M.A. Boderie, and H.W.G. Booltink. 1989. Influence of airborne ammonium sulfate on soils of an oak woodland ecosystem in the Netherlands: Seasonal dynamics of solute fluxes. In D.C. Adriano and M. Havas (eds.), *Acid Precipitation.* Vol 1. Case Studies. Springer-Verlag.

van Breemen, N., P.A. Burrough, E.J. Velthorst, H.F. van Dobben, T. de Wit, T.B. Ridder, and H.F.R. Reijinders. 1982. Soil acidification from atmospheric ammonium sulphate in forest canopy throughfall. *Nature* 299:548-550.

Vollenweider, R.A. 1975. Input-output models with special reference to the phosphorus loading concept in limnology. *Schweiz. Z. Hydrol.* 37(1):53-84.

Wallingford, G.W., W.L. Powers, and L.S. Murphy. 1975. Present knowledge on the effect of land application of animal waste. p. 580-582, 586. In: *Managing Livestock Wastes,* Proc. 3rd Int. Sym. on Livestock Wastes. Urbana-Champaign, IL. 21-24 April 1975. Am. Soc. Agric. Eng., St. Joseph, MI.

Wallner-Pendleton, E., D.P. Froman, and O. Hedstrom. 1986. Identification of ferrous sulfate toxicity in a commercial broiler flock. *Avian Diseases* 30:430-432.

Weil, R.R., W. Kroontje, and G.D. Jones. 1979. Inorganic nitrogen and soluble salts in a Davidson clay loam used for poultry manure disposal. *J. Environ. Qual.* 8:387-392.

Wilkinson, S.R. 1979. Plant nutrient and economic value of animal manures. *J. Animal Sci.* 48:121-133.

Wilkinson, S.R. and J.A. Stuedemann. 1990. Fate of copper in broiler litter applied to Coastal bermudagrass (*Cymodon dactylon* L. pers.) and Tall Fescue (*Festuca arundinacea* Shreb). *Agron. Abstracts* 1990 Annual Meetings, San Antonio, TX, 21-26 October, 1990, p. 17.

Wilkinson, S.R., J.A. Stuedemann, D.J. Williams, J.B. Jones, Jr., R.N. Dawson, and W.A. Jackson. 1971. Recycling broiler litter on tall fescue pastures at disposal rates and evidence of beef cow health problems. In: *Livestock Management and Pollution Abatement* Proc. Int. Symp. on Livestock Wastes. ASAE, St. Joseph, MI, pp. 321-328.

Williams, P.E.V. 1992. Socio constraints on poultry production - Addressing environmental and consumer concerns. pp. 14-29. In: *1992 Proc. Ark. Nutrition Conference*, Fayetteville, AR.

Witter, E. and H. Kirchmann. 1989. Effects of addition of calcium and magnesium salts on ammonia volatilization during manure decomposition. *Plant and Soil* 115:53-58.

Wood, C.W. 1992. Broiler litter as a fertilizer: Benefits and environmental concerns. p. 304-312. In: *Proc. National Poultry Waste Management Symposium*, Birmingham, AL. Auburn University, AL.

Cattle Feedlot Manure and Wastewater Management Practices

John M. Sweeten

I. Introduction

Regulatory requirements, research, and waste-management system operating practices have been developed in the United States over the last 25 years to address environmental concerns regarding cattle feedlots. To prevent potential pollution problems to both water and air, feedlot operators must be proactive to establish technically sound programs and systems to: control rainfall runoff, manage solid manure, maintain the feedlot surface, operate wastewater handling facilities, and utilize manure and wastewater on agricultural land at beneficial rates.

Table 1. Annual yield and concentrations of nitrogen and phosphorus in land runoff

Source	-----Total nitrogen-----		----Total phosphorus----	
	ppm	kg ha^{-1}	ppm	kg ha^{-1}
Grazed pastures	---	0.6-9.7	---	0.4-4.6
Precipitation	1.2-2.3	5.6-10.0	0.02-0.04	0.04-0.06
Forested land	0.3-1.8	3.0-13.0	0.01-0.11	0.03-1.0
Cropland runoff	9	0.1-13	0.02-1.7	0.06-2.9
Irrigated cropland in western U.S.	0.6-2.2	3.0-27.0	0.2-0.4	1.0-4.4
Urban land drainage	3	7.1-9.0	0.2-1.1	1.1-5.6
Feedlot runoff	920-2,100	100-1,600	290-360	10-620

(Adapted from Loehr, 1974 and Doran et al., 1981.)

Water pollution potential from cattle feedlots needs to be placed in perspective relative to the water quality effects of livestock grazing operations. In many cases, the effects of unconfined cattle on range or pasturelands cannot be discerned from natural or background levels of water pollution without livestock. The annual amount of nutrients transported off of grazed pastureland (Table 1) ranges from 0.56 to 9.7 kg nitrogen/ha/year, and 0.04 to 4.6 kg phosphorus/ha/year (Doran et al., 1981). These nutrient levels are comparable to values for cropland and input from forests and rainfall (Loehr, 1974). Runoff from livestock pastures often does not exceed nutrient levels in runoff from ungrazed pasturelands, forests, or dryland farms (Saxton et al., 1983). Nutrient losses and runoff amounts are greater for overgrazed pastures than for properly managed grazing systems (Smeins, 1977). Detectable water pollution from unconfined cattle operations does not appear to be related to cattle numbers or manure quantity, but rather to conditions that contribute to rapid surface runoff for sediment movement (Dixon, 1983a). Unconfined livestock may decrease vegetative cover and increase runoff, erosion, transport of sediment, plant nutrients and oxygen demand.

The most common change in stream water quality from unconfined livestock production is elevated concentrations of bacteria and sediment (Milne, 1976; Saxton et al., 1983). Chemical pollutants are sometimes increased slightly but do not usually exceed federally-approved stream quality standards. Watersheds containing cattle grazing sometimes show increased concentrations of bacterial indicator organisms, primarily coliforms and streptococcus, in the adjacent streams (Dixon, 1983a and 1983b; Milne, 1976). Reported effects are erratic and detectable only for short distances downstream. Fecal deposits along drainage ways may contribute a disproportionate share of the bacteria from grazed watersheds.

At high-impact feeding and watering sites, the sediment load can be minimized by riparian zone management practices, which include protecting fragile stream banks, maintaining vegetative cover, restricted or controlled grazing, distributing salt and water, providing feed, salt or water away from streams (Sweeten and Melvin, 1985).

II. Runoff from Cattle Feedlots

A. Water Quality Management

Runoff from cattle feedlots contains high concentrations of nutrients, salts, pathogens and oxygen-demanding organic matter, measured as BOD (biochemical oxygen demand) or COD (chemical oxygen demand) (USEPA, 1973; Reddell and Wise, 1974). Some typical concentrations of cattle feedlot runoff are shown in Table 2 (Clark et al., 1975a) and Table 3 (Sweeten et al., 1981; Clark et al., 1975b). Other researchers (Wells et al., 1969; USEPA, 1973) showed that BOD_5 concentrations of around 2,000 mg/L (about 8 times the concentration in raw domestic sewage) were commonplace. Feedlot runoff can contain 100 times more nitrogen and phosphorus than runoff from grazing land (Table 1). When feedlot runoff enters streams, the excess organic matter and nutrients can cause oxygen depletion and eutrophication, which leads to fish kills (Paine, 1973).

Feedlots in the Great Plains and Southwestern United States, beginning the late 1960s and 1970s, have had to control discharges and meet state and/or federal regulations that do not allow any discharge of wastewater from off the feedlot property. Several cattle feeding states including Texas had instituted individual permit programs by the early 1970s that are still in effect.

The U.S. Environmental Protection Agency (USEPA) adopted feedlot effluent guidelines (1974) requiring no-discharge and a federal permit system (1976) for feedlots over 1,000 head that discharge from less than a 25-year, 24-hour duration storm event. A feedlot/concentrated animal feeding operation (CAFO) is defined as a (1) manure-covered surface that does not sustain the growth of forage, crops or other vegetation, (2) feeding facility within a fence or enclosure, and (3) cattle feeding operation maintained for 45 days or more per year.

The Texas Natural Resource Conservation Commission (TNRCC), formerly the Texas Water Commission, developed and implemented a set of state regulations (TWC, 1987) with active participation and operation of livestock commodity groups and agricultural agencies. This Subchapter B regulation stated that it is the policy of the State of Texas that there shall be no discharge from livestock feeding facilities, but rather the animal waste materials must be collected and utilized or disposed of on agricultural land. This regulation has three major types of requirements: (1) surface water protection, (2) groundwater protection, and (3) proper land application of manure and wastewater. Beef cattle feedlots with over 1,000 head of cattle on feed have to get a state permit (comparable numbers apply to other animal species). With less than 1,000 beef cattle on feed, feedlots still have to meet the no-discharge requirements for water pollution control but do not have to obtain a permit. Local governmental involvement in permits includes the opportunity to respond to public notification, review and comment.

Table 2. Average chemical characteristics of runoff from beef cattle feedyards in the Great Plains

Location	Total solids	Chemical oxygen demand	Total nitrogen	Total phosphorus	Potassium	Sodium	Calcium	Magnesium	Chloride	Electrical conductivity
					ppm					mmhos cm^{-1}
Bellville, TX	9,000	4,000	85	85	340	230	--	--	410	--
Bushland, TX	15,000	15,700	1,080	205	1,320	588	449	199	1,729	8.4
Ft. Collins, CO	17,500	17,800	--	93	--	--	--	--	--	8.6
McKinney, TX	11,430	7,210	--	69	761	408	698	69	450	6.7
Mead, NE	15,200	3,100	--	300	1,864	478	181	146	700	3.2
Pratt, KS	7,500	5,000	--	50	815	511	166	110	--	5.4
Sioux Falls, SD	2,990	2,160	--	47	--	--	--	--	--	--

(Adapted from Clark et al., 1975a.)

Table 3. Average concentration of nutrients, salts, and other water quality parameters from stored cattle feedlot runoff in Texas[a]

Water quality parameters	Texas High Plains[b]			South Texas[c]
	Fresh runoff	Holding ponds	Playa basin	Holding ponds
Nitrogen, ppm	1,083	145	20	180
Phosphorus, ppm	205	43	12	--
Potassium, ppm	1,320	445	60	1,145
Sodium, ppm	588	256	54	230
Calcium, ppm	449	99	55	180
Chloride, ppm	199	72	30	20
COD, ppm	1,729	623	86	1,000
Total solids, ppm	--	--	--	1,100
Sodium absorption ration (SAR)	--	--	--	2,470
	5.3	4.6	1.4	4.2
Electrical conductance (mmhos/cm)	8.4	4.5	1.0	4.5

[a]Quality of runoff after storage in the runoff holding pond for several weeks; playa basins typically catch runoff from areas other than the feedlot; thus, there is greater dilution effect.
[b]From Clark et al., 1975b.
[c]From Sweeten et al., 1981.

To obtain a permit, feedlot operators have to estimate daily and annual production of manure, wet and dry basis, and the major constituents including volatile solids, nutrients (N,P,K), salts (Na, Cl, etc.) and oxygen demand. The calculations involve a nutrient balance from manure collection through land application. The conventional basis for these calculations is a set of standards "Manure Production and Characteristics," D384.1 (ASAE, 1988). In Texas, the mean plus one standard deviation is used for all manure production and constituent calculations. In a confinement building, semisolid or liquid manure and wastewater need to be collected and structures provided with a minimum capacity that increases from west to east across the state according to annual rainfall and evaporation.

Surface water protection measures for open, dirt-surfaced feedlots include diverting the clean water around the feedlot, providing pen drainage, collecting the rainfall runoff, and dewatering of holding ponds or lagoons. Man-made runoff retention facilities must be built above the 100-year flood plain and must be dewatered within 21 days after they are half full or more with rainfall runoff. Irrigation is the most effective and popular means of dewatering and provides for beneficial use of wastewater. In dry areas, evaporation is an alternative. Numerous large cattle feedlots were built adjacent to natural playa basins which are used for runoff collection, evaporation and/or irrigation.

The USEPA (1993) adopted a general permit for Concentrated Animal Feeding Operations (CAFOs) in Texas, Louisiana, Oklahoma, and New Mexico. The general permit requires CAFO's with more than 1,000 animal units (feeder cattle equivalents) to come under the general permit. As well, those operations with 300 or more animal units and that discharge wastewater through a man-made convey-ance structure should come under the general permit. The general permit requires design, implementation, and maintenance of best management practices (BMPs) for control of rainfall runoff, manure, and process wastewater including overflow cattle drinking water; prevention of a hydrologic connection to surface waters; and application of manure and wastewater onto land at nutrient loading rates compatible with agronomic considerations. The general permit requires the operator to develop, follow and maintain an accurate pollution prevention plan (PPP) that includes a well-designed nutrient management plan, certification of lagoon or holding pond liners, inspection and maintenance records, dewatering records, employee training logs, spill prevention, monitoring and self-reporting of discharges that could enter waters of the United States.

Appreciable movement of nutrients and salts beneath typical playas used as cattle feedlot wastewater catchment, storage and evaporation ponds has not occurred (Smith et al., 1994). Legislation was adopted in Texas in 1993 that prohibited the future conversion of playa basins for wastewater retention or treatment facilities. This legislation had the effect of eliminating construction of feedyards on playa basins if the basin is to be used as the runoff retention facility.

B. Runoff Collection

Runoff holding ponds must be designed to collect and store all runoff from a 25-year frequency, 24-hour duration storm. The design rainfall event is approximately 125 mm in the main cattle feeding regions of the Southern High Plains. To calculate runoff from the design storm, the TNRCC regulation (TWC, 1987) requires using the USDA-Natural Resources Conservation Service (NRCS) soil cover complex curve number 90. For a design rainfall of 125 mm from the 25-year, 24-hour storm, the design runoff is about 97 mm, which is between 75 and 80% of the rainfall. Research was conducted to develop feedlot runoff vs. rainfall relationships in the 1970s by state Agricultural Experiment Stations and the Agricultural Research Service-USDA when the feedlots were faced with controlling water pollution (Gilbertson et al., 1981; Clark et al., 1975a). Researchers determined that it takes about 13 mm of rainfall to induce runoff from a cattle feedlot (Gilbertson et al., 1980). Thereafter, the rainfall versus runoff relationships predict less runoff per inch of rainfall in dry climates than in wetter climates (Clark et al., 1975a). Nevertheless, holding ponds should be designed using the USDA-NRCS Curve 90 as it provides a more conservative estimate.

Lott et al. (1994b) found that rainfall retained on the feedlot surface varied from 1 to 19 mm per event. Lott (1995) measured runoff yields, expressed as a percentage of rainfall, of 50%, 39%, and 22% for high, medium, and low annual rainfall sites near Toowoomba, Queensland, Australia. The annual amount of runoff expected is about 20 to 33% of rainfall in the Great Plains cattle feeding regions according to calculations of Phillips (1981), based on NRCS Curve 90 and weather records. Therefore, a 40-ha feedlot in a 450 mm rainfall area will produce an average of 42,000 m³ of runoff per year to be disposed of by irrigation.

For a 40-ha feedlot module, multiplying the runoff volume of 42,000 m³ per year by concentration data for fresh runoff at Bushland in Table 2 (Clark et al., 1975a and 1975b) yielded an estimate of mass loading rates into a holding pond or playa of approximately 630 Mg per year of total solids, 660 Mg per year of chemical oxygen demand (COD), 46 Mg per year of total nitrogen, 9 Mg per year of total phosphorus, 56 Mg per year of potassium, 25 Mg per year of sodium, and 73 Mg per year of chloride (Sweeten, 1994). The total solids loading rate represents less than 3% of the total solids generated by cattle in the feedlot.

C. Seepage Control

Groundwater quality is protected by feedlots in accordance with state regulations. Standards for seepage control at existing feedlots require that a runoff holding pond or lagoon be built in, or lined with, at least 30 cm compacted thickness of soil material with 30% or more passing a No. 200 mesh sieve, a liquid limit (LL) of 30% or more and a plasticity index (PI) of 15 or more (TWC, 1987). These three criteria basically require a sandy clay loam, clay-loam or clay soil and together are consistent with attaining a coefficient of hydraulic conductivity of around 1×10^{-7} centimeters per second (cm/sec) which has usually been stipulated in permits. USEPA (1993) specified a clay liner of 45 cm thickness with materials having a hydraulic conductivity of 1.0×10^{-7} cm/sec as one of the methods for establishing "no hydrologic connection" to waters of the U.S. New feedlots in Texas need to be built with a similar standard (TNRCC, 1995), and embankments should be constructed according to Technical Note 716 of the USDA Natural Resources Conservation Service (Walker, 1995) or other equivalent means.

Holding ponds and manure treatment lagoons have a tendency to be partially self-sealing (Sweeten, 1988b). For example, research from California involving an unlined cattle manure storage pond showed the initial seepage rate was 1.3×10^{-4} cm/sec but after six months, the seepage rate was reduced nearly a hundred-fold (i.e., two orders of magnitude) to 3.5×10^{-6} cm/sec (Robinson, 1973). Research in Canada showed that clogging of soil pores by bacterial cells and organic matter is the mechanism responsible for partial self-sealing (Barrington and Jutras, 1983). The initial fresh water infiltration rate in ten-foot deep holding ponds was 10^{-2}, 10^{-3} and 10^{-4} cm/sec for sand, clay and loam, respectively. After only two weeks of storage, the infiltration rates of dairy lagoon effluent were reduced to only 10^{-6} cm/sec in loam and sandy soils as compared to $0-1.8 \times 10^{-6}$ cm/sec after a year for all three soils. It should be noted that the coefficients of hydraulic conductivity finally attained are still well short of USEPA (1993) and TNRCC (1995) requirements. Clay liners are important to reduce the movement of chemicals beneath manure storage ponds. Beneath a dairy manure storage pond in Canada, nitrate concentrations at five to ten feet soil depth below the pond were 0.4 mg/l for a clay soil, 1.2 mg/L for a loam soil and 17 mg/L for a sandy soil (Phillips and Culley, 1985). Seepage beneath feedlot runoff holding ponds has shown reduced seepage (Lehman and Clark, 1975) and very little nitrate or chloride movement occurs (Lehman et al., 1970; Clark, 1975). After five years, nitrates had leached less than three feet below the bottom of a natural playa lake with about 1 m thickness of montmorillonite clay bottom used to collect feedlot runoff. This study indicated minimal groundwater hazard from nitrate contamination. In a subsequent field research project, Smith et al. (1994) found that nitrate did not leach below 3 m on 3 playa basins in the Amarillo area.

Miller (1971) found measurable nitrate in the groundwater beneath 22 feedlots out of more than 80 sampled from monitoring wells into the Ogallala Aquifer in the Texas High Plains, where groundwater depth was typically 30 to 90 m. Nitrate-nitrogen concentrations ranged from less than one to twelve mg/L and were higher below the holding ponds than beneath the feedlot surface. Sweeten et al. (1990 and 1995a) found nitrate-nitrogen concentrations of 0.25 to 9.1 mg/L in wells within and around 26 cattle feedlots in 4 counties southwest of Amarillo. Wells downgradient from two feedlots that used playa basins as feedlot runoff holding ponds did not show higher concentrations of nitrate or salinity than upgradient wells or feedlot water supply wells on premises (Sweeten et al., 1995a).

D. Managing Settled Solids in Holding Ponds

Feedlot runoff control systems usually consist of a series of small settling basins that overflow into one or more large runoff retention ponds that need to be pumped down rapidly after each runoff event (Shuyler et al., 1973; Swanson et al., 1973; Sweeten, 1985b; Lott, 1995). Settling basins should be cleaned out promptly to restore capacity, remove wet sediment; and reduce odor and fly breeding sites. Settling basins and channels should be shallow, broad, and free-draining if possible with an outlet that does not clog with debris so that sediment is spread in shallow layers over a wide area to promote drying conditions and facilitate removal.

Holding pond sediment is part manure and part soil with nutrient values of N, P, and K that are similar to or slightly less than feedlot manure on a dry matter basis (Sweeten and Amosson, 1995; Sweeten, 1990b). After the holding pond effluent is removed by pumping, and after several months of drying, the sediment will form a dried crust leaving wet or semisolid material (for example, 80 percent moisture content) in the interior that is difficult to dry further or remove efficiently. Methods of sediment removal from sediment basins and runoff holding ponds include dragline, dozer, wheel loader, elevating scraper, floating dredge and slurry agitation/pumping (Sweeten et al., 1981; Lindemann et al., 1985; Sweeten and McDonald, 1979). Because hauling and spreading is usually a slow process due to limited numbers of dump trucks, spreader trucks or wagons available at one time, the sediment that cannot be hauled immediately should be placed or stacked in a drying area that drains back into the basin or holding pond for runoff control. After further drying, which may require several weeks, it needs to be reloaded, hauled and spread. Composting in place using windrow equipment is another good management tool to speed drying and spreadability. Wet sediment may require several turnings before aerobic conditions are established, after which rapid heating and moisture loss will occur.

In a field study, runoff sediment sampled from the dragline bucket contained only 17% total solids content wet basis (85% moisture) (Sweeten et al., 1981). About 55 percent of the total solids was volatile (biodegradable) solids and 45 percent was ash. The nutrient content in runoff holding pond sediment was substantial with more than three percent total nitrogen and over one percent phosphorus. Potassium and sodium were usually low because they are leached out with the liquid fraction, so the potential salt hazard was less than or similar to the runoff itself (Sweeten et al., 1981; Sweeten, 1990b).

One of the better ways to remove sediment from settling basins that are not free-draining or from small holding ponds is to use a commercially-available propeller agitator on a 6 to 7.3 m shaft (PTO driven) that will agitate and homogenize the sediment into a slurry for irrigation through a traveling or portable big-gun sprinkler. Field demonstrations have been conducted in which sediment with 3 to 14% solids content was applied at a depth of 12 to 19 mm, which provided adequate fertilization of coastal Bermuda grass (Lindemann et al., 1985).

To avoid or postpone having to clean out large holding ponds, feedlots should use sedimentation basins or traps that allow solids to settle in smaller, more accessible locations, which are conducive to rapid drying and frequent collection when using conventional equipment such as a wheel loader and spreader truck. Types of settling basins include shallow earthen basins or concrete pits 0.5 to 1.0 m deep with a grooved concrete entry ramp for solids removal. Outlets consist of a buried culvert with vertical riser pipe with perforations or vertical slot openings 19 to 25 mm wide (Loudon et al., 1985) protected by an expanded metal trash rack or weirs. Another method of solids settling is an earthen channel with less than 1% slope placed just outside feedpens. The settling channel should discharge supernatant to a holding pond through a screen-wire baffle, vertical slot inlet pipe, vertical planks, highway guard rail (horizontal), or other similar types of baffle outlets that allow complete drainage rather than ponding of effluent. Settling systems on research feedlots provided 70 to 80% reduction in total suspended solids using a relatively flat channel with a simple screen-wire baffle attached to a wooden frame (Swanson et al., 1977). The outlet section should be concrete-lined to anchor the outlet devices and allow mechanical cleaning (Loudon et al., 1985).

Runoff settling channels should be designed with 0.5 hour detention time for runoff from the 10-year, one-hour storm (Loudon et al., 1985). For example, if the design storm is 63 mm per hour, the debris basin or settling channel should have a capacity equivalent to 12 to 30 mm depth over the contributing watershed. The flow velocity should be less than one foot/second to deposit suspended solids.

As a general rule, a shallow broad basin with long flow path will be more effective than a short, deep, narrow settling basin. A shallow design will reduce velocity and reduce the vertical distance of particle fall into the sediment layer where it is captured. For example, a study was performed on a 330 m^3 settling basin on an open lot dairy farm in Texas with two concrete-lined parallel settling basins about 12 m wide by 24 m long and about 1 m deep. Flow of both liquid manure and open lot runoff was diverted alternatively into one half of the settling basin while the other side drained. A 15 to 60% reduction (45% average) in volatile solids concentration was measured (Sweeten and Wolfe, 1993).

Gilbertson et al. (1979a) concluded that a runoff sediment basin adjacent to the feedlot should have a volume equivalent to 32 mm runoff depth (32 mm ha^{-1} of feedlot surface). If the settling basin is remote from the holding pond, the volume should be as much as 75 mm-ha per ha (i.e., 75 mm equivalent runoff depth). The conditions under which this research was conducted were rather severe, for example, 15% slope, cattle spacing 28 to 37 square feet per head, and humid climate with cold winter weather. Thus, sediment basins designed for 32 to 38 mm should be adequate for most feedlot conditions. In the Nebraska studies, the amount of sediment removed by debris basins located outside of feed pens was 0.014 to 0.04 Mg of dry sediment per m^3 of runoff processed. On a average annual basis, this sediment volume was equivalent to a depth of around 11 to 41 mm across the feedlot.

Building sediment basins inside feed pens is not recommended. Instead water should be rapidly drained out of pens, to minimize problems with mud and odor (Watts and Tucker, 1993a).

E. Land Application of Feedlot Runoff

Feedlot runoff collected in holding ponds needs to be disposed of by land application and/or evaporation. Feedlot runoff characteristics are illustrated in Tables 2 and 3. Sprinkler irrigation is the preferred approach to land application of feedlot runoff. Sprinklers can apply as little as 12 mm per application, if necessary, to prevent runoff. With furrow irrigation, it is difficult to apply less than 75 to 100 mm to get complete coverage of a field, and this usually creates a tailwater problem in fine textured soils, resulting in high application rates for nutrients and salts. Level borders are a good way to apply feedlot effluent (Sweeten et al., 1995b). The application rate can be controlled on laser-leveled borders to 75 to 100 mm per irrigation with uniform distribution and prevent tailwater.

Serpentine waterways have been shown to be effective in removal of up to 80% of the solids and chemical oxygen demand from feedlot runoff for relatively small feedlots, e.g., below 1,000 head (Swanson et al., 1977). The runoff flowed through a settling basin to reduce the solids content and then entered the vegetated serpentine waterway which reduced runoff concentrations and volume discharged through a baffle or weir into the next terrace and similarly for succeeding terraces. Total size of the vegetated area should be one or two ha per ha of feedlot surface (Swanson et al. 1974 and 1977; Dickey et al., 1977).

Feedlot runoff application rates on crop or pasture land are usually limited either by nitrogen, salinity or sodium content (Butchbaker, 1973). Nitrogen concentrations of 89 to 364 mg/L, with 80 percent or more in the form of ammonium have been reported (Sweeten, 1990a and 1990b). Feedlot runoff stored in holding ponds generally has an electrical conductivity (EC) of one to ten millimhos per centimeter (mmhos/cm), depending on factors such as cattle ration and degree of evaporation. Clark et al. (1975b) determined a mean value of 4.5 mmhos/cm for feedlot runoff stored in holding ponds in the Texas High Plains, and Butchbaker (1973) found a mean value of 5.5 mmhos/cm for Kansas feedlots. Most of the salinity is in the forms of potassium and chloride, although sodium and ammonium are also important parameters. A relationship between EC and soluble sodium percentage (SSP) for feedlot runoff can be used as a guideline for sodium and salinity hazard in the soil (Butchbaker, 1973).

Runoff held in evaporation ponds has shown extremely high salt concentrations with electrical conductivity of over 20 mmhos/cm, which equates to around 12,000 to 15,000 ppm of total dissolved solids (Sweeten, 1990b). Evaporation pond effluent may not be suitable for irrigation.

It is an asset for a feedlot to have irrigation water available and to add dilution water to reduce salt content when irrigating with feedlot runoff (Powers et al., 1973). Dilution ratios of 3:1 up to 10:1 may be needed depending on soil texture and characteristics of effluent and irrigation water (Sweeten, 1976).

Salt tolerance has been established for most crops (FAO, 1985; Stewart and Meek, 1977). Salinity levels in soil and applied effluent that will cause 10, 25 and 50% reduction in yields have been determined (Stewart and Meek, 1977). Salt tolerant crops such as sorghum, barley, wheat, rye and Bermuda grass are good choices for feedlot runoff application (Butchbaker, 1973), while corn is less salt tolerant but is a high nitrogen user.

Runoff may not be low enough in concentration to use without dilution on corn. Research in Kansas showed that about 250 mm of undiluted feedlot runoff applied per year for three years produced peak yields of corn forage, but beyond that level it began to reduce crop yield (Wallingford et al., 1974). By comparison cattle feedlots in Texas reported using 50 to 150 mm per year of undiluted runoff (Sweeten and McDonald, 1979).

Research on germination of crops that received feedlot runoff showed soybean germination was zero to 30%; and hence is a poor crop to choose for runoff disposal (Coleman et al., 1971). Cotton and grain sorghum germinated much better, especially for 50 to 150 mm applications of dirt lot runoff. Yields of cotton, grain sorghum and Bermuda grass were increased with feedlot runoff applications as compared to groundwater. Treatments were 25 to 50 mm of runoff from dirt or concrete surface feedlots or groundwater every two weeks during a 14-week irrigation season (Coleman et al., 1971), resulting in total applications of 175 and 350 mm for the year.

Wells et al. (1969) showed very high reductions in most constituents in leachate when 442 mm of feedlot runoff was applied to cotton, sorghum and Midland Bermuda grass and percolated through 75 cm of soil. As compared to fresh runoff, removals of BOD, COD, total nitrogen and volatile solids in soil-column, leachate were 99.5%, 95-98%, -48 to +76%, and 77-82%, respectively.

Sweeten et al. (1995b) found that application of 100 to 235 mm per year of undiluted feedlot runoff in level borders maintained a good stand of winter wheat over a 4-year period in the Southern High Plains. Soil chemical changes were evident in elevated electrical conductivity to final levels of 1.4, 1.8, and 1.3 mmhos/cm for 100, 170, or 235 mm of annual effluent irrigation as compared to control treatments (i.e., no effluent irrigation) of 0.4 mmhos/cm. Deep chiseling the soil profile after the third year of effluent application, followed by 50% higher than normal rainfall in the ensuing year helped control soil salinity to levels that would maintain wheat production.

III. Land Application of Manure

Feedlot manure contains two to 2.5% nitrogen, 0.3 to 0.8% phosphorus (P) and 1.2 to 2.8% potassium (K) on a dry weight basis (Mathers et al., 1973; Arrington and Pachek, 1981; Sweeten and Amosson, 1995). Manure application rates depend on many factors including manure analysis, nutrient availability, physical and chemical soil characteristics, type of crop, yield goal, soil drainage, climate, groundwater depth and geology. The feedlot manager should work with a professional agronomist to determine the proper application rate. Overapplication of feedlot manure can depress yields, increase fertilization costs, and increase surface and groundwater pollution potential by applying nutrients that are not taken up by crops.

In most cases, manure application rates should be selected on the basis of plant-available nitrogen and phosphorus (Gilbertson et al., 1979b). Soil fertility guides are available for specific crops and yield goals. For grain sorghum with a yield goal of 7,800 lbs/acre, each year the crop will require about 168 kg/ha N, 38 kg/ha P, and 110 kg/ha K. To supply these requirements with feedlot manure in which about 40 to 50% of the nitrogen is available, an application of manure of about eight tons/acre dry basis would be needed.

Research at Texas A&M University determined yields of corn silage and grain sorghum that resulted from feedlot manure application rates (at 36 to 51% moisture) applied at rates of 0, 22, 56, 112, 224, 336, 670, 1,340, and 2,000 Mg/ha for two years (Reddell, 1974). Peak yield occurred at the 56 Mg/ha application rate for sorghum grain and at 22 and 56 Mg/ha for corn silage. The highest application rate of 2,000 Mg/ha, in which manure was applied about 30 cm deep before plowing, reduced yields to 33 to 38 percent of the peak yields with lower rates. However, two annual applications of 336 Mg/ha or less did not significantly reduce yields of either crop.

Crop yield data (relative to check plot yields) from three cattle feeding states in the U.S. used manure application rates of 0 to 600 Mg/ha/year (dry basis) (Stewart and Meek, 1977). The effects of soluble salt on crop yields was studied at Brawley, California (desert climate), where peak yields of sorghum grain occurred at 74 Mg/ha and yield reduction occurred at higher rates of 74 to 290 Mg/ha. Near Amarillo, Texas (Bushland), irrigated sorghum yield on a Pullman clay loam soil peaked at a manure application rate of only 22 Mg/ha tons/acre then decreased at higher application rates due to minimal leaching. For corn silage in Kansas, yields peaked at 100 to 200 Mg/ha and decreased at 200 to 600 Mg/ha. The harmful salinity effects of these high application rates generally do not persist (Stewart and Meek, 1977).

Research at USDA-Agricultural Research Service involved applying feedlot manure at various application rates to sorghum near Amarillo, Texas over a seven-year period (Mathers et al., 1975). Yield of sorghum grain with no manure applied

Table 4. Value of feedlot manure in grain sorghum production, 1969-1973, Bushland, Texas

Annual treatment	Avg. yield	Yield increase[a]	Incremental yield value	
	kg/ha/yr	kg/ha/yr	$/ha/yr	$/Mg
Check (no fertilizer)	5,030	---	---	---
N(270 and 135 kg N/ha)	7,220	2,190	219	---
N-P-K (270 and 135 kg N/ha)	7,180	2,150	215	---
Manure (Mg/ha)				
22	7,440	2,410	241	10.76
67	7,270	2,240	224	3.33
134	7,130	2,100	210	1.56
270	5,740	710	71	0.26
540(3 yr treatment and 2 yr recovery)	1,010/ 7,620	-1,380	-138	-0.26
540 (1 yr treatment and 4 yr recovery)	370/ 7,560	1,094	109	0.20

[a]Assumes price of sorghum grain is $0.10/kg.
(From Mathers et al., 1975.)

was 5,030 kg/ha (Table 4). Two different rates of commercial fertilizer produced 7,180 to 7,220 kg/ha of sorghum, an increase of over 40%. When feedlot manure was applied at annual rates of 22, 67, 134, and 269 Mg/ha with an average of about 40 percent moisture content, grain yields increased as compared to no fertilizer. Peak yields occurred at 22 Mg/ha/year, which produced an average of 7,440 kg/ha per year. Cost benefits of the sorghum yield vs. application rate results of Mathers et al. (1975) were calculated by Sweeten (1984). The lowest application rate of 22 Mg/ha was the most cost-effective, yielding nearly $11 in grain per Mg of manure applied as compared to $3.85 to $4.40/Mg for manure hauling and spreading in most cases. However, at the 134 Mg/ha application rate and above, the incremental yield increase of $1.56 per Mg of manure is less than the hauling and spreading costs. Again, the greatest benefits resulted from using relatively low application rates that matched the crop fertilizer requirements. In the same study, at the highest application rate of 270 Mg/ha, soil nitrate concentration increased to about 50 ppm in the soil (Mathers and Stewart, 1971). Successively lower nitrate concentrations occurred with lower application rates. The lowest application rate gave about the same soil nitrate concentration as when no fertilizer was applied. Excess nitrogen applied and not taken up by the crop has potential for leaching.

The effects of manure on sorghum, corn and wheat for various time intervals at Bushland, Texas are shown in Table 5 (Mathers and Stewart, 1984). Residual benefits of manure at high rates on corn and wheat were evident in terms of sustained yields after application of manure ceased.

John M. Sweeten

Table 5. Crop yields from feedlot manure application, Bushland, Texas, 1969-1980, USDA-ARS

Manure treatment (Mg/ha)	Other fertilizer	Manure applied, continuous (yrs)	Recovery period after application (yrs)	Average yields, kg/ha		
				Sorghum grain 1969-1973	Corn 1975, 1977, 1979	Wheat 1976, 1978, 1980
0	--	11	0	5,030	9,360	1,570
0	N	11	0	7,220	15,000	4,540
0	NPK	11	0	7,180	15,200	4,810
22	--	11	0	7,440	15,600	3,840
67	--	11	0	7,270	15,020	5,080
134	--	5	6	7,130	16,070	4,480
270	--	5	6	5,740	15,630	4,770
540	--	3	8	3,650	17,100	4,850
540	--	1	10	6,125	13,560	3,150

(From Mathers and Stewart, 1984.)

Residual nitrogen in the soil from using livestock manure was also shown to benefit crop production in experiments at Auburn University in Alabama (Lund et al., 1975; Lund and Doss, 1980). Solid dairy manure was applied to Bermuda grass on two different soil types. A commercial fertilizer (N-P-K) application of 448-94-390 kg/ha was compared with three rates of dry manure: 45, 90, and 134 Mg/ha manure per year on two sandy loam soils. On soil type A, during the three years that manure was applied, slightly higher hay yield occurred with 90 and 134 Mg/ha manure than with commercial fertilizer. However, when they ceased applying any fertilizer for the next three years, the yield on the commercial fertilizer plots dropped by 54%, while that from the manure-treated plots kept increasing because of residual nutrients. Similarly, on soil type B there was little difference among the treatments in terms of crop yield for three years during application of fertilizer and manure except that the 45 Mg/ha produced the lowest yields. However, for the three years after fertilizer and manure treatment ceased, the commercial fertilizer plots had almost no yield (reduced by 84%), but yields were sustained on plots that received manure. Hence, for the six-year program, the manure treatments produced 18 to 88% more hay than commercial fertilizer on soil type B, and 31 to 85% more yield for soil type A, which illustrated the benefits of residual nutrient value from manure.

Manure can also correct certain micronutrient deficiencies, such as iron chlorosis in sorghum, which is caused by iron deficiency (Mathers et al., 1980). In research near Lubbock, Texas where feedlot manure was applied to a calcareous soil (Arch fine sandy loam), grain sorghum yields increased from 2,600 kg/ha with no manure to 6,960 kg/ha with 11 Mg/ha manure and to 6,520 kg/ha with 34 Mg/ha manure. Commercial N and P fertilizer caused a 49% yield decrease versus the control. Manure is a good form of fertilizer on calcareous soils due to slow release of nutrients and chelating of ions (Mathers et al., 1980; Thomas and Mathers, 1979).

A technical guide for determining proper manure application rates based on nitrogen content was developed by the USDA Agricultural Research Service (Gilbertson et al., 1979b). This technical guide takes into account the slow rate of release of organic nitrogen in manure and the nitrogen concentration on a dry basis. For example, suppose cattle manure contains 2% nitrogen (dry basis). Fifteen Mg/ha of manure dry basis are required the first year to supply 112 kg/ha of available nitrogen. In succeeding years, release of residual organic nitrogen lowers the manure requirement to 13 Mg/ha in the second year and to 9.9 Mg/ha in the fifth year. Because of nitrogen losses after manure is applied to soil, application rates should be increased by approximately one-third if manure is to be surface-applied rather than incorporated into the soil.

Jones et al. (1995) investigated the effects of applying cattle feedlot manure (stockpiled vs. composted) and commercial fertilizer (18-46-0) to cultivated drylands at nitrogen application rates of 40 to 60 kg/ha at Bushland, Texas. Manure treatments were surface-applied and were incorporated into the soil following application on stubble mulch tillage plots, but were not soil-incorporated on no-tillage plots. The feedlot manure (stockpiled) and compost contained, respectively, 0.7 and 1.9% total N, 0.3 and 0.9% total P (as-received basis), and 39% and 51% moisture.

As compared to the control treatment, statistically higher wheat grain and total dry matter yields were obtained with the 18-46-0 fertilizer, followed in order by stockpiled manure and compost (Jones et al., 1995). On cultivated sorghum plots, soil N and P (0-15 cm depth) were similar at the end of fallow on cultivated crops except that the 18-46-0 treatments produced statistically higher levels of soil nitrate and total phosphorus than the unfertilized control treatment.

Runoff volume from dryland sorghum averaged 4.4 mm for stubble mulch tillage plots (1.5% of precipitation) and 15.5 mm or 5.4% of precipitation for the no-tillage plots (Jones et al., 1995). This indicated that greater infiltration occurred with stubble mulch tillage than with no-tillage with near-average precipitation. Nutrient losses from manure- or compost-treated plots were similar to check and commercially-fertilized (18-46-0) plots. Nutrient losses ranged from 0.09 to 0.12 kg/ha TKN and 0.02 to 0.04 kg/ha total P for both tillage systems. Measured sediment content of runoff was low. Nitrate-N concentrations in runoff from manure and compost plots averaged 0.2 to 0.3 mg/L, which was less than for the commercial fertilizer plots (which averaged 0.6 to 1.4 mg/L). Average runoff concentrations from manure and compost plots were 0.6-2.6 mg/L NH_4-N and 0.20 to 0.57 mg/L total phosphorus, which exceeded levels from the commercial fertilizer plots on no-till system. However, only runoff from the stockpiled manure treatment had water-soluble P and ammonium-N concentrations considerably above those measured on the unfertilized control treatments. Phosphorus concentrations in runoff were greater than desired in potable water but similar to values from similar unfertilized watersheds over a 10-year period at the same location. Conductivity values in runoff were 3.3 to 7.3 mS/m across manure treatments which exceeded EC values in runoff from check plots (4.6 mS/m). Jones et al. (1995) concluded that small amounts of manure nutrients required for dryland cropping can be surface-applied as commercial fertilizer, manure, or compost without a significant water quality effect.

Harman et al. (1995) evaluated the impacts of high-load single-frequency (HLSF) applications of feedlot manure using an irrigated corn/conservation-tilled wheat rotation in Moore County in the Texas High Plains during 1991-92 and 1992-93 crop rotation cycles. HLSF rates of manure averaged 47 and 38 Mg/ha for the first and second periods, respectively. Manure was applied one time prior to planting corn as the first crop of the 2-year rotation, followed by conservation-tilled wheat that received no supplemental fertilizer.

Corn yields averaged 220 and 250 kg/ha in the 1991-92 and 1992-93 test periods, respectively, whereas wheat yields averaged 4,840 and 4,170 kg/ha (Harman et al., 1994 and 1995). Estimated profits over variable costs averaged $465 per ha for corn and $242 per ha for wheat, totaling $707 per ha for the 2-year crop rotation cycles. After deducting fixed costs of $500 per ha, the 2-year net returns averaged $208 per ha. Breakeven manure prices (above which manure use would be discontinued) ranged from $5.30 to $9.10 per Mg , depending on cost of land rent and management fees.

Marek et al. (1995) assessed changes in soil infiltration rate in Sherm silty clay loam soil and water quality characteristics of runoff from furrow-irrigated cropping sequence of corn-wheat-soybean rotation in 2 years following HLSF manure treatments of 38 and 65 Mg/ha. Manure applications did not significantly alter the cumulative infiltration characteristics of Ogallala Aquifer water as tested by the double-ring infiltrometer methods after each of the sequential crops. Mean infiltration slope function values ranged from 0.61 with the inorganic fertilizer treatment following corn to 0.38 to 0.40 for the 67 Mg/ha manure treatment following soybeans at the end of the 2-year crop rotation cycle. However, these differences were not statistically significant.

The highest concentrations of nitrate, phosphate, and total phosphorus in furrow runoff occurred with the first (preplant) irrigation of corn and generally decreased through the corn and wheat growing seasons (Marek et al., 1995). Total phosphorus concentrations (with either ro-tilled or no-tillage treatment) were higher with the 67 Mg/ha manure application than with either the 45 Mg/ha or inorganic fertilizer applications. Water quality of runoff was found reusable for agricultural purposes (Marek et al., 1994 and 1995). With the 67 Mg/ha HLSF rate, total phosphorus concentrations of runoff decreased from 1.6 to 1.7 mg/L with the corn preplant irrigation to 0.5 to 0.6 mg/L from soybeans (final irrigation). Nitrate concentrations were slightly higher from the inorganically fertilized plots throughout the experiment than from either the 45 or 67 Mg/ha manure applications. Across all treatments, runoff nitrate ranged from 3.2 to 4.4 mg/L from the corn preplant to lower values of 0.8 to 1.6 mg/L midway through the ensuing wheat irrigation season. Soil conservation practices should be maintained to prevent phosphorus movement off the fields (Marek et al., 1995).

The best returns from using manure for fertilizer usually result from: applying manure on those soils and crops that need both nitrogen and phosphorus; using a low application rate (e.g., 11 to 22 dry Mg/ha/year); and applying manure on poor soils with a chemical imbalance, such as iron deficiency problem. Some benefit can be expected both from residual nutrients and micro-nutrients. Finally, over a period of years even at low application rates, improvement in soil physical properties, such as improved water infiltration rate, greater nutrient holding capacity and greater soil aggregate stability can be realized (Mathers and Stewart, 1981; Sweeten and Mathers, 1985).

IV. Air Quality and Manure Management

A. Feedlot Dust

In hot dry weather, feedlot cattle can create high dust concentrations especially for about two hours around dusk when cattle activity increases, but dust is usually minimum in early morning (Elam et al., 1971). Under calm conditions, feedlot dust can drift over nearby highways and buildings. In 1970, the California Cattle Feeders Association (CCFA) sponsored a study of dust emissions at 25 cattle feedlots (Algeo et al., 1972). Standard high volume samplers were stationed upwind and downwind of the feedlots to monitor dust concentrations. The range measured for total suspended particulates (TSP) in the CCFA research project was 54 to 1,268 $\mu g/m^3$, and the overall average for the 25 feedlots was 654 $\mu g/m^3$, more than four times the federal secondary standard 150 $\mu g/m^3$ (Alego et al., 1972; Peters and Blackwood, 1977). Some feedlot dust emissions were lower than the USEPA standards, due possibly to sprinkling or other management practices. Based on the above California data, the USEPA established an "emission factor" for beef cattle feedlots (USEPA, 1986; Peters and Blackwood, 1977) based on essentially "worst-case" assumptions. The EPA emission factor was 127 kg dust per day per 1,000 head of feedlot capacity.

A research project on feedlot dust emissions sponsored by the Texas Cattle Feeders Association measured feedlot dust concentrations at three feedlots and was replicated three times (Sweeten et al., 1988). A mean net increase was found in TSP concentration of 412 $\mu g/m^3$ (difference between upwind and downwind dust concentrations), and the range was 16 to 1,700 $\mu g/m^3$. These values were generally lower than found in the California dust emission studies, possibly because Texas receives more precipitation in its cattle feeding regions than California feedlots. However, the Texas values were still well above the state and federal TSP standards.

Lesikar et al. (1994) calculated a range of particulate emission factors for feedyards using field data from Sweeten et al. (1988). Calculated emission factors ranged from 0.04 to 35 kg per day per 1000 head, with a mean value of 4.5 kg per day per 1000 head. The latter figure represents only 3.5% of the USEPA (1986) emission factor for feedyards.

In July, 1987, the U.S. Environmental Protection Agency (USEPA, 1987) changed the basis for ambient air quality standards for particulate emissions to the PM-10 standard (i.e., median aerodynamic particle size of 10 microns). Accordingly, in the Texas studies (Sweeten et al., 1988), two different types of PM-10 monitors were used to sample feedlot dust and compare with USEPA's new PM-10 standard of 150 $\mu g/m^3$. In general, PM-10 dust concentrations were 19 and 40% below the TSP measurements for the same sampling sites. Mean particle sizes for feedlot dust were 8.5 to 12 microns (μm). However, only 2 to 4% of the total collected dust was respirable dust, which is considered to be two microns and below.

Dust control methods include watering unpaved roads, watering feedlot surfaces with mobile tankers or solid-set sprinklers, and controlling cattle stocking rate in relation to precipitation and evaporation (Sweeten, 1982; Sweeten and Lott, 1994). The amount of manure moisture generated by cattle varies directly with liveweight and inversely with stocking rates (Table 6). Large cattle on tight spacing can easily double the effective "precipitation" on the feedlot surface, which has implications for both dust and odor control.

Lott (1995) fitted a 3-stage drying model to the data for evaporation from pen surfaces at 3 feedlots. The first stage of manure surface drying represents a fast-rate linear function that is directly related to potential evaporation. In the second stage of drying, evaporation rate slows as manure restricts the release of water, and evaporation occurs at a very slow rate in the third stage. This drying rate function is analagous to drying curves for typical soils (Sweeten, 1982).

Table 6. Moisture in fresh manure produced in beef cattle feedlots versus stocking rate equivalent annual depth on feedlot surface

Animal size	Average animal spacing (m²/head)				
	9.3	13.9	18.6	23.2	27.9
(kg/head)	Annual moisture produced[a] (mm)				
180	350	235	175	140	117
270	530	350	260	210	175
360	710	470	350	280	235
450	880	590	440	350	290
540	1,060	710	530	420	350
725	1,410	940	710	560	470

[a]Calculated from manure characteristics data in ASAE, 1988.
(From Sweeten, 1990a.)

B. Odor Management

Feedlot odor research conducted using state-of-the-art odor measurement equipment determined that odor concentrations from feedlot manure pads were approximately 50 to 100 times greater than those of dry pads; odor concentrations and offensiveness peaked 2 days after the manure became wet; wet pads produced more offensive odors than dry pads (Watts et al., 1992; Tucker, 1992; Watts and Tucker, 1993a; Watts and Tucker, 1993b). Therefore, a small area of wet feedlot pad can be responsible for the generation of a significant amount of odor, and poorly drained feedlots with manure build-up can create far more odor than well-drained feedlots which are regularly cleaned and dry rapidly.

V. Manure Collection

The proper approach for manure collection should be that of "manure harvesting" rather than "cleaning pens." Methods designed for "manure harvesting" should maintain a surface seal, to promote good drainage, and to collect a high quality product (Lott et al., 1994a). Solid manure can be collected in several ways, as studied several years ago (Sweeten and Reddell, 1979; Sweeten, 1979). Where manure is collected once a year, the most efficient method is to use a chisel plow to reduce particle size, followed by a wheel loader to stack in the pens and load into a truck. Collection rates were measured, normalized to 100% operating efficiency, at 145 Mg/hour with an overall energy requirement of 1.07 kW-hr/Mg. However, the chisel plow/wheel loader method produces a strong risk of disturbing the compacted manure/soil interfacial seal that should be left to protect against leaching. The wheel loaders provide collection rates of 96 Mg/hour and energy requirements of 1.09 to 1.41 kW-hr/Mg.

Powell (1994) measured feedlot manure harvesting rates of 48 Mg per hour for a box-scraper and 79 Mg per hour with wheel loaders. Box scrapers (tractor drawn) are frequently used for collection of loose surface manure for dust and odor control, for maintaining a smooth surface for drainage, and for mound building.

In connection with characterizing feedlot manure as a fuel feedstock for combustion or gasification for electricity generation, data has been published on feedlot manure characteristics resulting from manure collection (Sweeten et al., 1985). Ash content in the collected manure was only half that of the uncollected manure that was left above the soil interface (32% vs. 61% ash, respectively). Nitrogen content and heat of combustion were also higher in the harvested manure.

Solid manure from open dirt unpaved feedlots in Texas are collected mainly by private contractors. Each contractor usually serves several feedlots and may also be involved in silage hauling or other activities. Manure is collected and loaded into a fleet of trucks that haul and spread manure at typical application rates of 11 to 56 Mg/ha wet basis.

In 1985, 13 contractors who collected manure from 61 Texas cattle feedlots having 1.5 million head of cattle on feed were surveyed (Sweeten, 1985a), to obtain information concerning their payments to feedlots, charges to farmers and manure application rates on crops. At that time contractors were paying feedlots as much as $1.10 per Mg (or in some cases were being paid as much as $0.55 per Mg) for manure collected, loaded and weighed at the feedlot scales. The contractor then sells the manure to farmers on a haul and spread basis. As of 1985, the farmers were paying contractors an average of $3.00 per Mg plus $0.085 per Mg per km.

The survey was repeated in 1989 and manure costs were even lower: $2.37 per Mg plus $0.08 per Mg-km. So for a 16 km one-way haul distance, the cost to the farmer averages $3.70 per Mg of applied manure. Thus, for a 22 Mg per ha application rate, the cost is $83 per ha for manure fertilizer. Most manure is being hauled 8 to 24 km, although in a few places where vegetable crops are grown on sandy land, for example, manure is occasionally hauled 50 km or more.

For haul distances over about 16 km, it often becomes more practical to haul manure using a larger truck (i.e., 18 wheel, 30 m^3), which is unloaded at the farmer's field and later spread. An increasing amount of manure is being hauled in this manner so that feedlots are less dependent on crop cycles to have manure harvested from pens and hauled. Temporary stockpiles need to be carefully located to avoid surface and groundwater pollution problems, however, and in some cases temporary holding ponds should be considered.

The 1985 contractor's survey (Sweeten, 1985a) showed that crops were being fertilized with manure at an average rate of 25 Mg per ha per year wet or as received basis on irrigated land. The range was 18 to 34 Mg/ha per year on grain and vegetable crops and 11 to 22 Mg ha^{-1} per year on cotton. Application rates for dryland crops were about half those for irrigated crops.

VI. Manure Storage and Processing

A. Stockpiling

Solid manure is frequently placed in stockpiles outside feedpens while awaiting reloading, hauling and spreading. The purpose of a stockpile should be to allow manure to be harvested from pens regularly even though spreader trucks or cropland are not available. The volume of stockpiles should be kept to a minimum since manure is a perishable commodity and should be used promptly for maximum benefit. Manure stacked in feedpens should be promptly hauled to a stockpile unless properly shaped and used for a mound. Powell (1994) measured weight reductions of 19-38% per year in in-pen mounds as compared to 29-38% reduction in stockpiles.

Manure stockpiles are subject to smoldering or spontaneous combustion (which produces air pollution emissions). If manure is stockpiled deeper than about 2 m, it should be compacted to partially exclude oxygen. Generally, the hottest zone in a stockpile is 0.6 to 0.9 m below the crown. Relatively wet manure is another cause of fire in manure stockpiles. Fires in manure stockpiles are difficult to extinguish. Consequently, small, discrete stockpiles are preferable to large stockpiles, especially for wet manure. State water pollution abatement regulations require stockpiles to be isolated by dikes or diversion terraces to prevent entry of rainfall runoff. Likewise, runoff from stockpiles has to be collected in holding ponds.

B. Composting Manure

Many feedlots and contractors practice composting (aerobic thermophilic treatment) by placing the manure in windrows 1 to 2 m high and turning it one or more times. To speed up composting, manure should be turned frequently, an average of once per week for six to eight weeks. Different methods of turning windrows include wheel loaders and mobile units with rotary-spiked drums or augers (Sweeten, 1988a). The purpose of turning windrows is to aerate, increase temperature, release excess heat, and promote evaporation of excess moisture. The second approach to composting consists of aerated bins in which air may be blown or suctioned through a relatively shallow bed of composting material, for example, one meter deep in a concrete bin. Turning is done either by a spiked drum or flighted conveyor mounted on rails. Stable, good quality compost can be produced in four weeks with daily turning in an aerated bin system. Important variables for composting include moisture content (40 to 60%), particle size or structure, and aeration (Sweeten, 1988a). The carbon-nitrogen ratio should be about 30:1 for ideal composting, as compared to only 10 to 15:1 for fresh manure. Addition of crop residues or finished compost can raise this ratio in fresh manure. Temperatures should range from 45 to 79°C, with an ideal temperature of 60 to 71°C. These temperatures are sufficient to reduce pathogenic organisms by several hundred or thousand-fold and kill more than 95% of the weed seeds (Wiese et al., 1977).

VII. Managing the Feedlot Surface

The feedlot profile usually contains a compacted interfacial layer of manure and soil that provides a biological seal that reduces water infiltration rate to less than 0.05 mm per hour (Mielke et al., 1974; Mielke and Mazurak, 1976). This zone of low infiltration restricts leaching of salts, nitrates and ammonium into the subsoil and potentially to groundwater (Schuman and McCalla, 1975).

Bulk density of the manure layer was measured at 0.60 to 0.74 kg/m^3 (Mielke et al., 1974), which is less dense than water. Immediately below this manure layer, the compacted manure/soil interfacial layer had a density of 0.77 to 1.36 kg/m^3. The density of the underlying soil was 0.96 to 1.28 kg/m^3.

Since feedlot manure has about 50 to 75% of the density of the underlying soil, less energy should be needed when collecting just the organic matter and leaving the soil. The manure itself usually has a shear plane which facilitates manure collection above the interfacial layer.

Leaving a compacted layer of manure on the feedlot surface reduces the leaching of nutrients and salts into the underlying soil profile. In California feedlots, Algeo et al. (1972) found that soil nitrate levels at zero to 0.6 m below the feedlot surface were only slightly higher (60 to 180 ppm) than in adjacent fields. Below four feet of soil depth, soil nitrate levels were the same beneath the feedlot and the cropland (20 to 40 ppm NO_3-N). Chloride concentrations showed similar trends.

Beneath a Nebraska feedlot, nitrate concentrations were 7.5 ppm in the top 100 mm of soil depth (Schuman and McCalla, 1975). Below 200 mm, however, nitrate dropped to less than one ppm NO_3-N due to dentrifying conditions and the presence of a manure/soil seal. Ammonia content also decreased with soil depth from 35 ppm at zero to 50 mm to less than two ppm below 100 mm.

Norstadt and Duke (1982) measured soil nitrate levels that decreased from 80 ppm at the top of feedlot soil profiles to less than 10 ppm at 1 to 1.5 m depth. Nearly the same result was obtained both for a clay loam soil and a layered soil that consisted of 0.75 m of sand over 0.75 m of clay loam.

In Florida, Dantzman et al. (1983) measured salt concentrations in soil profiles (fine sand) 10 and 15 years after operation of cattle pens. After 15 years, organic matter in the soil tripled to 15% and the soluble salts accumulated to 4,000 ppm in the top 25 to 30 cm but were less than 500 ppm below 50 cm, which is a relatively low soluble salt level. By comparison, at a nearby control location, soluble salt concentration was found throughout 1.5 m of the soil profile. Phosphorus levels rose from less than 4.6 ppm in check fields to over 1,500 ppm after 10 and 18 years.

Elliott et al. (1972) collected soil water samples at 0.45, 0.70, and 1.1 m beneath a level cattle feedlot surface on a silt loam/sand soil profile. Nitrate concentrations were generally less than 1 ppm as compared to 0.3 to 101 ppm in the top 75 mm. The low nitrate-nitrogen values below 75 mm indicate that dentrification takes place beneath the soil surface due to anaerobic conditions.

VIII. Cattle Spacing in Feedlots

Cattle spacing has an influence on manure moisture content and hence on dust, odor, runoff and muddy lot conditions. Large commercial cattle feedlots in the United States utilize cattle spacings of as little as 9.3 m^2 per head in arid climates such as Southern California (i.e., below 250 mm annual rainfall), to around 14 to 21 m^2/head in 500 mm rainfall zones, and 28 to 37 m^2/head or greater in humid climates of over 750 mm annual rainfall.

During dry seasons, a higher stocking density (i.e., lower pen spacing) is suitable due to high evaporative removal of excess moisture. In fact reducing the pen spacing in dry seasons can be a dust control strategy. But high stocking rates should be avoided in normally wet, cool seasons, due to low evaporation and prolongation of wet manure into warmer weather.

Nienaber et al. (1974) measured the effect of lot slope and animal density on beef cattle performance in Nebraska over a four-year period. Cattle spacings of 9.3 and 18.6 m^2/head did not have a significant effect on cattle performance, nor did pen slopes of 3, 6, and 9%.

IX. Summary

Major reasons for feedlot managers to focus attention on manure and wastewater management include: 1) to control air pollution (odor and dust); 2) to control surface and ground water pollution; 3) to maintain or increase cattle productivity by providing well-maintained feedlot conditions that provide all cattle with a similar production environment; 4) to recover nutrients in the form of fertilizer, feedstuffs or energy; and 5) to maintain or increase efficiency of the feedlot by avoiding operational obstacles such as muddy pen surfaces, excessive stockpiled manure, and full runoff holding ponds and settling basins that threaten illegal discharges. All five of these are good reasons why cattle feeders should be willing to focus greater attention on feedlot waste management and water and air pollution abatement.

References

Algeo, J.W., C.J. Elam, A. Martinez, and T. Westing. 1972. Feedlot air, water, and soil analysis. Bulletin D, *How to Control Feedlot Pollution*. California Cattle Feeders Association. Bakersfield, California. 75 pp.

American Society of Agricultural Engineers. 1988. Manure production and characteristics. ASAE D-384-1, American Society of Agricultural Engineers, St. Joseph, Michigan.

Arrington, R.M. and C.E. Pachek. 1981. Soil nutrient content of manures in an arid climate. p. 150-152. In: *Livestock Waste: A Renewable Resource*, Proceedings of the Fourth International Symposium on Livestock Waste, 1980. American Society of Agricultural Engineers, St. Joseph, Michigan.

Barrington, S.F. and P.J. Jutras. 1983. Soil sealing by manure in various soil types. Paper 83-4571, American Society of Agricultural Engineers, St. Joseph, Michigan. 28 pp.

Butchbaker, A.F. 1973. Feedlot runoff disposal on grass or crops. L-1053, Texas Agricultural Extension Service, Texas A&M University. DPE-7521, Great Plains Beef Cattle Feeding Handbook. 4 pp.

Clark, R.N. 1975. Seepage Beneath Feedyard Runoff Catchments. p. 289-295. In: *Managing Livestock Wastes*, Proceedings of the Third International Symposium on Livestock Wastes, American Society of Agricultural Engineers, St. Joseph, Michigan. pp.289-295.

Clark, R.N., C.B. Gilbertson, and H.R. Duke. 1975a. Quantity and Quality of Beef Feedyard Runoff in the Great Plains. p. 429-431. In: *Managing Livestock Wastes*. Proceedings of the third International Symposium on Livestock Wastes, Proc-275, American Society of Agricultural Engineers, St. Joseph, Michigan.

Clark, R.N., A.D. Schneider, and B.A. Stewart. 1975b. Analysis of runoff from Southern Great Plains feedlots. *Transactions ASAE* 15(2):319-322.

Coleman, E.A., W. Grub, R.C. Albin, G.F. Meenaghan, and D.M. Wells. 1971. Cattle feedlot pollution study. Interim Report No. 2, WRC-71-2, Water Resources Center, Texas Tech University, Lubbock, Texas.

Dantzman, C.L., M.F. Richter, and F.G. Martin. 1983. Chemical Elements in Soils Under Cattle Pens. *Journal Environmental Quality* 12(2):164-168.

Dickey, E.C., D.H. Vanderholm, J.A. Jackobs, and S.L. Spahr. 1977. Vegetative filter treatment of feedlot runoff. ASAE Paper No. 77-4581, American Society of Agricultural Engineers, St. Joseph, Michigan.

Dixon, J.E. 1983a. Comparison of runoff quality from cattle feeding on winter pastures. *Transactions ASAE* 26(4):1146-9.

Dixon, J.E. 1983b. Controlling water pollution from cattle grazing and pasture feeding operations. p. 107. In: J.H. Smits (ed.), *Profit Potential of Environmental Protection Practices of Cattlemen*. National Cattlemen's Association, Englewood, Colorado.

Doran, J.S., J.S. Schepers, and N.P. Swanson. 1981. Chemical and Bacteriological Quality of Pasture Runoff. *J. Soil and Water Conservation* 36(3):166-171.

Elam, C.J., J.W. Algeo, T. Westing, et al. 1971. Measurement and control of feedlot particulate matter. Bulletin C, *How To Control Feedlot Pollution*. California Cattle Feeders Association, Bakersville, California. January.

Elliott, L.F., T.M. McCalla, L.N. Mielke, and T.A. Travis. 1972. Ammonium, nitrate and total nitrogen in the soil water of feedlot and field soil profiles. *Applied Microbiology* 23:810-813.

FAO. 1985. Water quality for agriculture. Irrigation and Drainage Paper 29:34-35. Revision 1, Food and Agriculture Organization of the United Nations, Rome.

Gilbertson, C.B., R.N. Clark, J.C. Nye, and N.P. Swanson. 1980. Runoff control for livestock feedlots: state of the art. *Transactions ASAE* 23(5):1207-1212.

Gilbertson, C.B., J.A. Nienaber, J.L. Gartrung, J.R. Ellis, and W.E. Splinter. 1979a. Runoff Control Comparisons for Commercial Beef Feedlots. *Transactions ASAE* 22(4):842-849.

Gilbertson, C.B., F.A. Nordstadt, A.C. Mathers, R.F. Holt, A.P. Garnett, T.M. McCalla, C.A. Onstad, and R.A. Young. 1979b. p. 32-34. In: Animal waste utilization on cropland and pastureland: A manual for evaluating agronomic and environmental effects. USDA Report No. URR-6, U.S. Department of Agriculture, Science and Education Administration, Hyattsville, Maryland.

Gilbertson, C.B., J.C. Nye, R.N. Clark, and N.P. Swanson. 1981. *Controlling Runoff From Feedlots--A State of the Art.* Ag. Info. Bulletin 441, U.S. Department of Agriculture, Agricultural Research Service, Washington, DC 20250. 19 pp.

Harman, W.L., G.C. Regier, T.H. Marek, and J.M. Sweeten. 1995. Profits using feedlot manure in high-load single-frequency (HLSF) applications with conservation tillage systems, Texas High Plains. p. 134-139. In: *Innovations and New Horizons in Livestock and Poultry Manure Management.* Proceedings (Volume 1) September 6-7, 1995. Austin, Texas. Texas Agricultural Extension Service and Texas Agricultural Experiment Station. College Station, Texas.

Harman, W.L., T.H. Marek, G.C. Regier, and J.M. Sweeten. 1994. Economics of Using High-load Single Frequency (HLSF) Manure Applications with Conservation Tillage. p. 259-266. In: *Balancing Animal Production and the Environment.* Proceedings, Great Plains Animal Waste Conference on Confined Animal Production and Water Quality. GPAC Publication No. 151, Great Plains Agricultural Council, Denver, Colorado.

Jones, O.R., W.M. Willis, S.J. Smith, and B.A. Stewart. 1995. Nutrient cycling of cattle feedlot manure and composted manure applied to Southern High Plains drylands. p. 265-272. In: K. Steele (ed.), *Animal Waste and the Land-Water Interface.* Proceedings of Animal Waste in the Land-Water Interface Conference, Fayetteville, Arkansas. July 16-19. Lewis Publishers, Boca Raton, FL.

Lehman, O.R. and R.N. Clark. 1975. Effect of cattle feedyard runoff on soil infiltration rates. *Journal Environmental Quality* 4:437-439.

Lehman, O.R., B.A. Stewart, and A.C. Mathers. 1970. Seepage of feedyard runoff water impounded in playas. MP-944, Texas Agricultural Experiment Station, Texas A&M University, College Station, Texas. 7 pp.

Lesikar, B.J., S.A. Parnell, and J.M. Sweeten. 1994. Dispersion modeling for prediction of emission factors from cattle feedyards. Final Report. National Cattlemen's Association, Englewood, Colorado. November 17. 121 pp.

Lindemann, E.R., J.M. Sweeten, and J.P. Burt. 1985. Sludge removal from dairy lagoons. p. 653-659. In: *Agricultural Waste Utilization and Management.* Proceedings of the Fifth International Symposium on Agricultural Wastes, Chicago, Illinois. American Society of Agricultural Engineers, St. Joseph, Michigan.

Loehr, R.C. 1974. Characteristics and comparative magnitude of nonpoint sources. Journal Water Pollution Control Federation 46:1849.

Lott, S.C., E. Powell, and J.M. Sweeten. 1994a. Manure collection, storage and spreading. In: P.J. Watts and R. Tucker (eds.), *Designing Better Feedlots.* Queensland Department of Primary Industries, Toowoomba, Queensland, Australia. 15 pp.

Lott, S.C. 1995. Australian feedlot hydrology, Part I (Data). In: Proceedings, Feedlot Waste Management Conference, Queensland Department of Primary Industries, Gold Coast, Queensland, Australia. 29 pp.

Lott, S.C., P.J. Watts, and J.R. Burton. 1994b. Runoff from Australian cattle feedlots. p. 47-53. In: *Balancing Animal Production and the Environment.* Proceedings, Great Plains Animal Waste Conference on Confined Animal Production and Water Quality. GPAC Publication No. 151, Great Plains Agricultural Council, Denver, Colorado.

Loudon, T.L., D.D. Jones, J.B. Peterson, et al. 1985. *Livestock Waste Facilities Handbook.* MWPS-18 (2nd ed.), Midwest Plan Service, Iowa State University, Ames, Iowa. p. 2.1-2.2; 5.1-5.9.

Lund, Z.F. and B.D. Doss. 1980. Coastal bermudagrass yield and soil properties as affected by surface-applied dairy manure and its residue. *Journal Environmental Quality* 9:157.

Lund, Z.F., B.D. Doss, and F.E. Lowry. 1975. Dairy cattle manure--its effect on yield and quality of coastal Bermudagrass. Journal Environmental Quality 4:358.

Marek, T.M., W.L. Harman, and J.M. Sweeten. 1995. Infiltration and water quality inferences of high load, single frequency (HLSF) applications of feedlot manure. p. 162-169. In: *Innovations and New Horizons in Livestock and Poultry Manure Management.* Proceedings (Volume 1) Austin, Texas September 6-7, 1995. Texas Agricultural Extension Service and Texas Agricultural Experiment Station. College Station, Texas.

Marek, T.H., W.L. Harman, and J.M. Sweeten. 1994. Irrigation and runoff water quality implications of high load, single frequency (HLSF) applications of feedlot manure. p. 119-124. In: *Balancing Animal Production and the Environment.* Proceedings, Great Plains Animal Waste Conference on Confined Animal Production and Water Quality. GPAC Publication No. 151, Great Plains Agricultural Council, Denver, Colorado.

Mathers, A.C. and B.A. Stewart. 1971. Crop production and soil analysis as affected by application of cattle feedlot waste. p. 229-231, 234. In: Livestock Waste Management. Proceedings of the Second International Symposium on Livestock Wastes, American Society of Agricultural Engineers, St. Joseph, Michigan.

Mathers, A.C., B.A. Stewart, J.D. Thomas, and B. J. Blair. 1973. Effects of cattle feedlot manure on crop yields and soil conditions. Technical Report No. 11. USDA Southwestern Great Plains Research Center, Bushland, Texas.

Mathers, A.C. and B.A. Stewart. 1981. The effect of feedlot manure on soil physical and chemical properties. p. 159-162. In: *Livestock Waste: A Renewable Resource,* Proceedings of the Fourth International Symposium on Livestock Waste, 1980. American Society of Agricultural Engineers, St. Joseph, Michigan.

Mathers, A.C. and B.A. Stewart. 1984. Manure effects on crop yields and soil properties. *Transactions ASAE* 27:1022-1026.

Mathers, A.C., B.A. Stewart, and J.D. Thomas. 1975. Residual and annual rate effects of manure on grain sorghum yields. In: *Managing Livestock Wastes.* Proceedings of the Third International Symposium on Livestock Wastes, 1975. American Society of Agricultural Engineers, St. Joseph, Michigan.

Mathers, A.C., J.D. Thomas, B.A. Stewart, and J.E. Herring. 1980. Manure and inorganic fertilizer effects on sorghum and sunflower growth on iron-deficient soil. *Agronomy Journal* 72:1025-1029.

Mielke, L.N. and A.P. Mazurak. 1976. Infiltration of water on a cattle feedlot. *Transactions ASAE* 19:341-344.

Mielke, L.N., N.P. Swanson, and T.M. McCalla. 1974. Soil profile conditions of cattle feedlots. *Journal Environmental Quality* 13:14-17.

Miller, W.D. 1971. Infiltration rates and groundwater quality beneath cattle feedlots. Texas High Plains. Final Report No. 16060 EGS, Water Quality Office, U.S. Environmental Protection Agency, Washington, D.C. 55 pp.

Milne, C.M. 1976. Effect of a livestock wintering operation on a western mountain stream. *Transactions ASAE* 19:749-752.

Nienaber, J.A., C.B. Gilbertson, T.J. Klopfenstein, S.D. Parlin, and T.M. McCalla. 1974. Animal performance and lot surface conditions as affected by feedlot slope and animal densities. p. 130-137. In: Proceedings, International Livestock Environment Symposium, Lincoln, Nebraska.

Norstadt, F.A. and H.R. Duke. 1982. Stratified profiles: characteristics of simulated soils in a beef cattle feedlot. *Soil Science Society America Journal* 46:827-832.

Paine, M.D. 1973. Confined animals and public environment. GPE-7000, Great Plains Beef Cattle Feeding Handbook, Cooperative Extension Service, Oklahoma State University, Stillwater, Oklahoma. 4 pp.

Peters, J.A. and T.R. Blackwood. 1977. *Source Assessment: Beef Cattle Feedlots*. Monsanto Research Corporation, EPA-600/2-77-107, U.S. Environmental Protection Agency, Industrial Environmental Research Laboratory, Research Triangle Park, North Carolina. 101 pp.

Phillips, P.A. and J.L.B. Culley. 1985. Groundwater nutrient concentrations below small-scale earthen manure storage. p. 672-679. In: *Agricultural Waste Utilization and Management.* Proceedings of the Fifth International Symposium on Agricultural Wastes, American Society of Agricultural Engineers, St. Joseph, Michigan.

Phillips, R.L. 1981. Maps of runoff volumes from feedlots in the United States. p. 274-277. In: *Livestock Waste: A Renewable Resource.* Proceedings of the Fourth International Symposium on Livestock Waste, American Society of Agricultural Engineers, St. Joseph, Michigan.

Powell, E.E. 1994. Economic management of feedlot manure. Final Report, Part 2. Evan Powell Rural Consultants, Dalby, Queensland. Meat Research Corporation Contract M.087, Sydney, NSW, Australia. 35 pp.

Powers, W.L., R.L. Herpich, L.S. Murphy, D.A. Whitney, H.L. Manges, and G.W. Wallingford. 1973. Guidelines for land disposal of feedlot lagoon water. C-485, Cooperative Extension Service, Kansas State University, Manhattan, Kansas.

Reddell, D.L. 1974. Forage and grain production from land used for beef manure disposal. p. 464-483. In: *Proceedings and Management of Agricultural Wastes*, Cornell University, Ithaca, New York.

Reddell, D.L. and G.G. Wise. 1974. Water quality of storm runoff from a Texas beef feedlot. Texas Agricultural Experiment Station, PR-3224, College Station, Texas.

Robinson, F.E. 1973. Changes in seepage rates from an unlined cattle waste digestion pond. *Transactions ASAE* 16:95-96.

Saxton, K.E., L.F. Elliot, R.I. Papendick, M.D. Jawson, and D.H. Fortier. 1983. Effect of animal grazing on water quality of nonpoint runoff in the Pacific Northwest: Project Summary. EPA-600/S2-82-071. U.S. Environmental Protection Agency, Ada, Oklahoma.

Schuman, G.E. and T.M. McCalla. 1975. Chemical characteristics of a feedlot soil profile. *Soil Science* 119:113-118.

Shuyler, L.R., D.M. Farmer, R.D. Kreis, and M.E. Hula. 1973. Environment protecting concepts of beef cattle feedlot wastes management. National Environmental Research Center, Office of Research and Development, U.S. Environmental Protection Agency, Corvallis, Oregon. July.

Smeins, F.E. 1977. Influence of vegetation management on yield and quality of surface runoff. Final Report No. TR-84, Texas Water Resources Institute, Texas Agricultural Experiment Station, Texas A&M University, College Station, Texas.

Smith, S.J., A.N. Sharpley, B.A. Stewart, J.M. Sweeten, and T. McDonald. 1994. Water quality implications of storing feedlot waste in southern great plains playas. p. 267-270. In: *Balancing Animal Production and the Environment.* Proceedings, Great Plains Animal Waste Conference on Confined Animal Production and Water Quality. GPAC Publication No. 151, Great Plains Agricultural Council, Denver, Colorado.

Stewart, B.A. and B.D. Meek. 1977. Soluble salt considerations with waste application. p. 219-232. In: L.F. Elliott and F.J. Stevenson (eds.), *Soils for Management of Organic Wastes and Wastewaters.* Soil Science Society of America, Madison, Wisconsin.

Swanson, N.P., C.L. Lindemann, and L.N. Mielke. 1974. Direct land disposal of feedlot runoff. p. 255-257. In: *Managing Livestock Wastes.* Proceedings of the Third International Symposium on Animal Wastes, American Society of Agricultural Engineers, St. Joseph, Michigan.

Swanson, N.P., J.C. Lorimor, and L.N. Mielke. 1973. Broad basin terraces for sloping cattle feedlots. *Transactions ASAE* 16:746-749.

Swanson, N.P., L.N. Mielke, and J.R. Ellis. 1977. Control of beef feedlot runoff with a waterway. ASAE Paper No. 77-4580, American Society of Agricultural Engineers, St. Joseph, Michigan.

Sweeten, J.M. 1988a. Composting manure and sludge. L-2289, Texas Agricultural Extension Service, Texas A&M University, College Station, Texas. 4 pp.

Sweeten, J.M. 1985a. Cost of manure for fertilizer. Paper presented at Extension Seminars on Feedlot Manure Utilization for Fertilizer. Etter and Stratford, Texas, June 17-18. 5 pp.

Sweeten, J.M. 1976. Dilution of feedlot runoff. MP-1297, Texas Agricultural Extension Service, Texas A&M University, College Station, Texas. 8 pp.

Sweeten, J.M. 1982. Feedlot dust control. L-1340, Texas Agricultural Extension Service, Texas A&M University, College Station, Texas. 7 pp.

Sweeten, J.M. 1988b. Groundwater quality protection for livestock feeding operations. B-1700, Texas Agricultural Extension Service, Texas A&M University System, College Station, Texas. 10 pp. (revised 1992).

Sweeten, J.M. 1979. Manure Management for Cattle Feedlots. L-1094, Texas Agricultural Extension Service, Texas A&M University, College Station, Texas. 6 pp.

Sweeten, J.M. 1985b. Removal and utilization of feedlot runoff and sediment. p. G1-G16. In: Proceedings, Great Plains Cattle Feeders Conference and Oklahoma Cattle Feeders Seminar, Guymon, Oklahoma. May 8-9.

Sweeten, J.M. 1990a. Cattle feedlot waste management practices for water and air pollution control. B-1671, Texas Agricultural Extension Service, Texas A&M University, College Station, Texas. 24 pp.

Sweeten, J.M. 1984. Utilization of cattle manure for fertilizer. p. 59-74. In: F.H. Baker and M.E. Miller (eds.), *Beef Cattle Science Handbook*. Volume 20, Westview Press, Boulder, Colorado.

Sweeten, J.M., R.P. Egg, D.L. Reddell, F. Varani, and S. Wilcox. 1985. Characteristics of cattle feedlot manure in relation to harvesting practices. p. 329-337. In: *Agricultural Waste Utilization and Management*, Proceedings of the Fifth International Symposium on Agricultural Wastes, American Society of Agricultural Engineers, St. Joseph, Michigan.

Sweeten, J.M. and A.C. Mathers. 1985. Improving soils with livestock manure. *Journal Soil and Water Conservation* 40:206-210.

Sweeten, J.M. and R.P. McDonald. 1979. Results of TCFA environmental and energy survey--1979. Texas Cattle Feeders Association, Amarillo, Texas.

Sweeten, J.M. and S.W. Melvin. 1985. Controlling water pollution from nonpoint source livestock operations. p. 215-217. In: Perspectives on Nonpoint Pollution, EPA440/5-85-001, U.S. Environmental Protection Agency, Office of Water Regulations and Standards, Washington, D.C.

Sweeten, J.M., C.B. Parnell, R.S. Etheredge, and D. Osborne. 1988. Dust emissions in cattle feedlots. p. 557-578. In: J.L. Howard (ed.), *Stress and Disease in Cattle, Veterinary Clinics in North America; Food Animal Practice*, Vol. 4, No. 3.

Sweeten, J.M. and D.L. Reddell. 1979. Time-motion analysis of feedlot manure collection systems. *Transactions ASAE* 22:138-143.

Sweeten, J.M., L.M. Safley, and S.W. Melvin. 1981. Sludge removal from lagoons and holding ponds: case studies. p. 204-210. In: *Livestock Waste: A Renewable Resource*. Proceedings of the Fourth International Symposium on Livestock Wastes, American Society of Agricultural Engineers, St. Joseph, Michigan.

Sweeten, J.M., T.H. Marek, and D. McReynolds. 1995a. Groundwater quality near two cattle feedlots in the Texas high plains. *Applied Engineering Agriculture* 11:845-850.

Sweeten, J.M. and S. H. Amosson. 1995. Manure quality and economics. In: *Total Quality Manure Management Manual*. Texas Cattle Feeders Association, Amarillo.

Sweeten, J.M., G.L. Sokora, R.M. Seymour, M.G. Hickey, and S.M. Young. 1995b. Irrigation of cattle feedlot runoff on winter wheat (Extended Abstract). p. 14-16. In: Proceedings, Animal Waste and the Land-Water Interface Conference, University of Arkansas Water Resources Center, Fayetteville, Arkansas.

Sweeten, J.M. and S.C. Lott. 1994. Dust management. p. 6.23-6.30. In: P.J. Watts and R. Tucker (eds.), *Designing Better Feedlots*. Queensland Department of Primary Industries, Toowoomba, Queensland, Australia.

Sweeten, J.M. 1994. Water quality associated with playa basins receiving feedlot runoff. p. 161-174. In: L.V. Urban and A.W. Wyatt (eds.), Proceedings of the Playa Basin Symposium, Texas Tech University, Lubbock, Texas.

Sweeten, J.M. and M.L. Wolfe. 1993. *The Expanding Dairy Industry: Impact on Ground Water Quality and Quantity with Emphasis on Waste Management Evaluation for Open Lot Dairies*. TR-155, Texas Water Resources Institute, Texas A&M University, College Station, TX. August. 142 pp.

Sweeten, J.M., H.D. Pennington, D. Seale, R. Wilson, R.M. Seymour, A.W. Wyatt, J.S. Cochran, and B.W. Auvermann. 1990. Well water analysis from 26 cattle feedyards in Castro, Deaf Smith, Parmer, and Randall Counties, Texas. Texas Agricultural Extension Service, Texas A&M University System, College Station, Texas. October 15. 16 pp.

Sweeten, J.M. 1990b. Feedlot runoff characteristics for land application. p. 168-184. In: *Agricultural and Food Processing Wastes*. Proceedings of the 6th International Symposium on Agricultural and Food Processing Wastes, Chicago, Illinois.

TNRCC. 1995. Subchapter K, Concentrated animal feeding operations. Chapter 321. Control of Certain Activities by Rule. Texas Natural Resource Conservation Commission. *Texas Register*, June 30. 20:4727-4742.

Texas Water Commission. 1987. Control of Certain Activities By Rule. Chapter 321, Subchapter B, Part IX, (31 TAC 321.31-321.38, 321-41). Austin, Texas. *Texas Register*, March 17, 1987. p. 904-909.

Thomas, J.D. and A.C. Mathers. 1979. Manure and iron effects on sorghum growth on iron deficient soil. *Agron. J.* 71:792.

Tucker, R. 1992. Odour measurements from simulated feedlot pads. p. 103-121. In: *Odour Update '92*. Proceedings of a Workshop on Agricultural Odours. MRC Report No. DAQ 64/24, Department of Primary Industries, Toowoomba, Queensland, Australia. April.

U.S. Environmental Protection Agency. 1973. Development document for proposed effluent limitations guidelines and new source performance standards for the feedlots point source category. EPA-440/1-73/004, USEPA, Washington, D.C. p. 59-64.

U.S. Environmental Protection Agency. 1986. Supplement A to compilation of air pollution emission factors, section 6.15 beef cattle feedlots. Volume I: Stationary Point and Area Sources. AP-42, Office of Air Quality Planning and Standards, Research Triangle Park, North Carolina. 3 pp.

U.S. Environmental Protection Agency. 1987. 40CFR50, revisions to the national ambient air quality standards for particulate matter and Appendix J--reference method for the determination of particulate matter as PM-10 in the atmosphere. *Federal Register*, 52(126):24634, and 24664-24669.

U.S. Environmental Protection Agency. 1993. National pollutant discharge elimination system general permit and reporting requirements for discharges from concentrated animal feeding operations. *Federal Register*, February. 8 pp. 7610-7644.

Walker, J. 1995. Seepage control from waste storage ponds and treatment lagoons. p. 70-78. In: *Innovations and New Horizons in Livestock and Poultry Manure Management Conference.* Proceedings (Volume 1) September 6-7, 1995, Austin, Texas. Texas Agricultural Extension Service and Texas Agricultural Experiment Station, Texas A&M University System, College Station, Texas.

Wallingford, G.W., L.S. Murphy, W.L. Powers, and H.L. Manges. 1974. Effect of beef feedlot lagoon water on soil chemical properties--growth and composition of corn forage. *Journal Environmental Quality* 3:74-78.

Watts, P.J., M. Jones, S.C. Lott, R.W. Tucker, and R.J. Smith. 1992. Odor measurement at a queensland feedlot. ASAE Paper 92-4516, Presented at International Winter Meeting of the American Society of Agricultural Engineers, Nashville, TN. December 15-17. 18 pp.

Watts, P.J. and R.W. Tucker 1993a. The creation and reduction of odour at feedlots. p. 3.1-3.14. Workshop on Agricultural Odours, Australian Water and Wastewater Association and Clean Air Society of Australia and New Zealand.

Watts, P.J. and R.W. Tucker. 1993b. The effect of ration on waste management and odour control in feedlots. p. 117-129. Recent Advances in Animal Nutrition in Australia 1993, University of New England, Armidale, NSW.

Wells, D.M., E.A. Coleman, W. Grub, R.C. Albin, and G.F. Meenaghan. 1969. Cattle Feedlot Pollution Study - Interim Report No. 1, WRC-69-7, Water Resources Center, Texas Tech University, Lubbock, Texas.

Wiese, A.F., D.E. Lavake, E.W. Chenault, and D.A. Crutchfield. 1977. Effect of composting and temperatures on weed seed germination. Abstract, Proceedings of Southern Weed Science Society 30:167.

Use of Manure on Grazing Lands

W.A. Phillips

I. Introduction

As livestock production systems become intensive and the production of animal manure concentrated, consideration must be given to the disposition of animal manure. Disposal of animal manure produced from intensively managed confinement livestock production systems is becoming a challenge for livestock producers because of potential negative environmental impact. In theory, returning captured nutrients to the land to be recycled through the soil and into plants, which were the original source of the nutrients as animal feeds, appears to be a logical and an ecologically friendly solution to the problem. Since grazing lands represent over 50% of the land mass used for agricultural production, are not densely populated, and have a populace with a mind set that is friendly to agricultural needs, it would seem reasonable to simply apply collected animal manure to grazing lands and let it be recycled back into plant matter.

Management of animal manure and its disposition have been the topic of symposiums and books that give nutrient composition of different types of animal manure as well as storage, handling, and consideration for disposal (Hore, 1975; Overcash et al., 1983a,b; Ross, 1979; Stewart, 1980; Van Der Meer et al., 1987). They are excellent reviews and cover this topic in great detail. I will not attempt to duplicate their work in this paper, but will focus on the following objectives: 1) to show the geographic relationship between grazing lands, animal distribution and the sources of animal manure, 2) to show the need for water quality research on grazing lands under different environments, plant communities, and grazing management to accurately predict the environmental impact of animal manure application, 3) to illustrate the impact of grazing on nutrient recycling and its implications under animal manure application, and 4) to list challenges for future research in applying animal manure to grazing lands.

II. Livestock Production Resources

A. Land Resources

The United States has 2.1 million farms that use 917 million ha of land for the production of food and fiber (USDA, 1992). Based on these data, the average U.S. farm has 440 ha of land, but the range in the size (105-1000 ha) of a farm is great across the nation and reflects different agricultural production systems that are determined by the climate and the adaptation of agricultural practices that best fit the land resources within the regional temperature and moisture ranges. Of the total land area available to agriculture, 187 million ha are used to grow crops. Cropping systems are located on the most productive land in the region requiring substantial investment in equipment to prepare the soil for planting, planting the crop, protecting the crop from insects, diseases, and weed encroachment, and harvesting the crop. Therefore, crop production requires a high yield per ha to recover the investment.

Land not suited for the production of a crop is dedicated to the production of plant biomass to be harvested by grazing livestock. USDA (1992) reported that 239 million ha of land are used to support grazing livestock. Grazing livestock are predominantly ruminant animals and can convert low value fibrous plant biomass to high value food (meat and milk) and fiber (wool and hair) products. In many cases the plant biomass is harvested directly by the animal, which eliminates the need for investment of fossil fuel and equipment for harvesting, storing, and feeding the forage. Grazing systems are based on renewable plant resources and tend to be extensive in nature, requiring larger land masses than crop production systems. Grazing livestock are also integrated with other agricultural production systems such as grazing forested areas.

Unique terms are used to describe the land, plant and animal components of a livestock-grazing system (Allen, 1991). First, grazing lands are any vegetated land that is grazed or has the potential to be grazed. This may include rangelands, prairie, meadows, and cropland. Rangelands can include grasslands, savannas, deserts, and tundra. Grasslands are dominated by grass, and savannas are grasslands with scattered trees. Savannas are often a transition zone between grasslands and forest-lands and were created by a climate that alternates between wet and dry seasons. The other two classes of rangelands are deserts and tundra, which are both areas where vegetation is absent or sparse due to a harsh growing environment. Prairies are nearly level or rolling grasslands that were originally treeless and usually characterized by fertile soils. Many of the original prairies are today's croplands,

Table 1. States in each of six agricultural regions

Region					
N. Atlantic	E. N. Central	W. N. Central	S. Atlantic	S. Central	Western
CT	IL	IA	DE	AL	AZ
ME	IN	KS	FL	AR	CA
MA	MI	MN	GA	KY	CO
NH	OH	MO	MD	LA	ID
NY	WI	NE	NJ	MS	MT
PA		ND	NC	OK	NV
RI		SD	SC	TN	NM
VT			VA	TX	OR
			WV		UT
					WA
					WY

which is land devoted to the production of cultivated crops. However, cropland may be used to produce a forage crop, which can be harvested and fed to animals or can be grazed directly by the animal. A forage is defined as the edible parts of plants, either consumed directly by the animal or mechanically harvested, transported to another area, and then fed to the animal.

Not all grazing lands are used exclusively to produce feed for animals. Owensby (1992) listed not only animals' use of rangelands, but hunting, fishing, outdoor recreation, vistas, forestry, and mineral development as uses of grazing lands. In some cases grazing animals are used as tools to impact the plant communities to make the area more productive for uses other than as a feed source for domestic animal production.

Land resources used for specific categories of agricultural production are regional, based on the soil resource and restricted by the climate. Using the same rationale as Overcash et al. (1983a,b), the 48 states were divided into six regions to describe the distribution of land resources for agricultural production (Table 1). The three regions that represent the central portion of the country (East North Central, West North Central and South Central) have over 157 million ha of cropland in use or idle (Table 2). This represents 84% of the total U.S. cropland resource. The implications of land resource allocation on animal manure production and utilization will be discussed later.

Regions that have soil and climatic limitations for crop production have smaller proportions of available land dedicated to crop production, but larger proportions of land used for forage production. This is also evident for land used for pasture, which in this case is defined as land suitable for crop production, but used for forage production. The Western North Central, South Central and Western regions

Table 2. Land resources found in each of six regions in the U.S.

Item	N. Atlantic	E. N. Central	W. N. Central	S. Atlantic	S. Central	Western
			------- million ha -------			
Cropland						
Crop	3.8	26.3	82.1	7.0	22.7	14.6
Idle	.5	4.8	14.9	2.0	6.6	2.4
Pasture	.8	2.6	10.3	2.3	10.5	2.2
Total	5.1	33.7	107.3	11.3	39.8	19.2
Grassland	1.0	4.5	35.7	4.8	54.6	106.0
Grassland + pasture	1.8	7.1	46.0	7.1	65.1	108.2

contain 196 million ha of grassland or 95% of the total ha in grassland (Table 2). When the ha in pasture is added to the ha in grassland, the total land area for forage production was increased to 219 million ha, but the percent of the total resource found in these regions was decreased slightly to 93%.

B. Animal Resources

Distribution and type of livestock across the six regions are a reflection of the land resource (Table 3). Livestock that are typically grazing animals (beef cattle, dairy cattle and sheep) are more prevalent in regions that have grassland resources than in regions dominated by cropland. Conversely, the population of nongrazing species, poultry and swine, are more numerous in regions with a large proportion of cropland, than in regions that are predominantly grassland. The animal distribution pattern revolves around a simple principle: it is more efficient to move the animal to the feed supply than the feed supply to the animal. This is especially true for low bulk density forage feeds. The three regions described in the previous paragraph as the major grassland resource contain 84% of the beef cattle, 48% of the dairy cattle and 95% of the sheep found in the U.S. From these data it is apparent that the lack of proximity of the animal production systems that produce manure and grasslands that could receive the manure is a major restriction in the use of animal manure on grasslands.

III. Grazing Lands and Water Quality

Grasslands are considered important in protecting surface water quality, and any surface management decision (application of animal manure, plant biomass

Table 3. Distribution of animals within the six regions of the U.S. on January 1, 1992

Item	N. Atlantic	E.N. Central	W.N. Central	S. Atlantic	S. Central	Western
			Region			
	---------------------------- 1,000 hd -------------------------------					
Beef cattle						
No. of hd	326	1602	9015	3528	12471	6811
% of total	1.4	4.8	26.7	10.5	36.9	20.2
Dairy cattle						
No. of hd	1716	2677	1605	714	1110	2069
% of total	17.3	27.1	16.2	7.2	11.2	20.9
Swine						
No. of hd	869	13793	28484	5553	3533	1181
% of total	1.6	25.8	53.3	10.4	6.6	2.2
Poultry						
No. of hd	141	69361	52763	78832	73582	70491
% of total	0.1	20.1	15.3	22.8	21.3	20.4
Sheep						
No. of hd	160	350	2546		2374	5293
% of total	1.4	3.3	23.7		22.1	49.5

(USDA, 1992.)

removal, tillage), changes in plant community composition, and nutrient load must be considered in light of their effect on water quality. In most watershed research, grasslands are used as check plots for water quality measurements because of their low erosion potential. However, few watersheds are dedicated to dealing with the impact of grazing animals in concert with surface management on water quality. Menzel et al. (1978) reported that yearly nutrient concentration and the amount of discharge from rangelands are as variable as those observed from cropland, but rangelands have lower average and maximum annual sediment discharge than croplands. Rangelands also have lower concentrations of soluble P than croplands, but any land use system must be studied over a long period of time to accurately record its effect on water quality. It can be assumed that the low sediment output from grassland watersheds can be attributed to the protection of the soil integrity by the undisturbed cover of plant material.

In well-managed grazing systems, runoff from rangelands is excellent quality. Neff (1982) reported that runoff from fine textured soils in southeastern Montana had an average total dissolved solid concentration of <300 mg/liter and that the annual erosion losses from these sites were <1500 kg/ha. Phillips et al. (1991)

reported annual sheet and rill erosion of native pastures to range from 2.5 to 9.9 Mg/ha on class 1 through 4 land, but these values were 50% lower than those reported for cropland. Soil erosion is highly correlated with nutrient losses and management practices that increase soil erosion and also increase nutrient losses (Swanson and Buckhouse, 1984). Animal manure applied to the soil surface of grasslands and not incorporated into the soil has the potential to move along the surface and to flow with the discharge.

Smith et al. (1984) concluded that the Modified Universal Soil Erosion Equation (MUSLE), which was developed for cropland, could be used to predict sediment yields from grasslands as well. The six components of the model are runoff volume, peak runoff rate, soil erodibility, slope length, erosion control practices, and crop management. Of these, only the latter two are under control of the land manager. In grazing systems the land manager can control agronomic inputs (fertilizer, herbicides, pesticides and animal manure), which determine the plant growth rate within climatic restrictions, the amount and timing of soil disruption, and the timing and intensity of plant biomass removal either by mechanical harvesting or by grazing, all of which affect soil erosion. Warren et al. (1986b) determined that infiltration rate was positively affected by surface litter, vegetative cover, and soil aggregate stability. If the amount of bare ground, rock cover, soil bulk density and mircorelief were increased, then infiltration rate was decreased. They also noted that as infiltration rate decreased, sediment production increased and as soil organic matter increased, sediment production decreased. From this perspective, applying solid animal manure to grasslands could provide nutrients for increased plant growth, which in turn would increase surface cover, soil organic matter content and eventually infiltration rate, which would result in a decrease in sediment production.

It appears that ground cover serves as an obstruction to overland sediment transport and as a protection from aggregate breakdown by direct raindrop impact (Thurow et al., 1986). While the type of ground cover is important from a land use standpoint, the amount of ground cover is more important than the type of cover in terms of soil stability (Thurow et al., 1988b). Raindrop impact energy is also affected by plant canopy structure, which intercepts rainfall and redirects stream flow (Thurow et al., 1987). Nutrient load and animal grazing strategies can alter plant community composition and affect sediment production (Thurow et al., 1988a; McCalla et al., 1984).

Application of animal manure to grazing lands can be used as part of a plant nutrient management strategy to achieve greater forage production. However, to utilize the increased forage produced, grazing systems must be more intensively managed. The intensity and duration of the grazing event is adjusted by the land manager to achieve maximum forage utilization or a desired animal performance goal. Grazing systems do impact infiltration rate and sediment production. In particular, intensive grazing systems tend to reduce soil hydrologic stability and increase sediment production (Warren et al., 1986a,b,c; McCalla et al., 1984; Takar et al., 1990; Wood and Blackburn, 1984; Olness et al., 1975).

IV. Impact of Grazing on Nutrient Recycling

A. Grazing Systems and Nutrient Balance

Grassland managers have two options for the harvesting of forages produced on grasslands. Forage can be 1) mechanically harvested as hay, haylage or silage, removed from the land, and stored for later use on this same site or at another location, or 2) harvested directly by grazing animals. The former option allows for a more even annual distribution of forage by harvesting excess forage during the growing season for feeding during low forage availability periods, but harvesting forages requires a large investment of capital and fossil fuel. Rotz and Muck (1994) estimate that 15 to 100% of the initial forage dry matter (DM) can be lost during the hay making process. Of these losses only a small portion occur during storage. In contrast, harvesting the forage for silage can reduce DM losses in the field, but increase the losses during storage (Phillips and Pendlum, 1984; Waldo, 1977). Overall, either process can result in an average of 15 to 25% of the initial forage DM being lost during harvesting or storage. While grazing does not utilize fossil fuel, harvesting efficiency of plant DM is low. Forage utilization under grazing may be as low as 45% or as high as 70% depending on animal management (Volesky et al., 1994). Under grazing management forage, DM is lost by the decay of dead plant parts, trampling of edible parts, and avoidance of certain areas due to plant structure or contamination by urine and feces (Cosgrove, 1992).

   In considering the use of animal manure on grasslands, the method of harvesting the forage must be factored into the decision process because it affects the recovery of applied nutrients. Ryden et al. (1984) noted ryegrass managed as a grazing system had greater annual nitrate losses than ryegrass managed under a hay harvesting system. However, the positive effects of haying were mitigated if the harvested hay was fed back to the animals on the same site at a later date. Under grazing, 80 to 90% of the nitrogen in the herbage consumed was returned to the land in the feces and urine.

B. Manure Production

Within a grazing system, 100% of the animal manure generated by the grazing animal is applied to the land daily. Thus, grazing lands are already receiving daily applications of animal manure. The chemical composition of feces and urine excreted is a function of diet, but the distribution is a function of stocking rate (hd/ha: Petersen et al., 1956; Russelle, 1992). Fecal and urinary N concentration as influenced by diet are presented in Table 4. Data were generated by conducting total collection of feces and urine using yearling steers (Phillips, unpublished data). The high quality diet was alfalfa pellets and the low quality diet was mature tall grass prairie hay. Diet N concentration was 3.1 and 0.9% for the high and low quality diet, respectively, and dry matter intake of the high quality diet was limited to 1.8% of body weight. Fecal output was similar among the two diets, but the steers fed the low quality diet had ad libitum access to the forage. This indicates that for

Table 4. Dry matter intake, fecal and urinary output and N retention by steers fed low and high quality forage diets

	Diet Quality	
	High	Low
Dry matter intake		
% body weight	1.8	1.8
kg/d	4.9	5.4
Fecal output, kg/d	2.2	3.3
Urine output, kg/d	9.3	2.6
N output, g/d		
Feces	50	36
Urine	77	13
N retention		
g/d	25	1
% of intake	16	2

grazing animals, a limit may exist on the amount of feces that can be processed each day. As diet quality increases, DM intake regulation shifts from physical to physiological controls, but fecal output will be similar among a wide range of diets (Hyer and Oltjen, 1986). The driving forces behind dry matter intake are physiological demand and gastrointestinal capacity (NRC, 1987).

In the example given in Table 4, urine output was greater when the steers consumed a high quality diet as compared to a low quality diet, and N concentration was higher, so N excretion was greater. The amount of N excreted in the feces is not as variable as the amount of N excreted in the urine. Typically, as N concentration in the forage increases, the amount of N excreted in the urine will also increase (Russelle, 1992). Ruminants are adept at extracting nitrogen from plants and excreting excess N through the urine. The amount of N retained by the animal is expressed as grams of N and is a reflection of energy:nitrogen and the physiological demand of the animal. In the present example, steers consuming the high quality diet retained 16% of the N consumed. This is the upper range of N retention for grazing animals and is observed only when the forage has a high DM digestibility. Of the remaining N, 39% was excreted in the feces and 61% was excreted in the urine. Steers consuming a low quality diet were almost in negative N balance. Each day some animals in this group excreted more N than they consumed. This N was derived from body stores and over an extended period, these animals would lose weight. Under these two schemes, one system was harvesting 16% of the N consumed while another was contributing N to the system. Under the latter scenario, supplemental energy and N would be provided to correct the nutritional deficiency, if the manager wanted the animals to gain weight. However, the majority of the extra nutrients provided by the supplemental feed would be added to the nutrient load of the pasture via the feces and urine.

Ball et al. (1991) described beef cattle systems for mature cows for the Southern Region. These systems were a combination of different forage types and land resources and each required a different supplemental program to offset any animal nutrient deficiencies. Similar systems could be described for each region of the country, but one factor that would be common to all systems is that some supplemental feed must be provided. Supplements are formulated and fed on a per animal basis. So as the stocking rate decreases, the added nutrients from the supplement are distributed over a larger area. However, if more land area is needed to support an animal unit, the forage production capacity is probably low and the capacity to recycle nutrients is also limited. The mixture of plants growing in an ecosystem is a function of the environment and management, and forage quality is also impacted by environment (Buxton and Fales, 1994; Nelson and Moser, 1994). Therefore, environment dictates plant community composition, forage quality, stocking rate, and nutrient recycling.

Church (1976) describes the chemical composition of feces as 50 to 69% moisture, 1.6 to 2.6% N and 21 to 29% fiber. The majority of fecal N is organic N and degrades slowly (Russelle, 1992). Nitrogen in the urine is predominantly urea (71%) with small amounts of ammonia. Because much of the deposited urinary N is lost through volatilization and leaching, only 30% is recovered by the plant (Ball et al., 1979: Russelle and Buzicky, 1988). Nitrogen cycling in dung pats is slower than urine spots and is also dependent upon climate.

C. Nutrient Distribution

Grazing increases heterogeneity in N distribution in pastures (Russelle, 1992). Nitrogen accumulated in the plant is relatively uniformly distributed across the surface, but grazers harvest plant N from a large area and then concentrate over 80% of the harvested N in localized spots (Petersen et al., 1956). Urine and dung spots are equivalent to applying 500 to 1000 kg of N /ha. Animals tend to avoid grazing areas contaminated by their own urine and feces, but more so for feces than urine. Cattle are more likely to graze areas where fresh dung has been deposited in comparison to areas where the decomposition process has begun (Latinga et al., 1987). Application of animal manure to grasslands would not affect the concentration of N in urine spots and dung pats any more than the use of inorganic N, but little research is available on the effects of solid animal waste application on grazing behavior.

Grazing systems harvest only a small portion of the N in the forage. Animal productivity and N harvested from three grazing systems based on different forages are compared in Table 5 (Phillips and Coleman). Native range (little bluestem, big bluestem and Indian grass) is a low input system using no N fertilizer, but produces less forage than systems based on introduced species such as Bermuda grass (*Cynodon dactylon* var. Midland) and Old World bluestem (*Bothriochloa* Spp.) that require N fertilization. Differences in forage production are reflected in the stocking

Table 5. Animal performance and nitrogen retention under three different forage systems

Item	Forage System		
	Native Range	Bermudagrass	Bluestem
Head/ha	0.6	3.7	4.8
Gain, kg/hd	78	59	65
Gain, kg/ha	49	220	314
N applied, kg/ha	0	140	110
N in gain/ha	0.84	3.8	5.4
N consumed/ha	6.94	40.8	52.9
N retained:consumed	0.121	0.093	0.102

(From Phillips and Coleman, 1995.)

rate and animal units/ha, but individual animal performance was similar among the three systems. Since animal gain per ha is a combination of individual animal performance and the stocking rate, gain/ha and the amount of N harvested/ha is greater for the introduced species than for native range. The Bermuda grass and Old World bluestem systems are more productive from an animal performance standpoint, but harvested less than 5% of the N applied. Lactating animals are more efficient in harvesting N from forages than growing livestock, but do not retain more than 20% of the N consumed. Wilkinson and Stuedemann (1991) concluded that grazing livestock remove 27 g of N/kg of body weight gain and 0.6 g of N/kg of milk produced. They estimated that a 250 kg steer consuming 6 kg of forage containing 3% N/d would retain only 12% of the N consumed. They also noted that mineral concentration of fecal material is higher than the original forage consumed.

Increasing the amount of N applied to pastures will in most cases increase plant growth but will not increase the recovery of applied N. Harvey et al. (1993) observed an increase in Bermuda grass forage production by doubling the amount of N applied in the form of swine effluent, but they did not increase the efficiency of harvesting applied N through grazing (Table 6). As the amount of applied N was increased, the recovery of that N decreased from 2.5 to 1.5%. Burns et al. (1990) used swine effluent to apply 335, 670 or 1340 kg of N/ha (Table 7). The amount of forage harvested as hay increased as the amount of N applied was increased, but the forage production response to additional N diminished as N application increased. At the highest level of N application, only 44% of the applied N could be recovered in the harvested forage. Not only is plant production affected by manure application, but excessive manure application can alter plant mineral composition and lead to health problems for animals that consume the forage (Ross, 1979).

Table 6. Animal performance and forage production of Bermuda grass pastures receiving swine lagoon effluent

	Amount of N Applied (kg/ha)	
	448	896
ADG, kg	0.54	0.65
Forage mass, kg/ha	1499	1484
Grazing days	1205	1231
Beef gain, kg/ha	651	800
N removed, kg/ha	11.1	13.6

(From Harvey et al., 1993.)

Table 7. Forage production and recovery of nutrients from Bermuda grass pastures receiving three levels of swine effluent over an 11-year period

	Level of Application		
	Low	Medium	High
N applied, kg/ha	335	670	1340
Forage produced, mg/ha	11.1	15.2	17.2
Forage DM:N	33	23	13
Recovery of nutrients, %			
N	72	74	44
K	72	69	42
P	44	41	27

DM = Dry matter.
(From Burns et al., 1990.)

V. Challenges for Future Research

Applying animal manure to grazing lands and harvesting the forage produced with grazing animals seems to be an ecologically sound practice, but there are a number of challenges that must be addressed before the practices can be integrated into sustainable livestock production systems. These are:

1) Transporting the manure from the site of production to the grazing lands.

2) Limiting the direct consumption of applied manure by grazing livestock.

3) Decreasing the effect of manure contamination of plants on forage intake.

4) Incorporating manure into the soil of grasslands to control odor and nutrient movement in surface and subsurface runoff.

5) Increasing the nutrient cycling capacity of the system to accept the added nutrients provided by the application of manure.

6) Developing plant communities that can tolerate grazing livestock and extract more available nutrients from the soil to prevent nutrient movement.

7) Determining the interactions of grazing livestock, plant growth dynamics, soil stability, and water quality.

References

Allen, V.C. 1991. *Terminology for Grazing Lands and Grazing Animals.* Commonwealth Press, Inc., Radford, VA.

Ball, D.M., C.S. Hoveland, and G.D. Lacefield. 1991. *Southern Forages.* Williams Printing Co., Atlanta, GA.

Ball, R., D.R. Keeney, P.W. Theobald, and P. Nes. 1979. Nitrogen balance in urine-affected areas of a New Zealand pasture. *Agron. J.* 71:309-313.

Burns, J.C., L.D. King, and P.W. Westerman. 1990. Long-term swine lagoon effluent applications on 'Coastal' bermudagrass. I. Yield, quality and element removal. *J. Environ. Qual.* 19:749-756.

Buxton, D.R. and S.L. Fales. 1994. Plant environment and quality. p. 155-199. In: G.C. Fahey (Ed.), *Forage Quality, Evaluation and Utilization.* ASA, CSSA and SSSA. Madison, WI.

Church, D.C. 1976. *Digestive Physiology and Nutrition of Ruminants.* Metropolitan Printing Co., Portland, OR.

Cosgrove, G.P. 1992. A comparison of grazing method effects on pasture and animal production. p. 305-329. In: Forrajes '92: *1ˢᵗ World Congress on Production, Utilisation and Conservation of Forages for Cattle Feeding*, Nov. 4-6, 1992. Buenos Aires, Argentina.

Harvey, R.W., J.P. Mueller, M.H. Poore, and B.W. Hogarth. 1993. Effect of nitrogen level from swine lagoon effluent on forage production, forage quality and steer performance. *J. Anim. Sci.*(Suppl. 1):23 (Abstract).

Hore, F.R. 1975. Managing livestock wastes. *Proc. 3ʳᵈ International Symposium on Livestock Wastes.* Urbana-Champaign, IL, April 21-24, 1975. ASAE. St. Joseph, MI.

Hyer, J.C. and J.W. Oltjen. 1986. Models of forage intake: A review. p. 226-231. In: F. Owens (ed.), *Feed intake by beef cattle.* November 20-22, 1986. Oklahoma City, OK. OSU, Stillwater, OK.

Latinga, E.A., J.A. Keuning, J. Groenwold, and P.J.A.G. Deenen. 1987. Distribution of excreted nitrogen by grazing cattle and its effects on sward quality, herbage production and utilization. p. 103-118. In: H.G. Van Der Meer (ed.), *Animal Manure on Grassland and Fodder Crops. Fertilizer or Waste?* Martinus Nijhoff Publ., Dordrecht, The Netherlands.

McCalla, G.R. II., W.H. Blackburn, and L.B. Merrill. 1984. Effects of livestock grazing on sediment production, Edwards Plateau of Texas. *J. Range Manage.* 37:291-294.

Menzel, R.G., E.D. Rhoades, A.E. Olness, and S.J. Smith. 1978. Variability of animal nutrient and sediment discharges in runoff from Oklahoma cropland and rangelands. *J. Environ. Qual.* 7:401-406.

Neff, E.L. 1982. Chemical quality and sediment content of runoff water from Southeastern Montana rangeland. *J. Range Manage.* 35:130-132.

Nelson, C.J. and L.E. Moser. 1994. Plant factors affecting forage quality. p. 115-154. In: G.C. Fahey, Jr. (ed.), *Forage Quality, Evaluation and Utilization.* ASA, CSSA and SSSA. Madison, WI.

NRC. 1987. *Predicting Feed Intake of Food-Producing Animals.* National Academy Press, Washington, D.C.

Olness, A., S.J. Smith, E.D. Rhoades, and R.G. Menzel. 1975. Nutrient and sediment discharge from agricultural watersheds in Oklahoma. *J. Environ. Qual.* 4:331-336.

Overcash, M.R., F.J. Humenik, and J.R. Miner. 1983a. *Livestock waste management.* Volume I. CRC Press, Inc., Boca Raton, FL.

Overcash, M.R., F.J. Humenik, and J.R. Miner. 1983b. *Livestock Waste Management.* Volume II. CRC Press, Inc., Boca Raton, FL.

Owensby, C.E. 1992. *Introduction to Range Management.* Kansas State University, Manhattan, KS.

Petersen, R.G., H.L. Lucas, and W.W. Woodhouse, Jr. 1956. The distribution of excreta by freely grazing cattle and its effect on pasture fertility. I. Excretal distribution. *Agron. J.* 48:440-444.

Phillips, W.A. and S.W. Coleman. 1995. Productivity and economic return of three warm season grass stocker systems for the Southern Great Plains. *J. Prod. Agric.* 8:334-339.

Phillips, W.A. and L.C. Pendlum. 1984. Digestibility of wheat and alfalfa silage with and without wheat straw. *J. Anim. Sci.* 59:476-482.

Phillips, J.M., H.D. Scott, and D.C. Wolf. 1991. *Environmental implications of animal waste application to pastures.* p. 30-36. In: Proc. 47th Southern Pasture and Forage Crop Improvement Conference, Mississippi State, MS. May 13-15, 1991.

Ross, I.J. 1979. *Animal waste treatment and recycling systems.* Southern Cooperative Series Bulletin 242. Project S-89. Univ. Kentucky, Lexington.

Rotz, C.A. and R.E. Muck. 1994. Changes in forage quality during harvest and storage. p. 828-868. In: G.C. Fahey, Jr. (ed.), *Forage Quality, Evaluation and Utilization.* ASA, CSSA and SSSA. Madison, WI.

Russelle, M.P. 1992. Nitrogen cycling in pasture and range. *J. Prod. Agric.* 5:13-23.

Russelle, M.P. and G.C. Buzicky. 1988. *Legume response to fresh dairy cow excreta.* p. 166-170. Proc. 1988 Forage and Grassland Conf., April 11-14, 1988. Baton Rouge, LA.

Ryden, J.C., P.R. Ball, and E.A. Garwood. 1984. Nitrate leaching from grassland. *Nature* 311:50-53.

Smith, S.J., J.R. Williams, R.G. Menzel, and G.A. Coleman. 1984. Prediction of sediment yield from Southern Plains Grasslands with the modified universal soil loss equation. *J. Range Manage.* 37:295-297.

Stewart, B.A. 1980. Utilization of animal manures on land: State-of-the-art. In: *Livestock Waste: A Renewable Resource.* 4th International Symposium on Livestock Wastes, Amarillo, TX, April 15-17, 1980. ASAE, St. Joseph, MI.

Swanson, S.R. and J.C. Buckhouse. 1984. Soil and nitrogen loss from Oregon lands occupied by three subspecies of big sagebrush. *J. Range. Manage.* 37:298-302.

Takar, A.A., J.P. Dobrowolski, and T.L. Thurow. 1990. Influence of grazing, vegetation life-form and soil type on infiltration rates and interrile erosion on a Somalion rangeland. *J. Range Manage.* 43:486-490.

Thurow, T.L., W.H. Blackburn, and C.A. Taylor, Jr. 1986. Hydrologic characteristics of vegetation types as affected by livestock grazing system, Edwards Plateau, Texas. *J. Range Manage.* 39:505-509.

Thurow, T.L., W.H. Blackburn, and C.A. Taylor, Jr. 1988a. Some vegetation responses to selected livestock grazing strategies, Edwards Plateau, Texas. *J. Range Manage.* 41:108-114.

Thurow, T.L., W.H. Blackburn, and C.A. Taylor, Jr. 1988b. Infiltration and interrile erosion responses to selected livestock grazing strategies, Edwards Plateau, Texas. *J. Range Manage.* 41:296-302.

Thurow, T.L., W.H. Blackburn, S.D. Warren, and C.A. Taylor, Jr. 1987. Rainfall interception by Midgrass, Shortgrass and Live Oak Mottes. *J. Range Manage.* 40:455-460.

USDA. 1992. *Agricultural Statistics.* United States Government Printing Offices, Washington, D.C.

Van Der Meer, H.G., R.J. Unwin, T.A. Van Dijk, and G.C. Ennik. 1987. *Animal Manure on Grassland and Fodder Crops. Fertilizer or waste?* Proc. International Symposium of the European Grassland Federation, Wageningen, The Netherlands, August 31-September 3, 1987. Martinus, Nijhoff Publishers, Dordrecht, The Netherlands.

Volesky, J.D., F. de Achaval O'Farrell, W.C. Ellis, M.M. Kothmann, F.P. Horn, W.A. Phillips, and S.W. Coleman. 1994. A comparison of frontal, continuous and rotation grazing systems. *J. Range Manage.* 47:210-214.

Waldo, D.R. 1977. Potential of chemical preservation and improvement of forages. *J. Dairy Sci.* 60:306-316.

Warren, S.D., W.H. Blackburn, and C.A. Taylor, Jr. 1986a. Effects of season and stage of rotation cycle on hydrologic condition of rangeland under intensive rotation grazing. *J. Range Manage.* 39:486-490.

Warren, S.D., W.H. Blackburn, and C.A. Taylor, Jr. 1986b. Soil hydrologic response to number of pastures and stocking density under intensive rotation grazing. *J. Range Manage.* 39:500-504.

Warren, S.D., T.L. Thurow, W.H. Blackburn, and N.E. Garza. 1986c. The influence of livestock trampling under intensive rotation grazing on soil hydrologic characteristics. *J. Range Manage.* 39:491-495.

Wilkinson, S.R. and J.A. Stuedemann. 1991. Macronutrient cycling and utilization in sustainable pasture systems. p. 12-18. In: Proc. 47th Southern Pasture and Forage Crop Improvement Conference. Mississippi State, MS, May 13-18, 1991.

Wood, M.K. and W.H. Blackburn. 1984. Vegetation and soil responses to cattle grazing systems in the Texas rolling plains. *J. Range Manage.* 37:303-308.

Impacts of Animal Manure Management on Ground and Surface Water Quality

A. Sharpley, J.J. Meisinger, A. Breeuwsma, J.T. Sims,
T.C. Daniel, and J.S. Schepers

I. Introduction

Since the late 1960s, point sources of water pollution have been reduced, due to their ease of identification and the passage of the Clean Water Act in 1972. However, water quality problems remain, and as further point source control becomes less cost-effective, attention is now being directed towards the role of agricultural nonpoint sources in water quality degradation. In 1994, the U.S. EPA reported that water quality problems in over 70% of surveyed rivers and lakes resulted from agricultural nonpoint sources (U.S. EPA, 1994).

Besides soil and pesticide loss from agriculture, most environmental concerns center on nonpoint transport of nitrogen (N) and phosphorus (P), which are essential inputs for optimum crop production. Due to their differing mobility in soil, N concerns revolve around nitrate (NO_3) movement through the soil profile to groundwater, while P concerns focus on soluble and particulate P transport in surface runoff.

Nitrate is a water quality concern because it has been linked to methemoglobine-mia in infants, to toxicities in livestock, and to increased eutrophication in both fresh and saline (e.g., estuaries) waters (Amdur et al., 1991; Sandstedt, 1990). The EPA has established a Maximum Contaminant Level (MCL) for NO_3-N in drinking water of 10 mg L^{-1} (45 mg $NO_3 L^{-1}$) to protect babies under 3-6 months of age. This segment of the population is most sensitive because bacteria that live in an infant's digestive tract can reduce NO_3 to nitrite causing hemoglobin to transform into methemoglobin, which interferes with oxygen-carrying ability of blood (Amdur et al., 1991).

Nitrate can also be toxic to livestock if reduced to nitrite where it can cause methemoglobinemia, which causes abortions in cattle. The tolerance level for livestock is about 40 mg NO_3-N L^{-1}, which is higher than for humans. Levels of 40-100 mg NO_3-N L^{-1} in drinking water are considered risky unless the feed is low in NO_3 and fortified with vitamin A (Sandstedt, 1990). Nitrate is also widely recognized as a contributing factor in eutrophication in estuaries where it can stimulate excess algal growth. This can lead to low dissolved oxygen concentrations as algae decompose and subsequently to adverse impacts on fish and shellfish populations.

Phosphorus in water is not considered to be directly toxic to humans and animals (Amdur et al., 1991). Because of this, no drinking water standards have been established for P (U.S. EPA, 1990a). However, advanced eutrophication of surface water restricts its use for fisheries, recreation, industry, and drinking, due to the increased growth of undesirable algae and aquatic weeds and oxygen shortages caused by their senescence and decomposition. Also, because of nonpoint N and P, many drinking water supplies throughout the world experience periodic massive surface blooms of cyanobacteria (Kotak et al., 1993). These blooms contribute to a wide range of water-related problems including summer fish kills, unpalatability of drinking water, and formation of trihalomethane during water chlorination in treatment plants (Kotak et al., 1994; Palmstrom et al., 1988). Consumption of cyanobacterial blooms or water-soluble neuro- and hepa-toxins, released when these blooms die, can kill livestock and may pose a serious health hazard to humans (Lawton and Codd, 1991; Martin and Cooke, 1994). Although N, carbon (C), and P are required for eutrophication, most attention has focused on controlling P inputs to fresh waters. Free air-water exchange of N and fixation of atmospheric N by blue-green algae means that N is often not a limiting factor in freshwater algal or bacterial growth.

In several states, confined animal operations are now the major source of agricultural income. For example, poultry production is the main agricultural industry in Alabama, Arkansas, Delaware, Georgia, and North Carolina. In Arkansas alone, it is a two billion dollar industry (National Agricultural Statistics Service, 1991), while in Delaware it accounts for over 65% of the total agricultural income. Large amounts of dairy manure are produced in California, Florida, Texas, and Wisconsin, and beef feedlot manure in California, Nebraska, and Texas. Swine manure is produced throughout the Midwest, especially in Indiana, Iowa, and North Carolina. In several other states, income from poultry and swine production has more than doubled in the last five years (e.g., MS, OK).

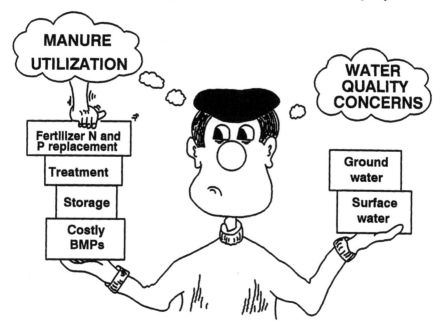

Figure 1. Given these factors, farmers often favor disposal rather than utilization of manure. (Adapted from Carton et al., 1993.)

The large amounts of manure produced in localized areas and the high cost of implementing effective Best Management Practices (BMPs) often favor disposal rather than utilization of manure (Figure 1). Continual application of manure at rates providing more N and P than removed by crops can increase soil N and P to levels that are of environmental rather than agronomic concern. Overapplication can enhance potential movement of N as NO_3 to groundwater and P in surface runoff. In many areas, land application of manure from confined animal operations is also of increasing concern as a source of N and P in agricultural drainage.

The extent to which NO_3 is causing problems in U.S. waters can be evaluated by reviewing U.S. Geologic Survey (USGS) and Environmental Protection Agency (EPA) data. A 1990 EPA national sample of 650 public system wells and 700 private wells found only 1.2% of the public wells exceeded the 10 mg NO_3-N L^{-1} standard while 2.4% of private wells were above the standard (USEPA, 1990b). The USGS data (USGS, 1988; USDA, 1991; Madison and Brunett, 1985) indicate the median level of NO_3-N in 288 of 316 principal U.S. aquifers (91%) did not exceed 3 mg L^{-1}, in 27 aquifers (8%) it was between 3-10 mg L^{-1} and in 1 aquifer it exceeded 10 mg L^{-1}. The same data base also shows there were 41 aquifers in 20 states for which 10% or more of the individual well samples had NO_3-N above 10 mg L^{-1}. These data indicate there are localized areas of high NO_3 concentrations within aquifers and point to a need for improved local management of N.

The USGS (Madison and Brunett, 1985) also analyzed samples collected over 25 years from nearly 124,000 wells and reported that in about 80% of wells, NO_3-N levels were at or below 3 mg L^{-1}, and that about 6% exceeded 10 mg L^{-1}. This estimate is considered to be biased upward due to heavier sampling in areas where NO$_3$ was suspected to be a problem. A survey of 1,430 randomly selected wells by the Monsanto company in 1988 and 1989 (Monsanto, 1990) reported that about 5% of them were above 10 mg L^{-1}, although for farmstead wells the value rose to nearly 10%.

Ritter and Chirnside (1987) reported that 34% of the 200 wells they sampled in southern Delaware, site of one of the most concentrated poultry producing areas in the U.S., exceeded the U.S. EPA drinking water standard. Similar reports of NO$_3$ contamination of groundwaters on the Delmarva Peninsula have been made by Robertson (1977), Weil et al. (1990), and most recently by Hamilton and Shedlock (1992) of the USGS. All studies have cited excess N from land application of poultry wastes as a major contributing factor to high NO$_3$ concentrations in shallow groundwater aquifers of the Delmarva Peninsula.

The NO$_3$ situation in southern Ontario (Gillham, 1989) is similar to the U.S. condition with NO$_3$ being ubiquitous in unconfined aquifers in areas of intensive agriculture (ranging from less than 10 to 50 mg NO$_3$-N L^{-1}), highly variable in concentration across a landscape, and highly variable in concentration vertically within an aquifer. Denitrification within aquifers can contribute to NO$_3$ variability and be an active NO$_3$ removal process under the proper conditions including shallow water tables (less than 2 m), transport of labile C to the water table, or microbially mediated reactions with sulfur compounds (Gillham, 1989; van Beek and Hettinga, 1989; Kolle et al., 1985).

There has been a rising trend for increased transport of N to coastal estuaries of the U.S. (Smith et al., 1987a; 1987b). The increased delivery rate is a concern because N is a common limiting factor for eutrophication in many bodies of water. It is generally thought that N is likely to be the limiting factor as salinity levels increase. Estuaries present a highly variable impact area because they are made up of mixtures of fresh and saline waters which may be P-limited in some areas and N limited in others. The extent to which agricultural loadings contribute to estuarian water quality is not well quantified, but agriculture's impact is known to be highly dependent on local conditions such as watershed land use (agriculture vs. forest vs. urban), hydrologic setting (precipitation, temperature, soils, etc.), and local estuarian factors (fresh vs. saline water mix, aquatic species, etc.).

The above brief overview indicates that NO$_3$ is a potential contaminant of our freshwater supplies and estuaries. The current level of contamination, however, is not widespread. Rather, NO$_3$ problem areas seem to be localized into relatively small domains with most contamination occurring in the upper part of the aquifer. These areas are usually associated with vulnerable water resources (shallow aquifers, coarse-textured soils, estuaries), excess N additions (high animal densities, excess fertilizer inputs, high mineralizing soils, etc.), and hydrologic/environmental conditions conducive to leaching (excess rainfall, excess irrigation, limited crop uptake, limited denitrification losses, etc.). Although NO$_3$ problem areas are not widespread, they are nonetheless important because they serve as a warning sign that remedial action (improved N management) is necessary in order to bring the agricultural-environmental system into closer harmony. The occurrence of NO_3 in groundwater may also signal other compounds could be leaching.

The potential for water resource contamination also strongly suggests that watershed or regional management of nutrients and animal manures must be instituted soon. Redistribution of excess nutrients away from land areas with contaminated aquifers or those rapidly approaching excessive N concentrations will require more than individual actions of farmers. The animal production industry as a whole (farmers, processors, integrators, distributors) and consumers of animal products must address the issue of improved N management, at a much larger scale, to avoid further degradation of groundwater quality in areas with high animal densities.

The number of soils with plant available P (soil test P) exceeding levels required for optimum crop yields has increased in recent years in areas of intensive agricultural and livestock production (Alley, 1991; Sims, 1992). In 1989, several state soil test laboratories reported the majority of soils analyzed had soil test P levels in the high or very high categories, which require little or no P fertilization (Figure 2). These categories vary among states and with soil test method (i.e., Bray vs. Mehlich), with high soil test P ranging from >10 to >75 mg P kg^{-1} and from >25 to >100 mg P kg^{-1} for very high. It is clear from Figure 2 that high soil P levels are a regional problem, with the majority of soils in several states testing medium or low (Figure 2). For example, most Great Plains soils still require P for optimum crop yields. Unfortunately, problems associated with high soil P are aggravated by the fact that many of these soils are located in lake-rich states and near sensitive water bodies such as the Great Lakes, Chesapeake, and Delaware Bays, Lake Okeechobee and the Everglades (Figure 2).

In Nordic countries (Denmark, Norway, Finland, and Sweden) the main reason for significant loss of P from agricultural land is considered to be the high net input of P to soil (calculated as 20 kg ha^{-1} yr^{-1} in recent decades) (Svendsen and Kronvang, 1991). Iserman (1990) reported P surpluses of between 55, 71 and 88 kg ha^{-1} yr^{-1} for West Germany, East Germany and The Netherlands, respectively. For agricultural soils in The Netherlands, Breeuwsma and Silva (1992) estimated that in 1990 about 43% of those in grass and 82% of those in maize were P-saturated. Dutch soils are considered saturated when more than 25% of its P sorption capacity is used (Breeuwsma and Silva, 1992). The potential for P loss from these P saturated soils is exacerbated by high water tables, tile drainage, and drainage channels. In the sand districts of central and southern Netherlands, leaching can contribute more to the transport of P to surface waters than runoff. For a watershed area with over 80% of P saturated soils and high total P concentrations in groundwater, the contribution by leaching and runoff was estimated to be 2.5 and 0.3 kg ha^{-1} yr^{-1}, respectively (Breeuwsma et al., 1989).

Animal manure can be a valuable resource if managed properly using cost-effective BMPs. In many areas, manure applications have improved soil structure and increased vegetative cover, thereby reducing runoff and erosion potential. However, in areas of intensive confined animal operations, where manure production exceeds local crop N and P requirements, agricultural, environmental and economical compromises are often opposed to one another. In most cases,

Figure 2. Percent of soil samples testing high or above for P in 1989. Highlighted states have 50% or greater of soil samples testing in the high or above range. (Adapted from Potash and Phosphate Institute, 1990; and Sims, 1993a.)

economic costs of remedial strategies are the crucial issues facing farmers trying to efficiently utilize manure (Figure 1).

Although we can develop BMPs for manure, should we expect farmers to bear the total cost of environmentally-sound management programs? If we do, we must expect lengthy delays in the implementation of new BMPs and only limited success in improving ground and surface water quality. There are several reasons for this. First, farmers cannot readily pass the cost of improved environmental practices directly to consumers, because they have little or no control over the price of their products. Therefore, they are unlikely to have additional funds to cover the costs of new practices. Second, farmers face severe labor shortages, and most improved BMPs require more labor inputs and more intensive management (e.g., additional testing, scouting, sampling, record-keeping). Farmers have little additional time and qualified, affordable technical assistance is often not available. Third, most farmers have little training in the rather complex field of solid waste management; hence, they must either use limited expertise to manage a difficult problem, or invest time and money in seeking either self-improvement through educational programs or advice from private consulting firms. Finally, it has been argued that expecting farmers to bear the entire cost of improved manure management practices for environmental purposes unfairly transfers the responsibility for an industry's waste management problem to those with the least resources and technical expertise.

This paper addresses issues involved in effective utilization of manure to minimize water quality impacts and draws on experience from the U.S. and Europe.

We attempt to answer the major questions of how water quality is impacted by land application of manure and what management strategies can minimize these impacts.

II. Groundwater Quality

A. Nitrogen Budgets

Reliable N budgets are more difficult to construct for manure than for fertilizer. The difficulties arise due to variability in initial manure composition, composition changes during storage, increased spatial variability due to nonuniform application, changes in soil bulk density, increased C availability which increases soil microbial activity, and the unpredictable nature of gaseous N losses by ammonia (NH_3) volatilization and denitrification. Gaseous N losses, either through volatilization of NH_3 or denitrification losses of N_2 and N oxides, are particularly difficult to quantify because of their dependence on environmental conditions and microbial activity. Anticipated NH_3 losses may not occur if little rainfall is received shortly after manure application. Manure application techniques that concentrate manure in smaller soil volumes (e.g., injection) may stimulate oxygen consumption and thus denitrification even in well-aerated soils. In many situations this can mean that the fate of as much as 30-40% of potentially-available N in manure cannot be accurately predicted.

Nevertheless, the general principle of conservation of mass used to construct N balances for soil and fertilizer N also applies to manures. This means that manure N additions will ultimately appear, to some degree, in one or more of the major N cycle outputs of NH_3 volatilization, denitrification, leaching beyond the root zone, crop uptake, or accumulation in the soil profile as organic N or mineral N.

Discussion of volatile NH_3 outputs in the context of a N budget usually focuses on direct losses from animal wastes or after application to soil. Recently, volatile N losses from living vegetation, presumably as NH_3, have been identified as another N loss mechanism. Francis et al. (1993) measured disappearance of 15N from corn (*Zea mays* L.) plants during productive growth stages and showed that volatile N losses from plants can be significant. They found these plant N losses increased with fertilizer N rate and reached 70 kg N ha^{-1} under conditions of near maximum corn yields. Volatile N losses from plant tissue are difficult to quantify without isotopic techniques because multiple metabolic processes (i.e., uptake, assimilation, remobilization, and volatile N loss) can occur simultaneously within a plant. This has been observed in a number of crops and is thought to be related to crop senescence (Holtan-Hartwig and Beckman, 1994). Unless specifically measured, plant volatile N losses are usually included in a category for "unaccounted for N." This category in a N budget is frequently assumed to be an estimate of NO_3 leaching because the other N sources and sinks can usually be measured or estimated. It is usually easier to calculate NO_3 leaching using the difference method than to measure the percolation rate and concentration of NO_3 in leachate. Improper assignment of volatile N losses from plants generally results in low estimates of crop N utilization from manure and fertilizer sources.

It is very difficult to quantitatively predict the fate of manure N due to the dynamic nature of the soil N cycle and the many interactions between manure-N and temperature, moisture, crop growth, and soil properties. However, the following discussion will develop some general precepts that apply toward managing manure N which can increase/decrease the odds of channelling manure N into one or more of the above mentioned outputs of the soil-crop N cycle.

1. Highly Instrumented Confined Micro-plots

It has not been feasible to date to utilize ^{15}N labelled manures due to the high cost of uniformly labelling all N compounds within the manure. This means that rather intensive, well-designed monitoring experiments must be used to study the fate and transformation of N under field conditions. Rolston and colleagues (Rolston and Broadbent, 1977; Rolston et al., 1978 and 1979) developed one of the most complete field N budgets using ^{15}N, illustrating the impact of manures on NO_3 transformations as well as the fickle nature of the soil N cycle. The study was conducted on 1 m^2 plots of well-drained Yolo loam soil and used two soil water levels corresponding to soil-water pressure heads of about -10 cm (about 90% saturation) and about -60 cm (about 80% saturation) that were maintained with an automatic traveling spray boom. The study was conducted in both summer and winter and included a manure treatment, a ryegrass (*Lolium perenne* L.) cropped treatment, and a nontreated control. The manured plots received the equivalent of 34 Mg ha^{-1} of beef feedlot manure (about 40% C) which was incorporated into the surface 10 cm of soil 2 weeks before the addition of $Ca(^{15}NO_3)_2$ which was added at 300 kg N ha^{-1}. No crop was grown on either manured or control plots. The plots were heavily instrumented with soil solution and atmosphere samplers, tensiometers, thermocouples, and neutron probe access tubes throughout the 1.2 m of undisturbed soil profile, in order to monitor soil temperature and estimate ^{15}N leaching. Temporary covers were also placed over each plot periodically to collect labelled N_2O and N_2 in order to directly estimate denitrification of labelled N. Following the 115-day study, eight soil cores (2.5 cm dia.) were taken to a depth of 1.2 m and labelled N was determined in the organic and inorganic fractions of soil. Ammonia loss was not a factor in this study since the labelled N was in the NO_3 form.

The summer data (Figure 3) show that manure markedly increased losses of labelled N due to denitrification (from 3% to 79%) in the warm summer months (average of 23° C) when the soil was very wet (-15 cm pressure head). Increased denitrification with manure was also associated with a decreased leaching loss of labelled N (from 87% to 12%). On the -70 cm pressure head plots, manure moderately increased denitrification, from 6% to 26%, increased soil immobilization from 8% to 21%, and caused a reduction of labelled residual NO_3 from 86% to 53% compared to the control. No leaching was observed on the -70 cm pressure head plots. Cropping the high moisture plots was not sufficient to prevent about

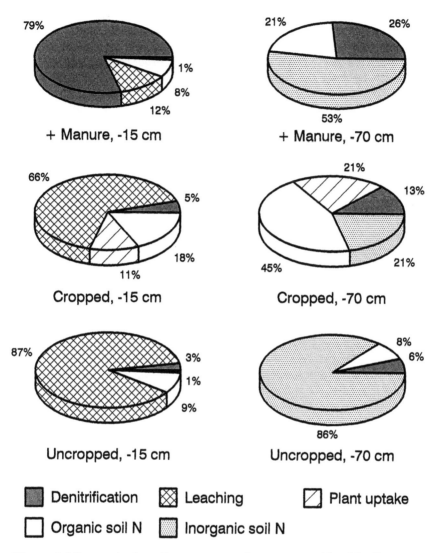

Figure 3. Nitrogen budgets for an uncropped, ryegrass, and beef feedlot manure treated (300 kg N ha⁻¹) Yolo loam in California during summer months at two soil moisture tensions (-15 cm and -70 cm). (Data adapted from Rolston and Broadbent, 1977; Rolston et al., 1978, 1979.)

66% of the $^{15}NO_3$ from leaching below the 60 cm root zone. Cool-season ryegrass was not an efficient N trap during warm months of the summer study.

Winter data for the high water treatment (Figure 4) show manure increased denitrification losses somewhat (from 0 to 22%) but major leaching losses occurred (about 77% leached). The fate of labelled N under wet soil moisture conditions after manure application was markedly different in the cool (average of 8° C) winter season where leaching dominated (Figure 4), compared to the warm summer season (Figure 3) where denitrification dominated. This was likely due to reduced microbial activity in the cool winter season. However, growing a cool-season grass with high water levels during winter reduced leaching to 39% due to N accumulation in the aboveground crop (35%) and immobilization of N in roots and soil organic matter (23%). When soil moisture was held at -50 cm, leaching losses were negligible and mineral N accumulated in the soil if uncropped. Denitrification losses increased with lower soil moisture due to a longer residence time of the labelled N, which gave the slower growing soil microbes more opportunity to denitrify the N. The winter data clearly show the benefit of growing a winter crop if one is seeking to reduce leaching losses and the level of residual NO_3.

Several precepts can be gleaned from the above study. One is that the fate of NO_3 is strongly affected by factors such as available C (which affects microbial activity and oxygen demand), soil water (which affects leaching and oxygen content), temperature (which affects microbial activity, crop growth, and water use), cropping practices (which create sinks for N and water), and soil properties (which interact with all of the above factors). If one is seeking to channel N through denitrification, then one should strive to juxtapose wet soil conditions (high soil moisture regimes, drainage management), high available C (recent manure additions), and high microbial activity (warm temperatures). If one is striving to minimize leaching, one should carefully control soil moisture (irrigation management if possible) and raise an actively growing crop. If one is interested in minimizing residual soil NO_3, then it is important to grow a crop with an N uptake pattern in close synchrony to availability of soil and added N, and immobilize N in above- and belowground biomass.

2. Large Long-term Field Plots

Pratt et al. (1976a, 1976b) reported N balance data for two types of dairy manure in the Chino-Corona Basin in California. Various rates of dry manure (open-air storage in corrals) and liquid manure (complete daily waste collection under confined animals) were applied to a well-drained Hanford sandy loam which had no textural discontinuities. Manure rates included levels suited to maximum forage production and high rates suited to manure application. Forage crops of winter barley (*Hordeum vulgare* L.) and summer sudangrass (*Sorghum bicolor sudanense* Moench) were grown annually over the 4- year study. Two irrigation regimes were studied, one with water inputs about equal to evapotranspiration (ET) and one about 33% greater than ET (1.33 x ET). Manures were applied just before crop establishment and were incorporated into the soil within 48 hours, which should

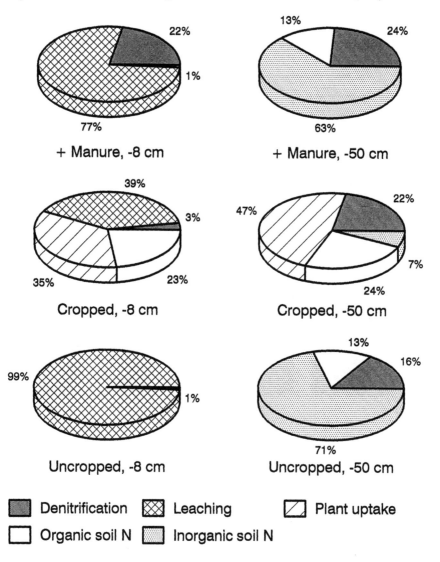

Figure 4. Nitrogen budgets for an uncropped, ryegrass, and beef feedlot manure treated (300 kg N ha⁻¹) Yolo loam in California during winter months at two soil moisture tensions (-8 cm and -50 cm). (Data adapted from Rolston and Broadbent, 1977; Rolston et al., 1978, 1979.)

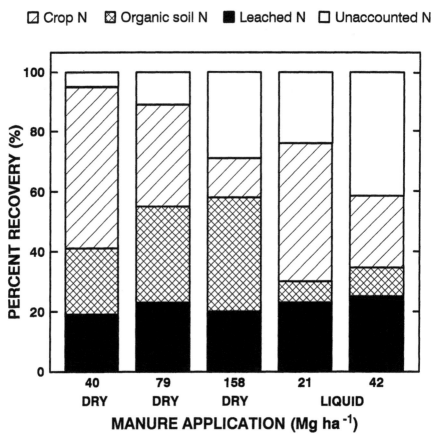

☑ Crop N   ☒ Organic soil N   ■ Leached N   ☐ Unaccounted N

Figure 5. Nitrogen budget for a Hanford sandy loam in California receiving dry or liquid dairy manure, where irrigational water inputs equalled evapotranspiration. (Data adapted from Pratt et al., 1976a,b.)

have reduced, but not eliminated, NH_3 loss. The N and Cl contents were determined in each manure, in all harvested crops, and in the soil (to 4.5 m) at the end of the replicated study. A chloride budget was constructed to estimate the fraction of the applied water that leached, and from the leaching fraction and soil water NO_3 concentration below the root zone, it was possible to estimate NO_3 leaching losses.

A summary of the irrigation equal to ET data (Figure 5) shows that dry manure behaved quite differently than liquid manure. Liquid manure was more biologically active, with more N lost to gaseous routes (denitrification plus NH_3 volatilization, i.e., unaccounted for N in Figure 5) and having a much smaller conversion into soil organic N. At the lowest rates of 40 Mg ha^{-1} dry manure (613 kg N ha^{-1} annually) and 21 Mg ha^{-1} liquid manure (922 kg N ha^{-1} annually) forage N uptake was the dominant process with leaching accounting for about 20% of the N. But as manure

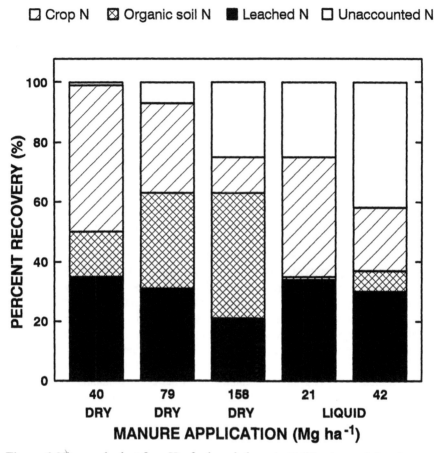

□ Crop N   ⊠ Organic soil N   ■ Leached N   □ Unaccounted N

Figure 6. Nitrogen budget for a Hanford sandy loam in California receiving dry or liquid dairy manure, where irrigational water inputs were 1.33 times evapotranspiration. (Data adapted from Pratt et al., 1976a,b.)

loading rates increased, the percentage going into crop uptake declined because crop N uptake was at a maximum for the site. Most of the extra N at the high rates was either converted into soil organic matter (dry manure) or was lost to the atmosphere (liquid manure).

With extra irrigation (Figure 6) the percentage lost to leaching increased from about 20% to 30%. The dry manure again gave a higher conversion into soil organic matter than the liquid, while gaseous losses were higher for the liquid. It is likely that NH$_3$ volatilization was higher with liquid manure since it was freshly exposed to the air, compared to the dry manure which had been stored in the open corrals and probably had lost much of its ammoniacal N before application (Safley et al., 1992).

These data illustrate some generalizations of the effect of manure type on soil N transformations. Liquid manure was much more biologically active per unit of applied N and resulted in greater gaseous N losses and lower increases in soil organic matter. Dry manure caused a significant increase in soil organic N which should be taken into account for future crops because it will increase soil N mineralization and reduce the need for supplemental fertilizer N. Crop N uptake was a major sink for manure N at low application rates, but declined in importance at high rates. These data also show that significant gaseous N losses can occur with manures, even on coarse textured soils, due to the large pool of available C added with manure. Nitrogen lost through leaching was affected more by irrigation practices than manure type. However, leaching losses of about 20% were encountered even with the lowest rate of manure, a continuously cropped soil, and careful irrigation management.

Other field N budget research with manure has been reported by Sherwood (1985) who applied liquid swine manure to grassland on two soil series and found that soil drainage had a marked effect on the partitioning of N between leaching and "unaccounted for N" (presumably denitrified N). A moderately well-drained loam soil had leaching losses of about 15% while a poorly-drained gley soil had leaching losses of only 2%. However, the poorly-drained soil had nearly twice as much N "unaccounted for" (16% vs. 9%). Thus, soil drainage interacted with manure to channel more N through leaching under well-aerated conditions and to favor denitrification under poorly-aerated conditions. Other manure N management research (Mathers and Stewart, 1971; Marriott and Bartlett, 1975) concluded that NO_3 leaching could be minimized only when the crop utilized most of the N; with excess manure N, NO_3 either accumulated in the crop, accumulated in the soil profile, or leached through the soil. Therefore, manure-derived NO_3 can contribute to leaching beyond the root zone, with potential impacts on groundwater quality, if it is not utilized by a crop or converted to gaseous N. However, enhancing gaseous losses as either NH_3 or N oxides can have negative environmental effects via damage to the ozone layer, increasing acid deposition, and increasing rainfall inputs of N to nearby surface waters.

B. Phosphorus Movement

High net inputs of P to soils do not immediately lead to enhanced leaching, because P is strongly adsorbed in most soils. However, the P sorption capacity of soils is not unlimited. Therefore, P may eventually reach groundwater depending on loading rates, sorption capacity, and depth to the groundwater table.

Figure 7 indicates the concentration that can be found in groundwater at very shallow depths (10-50 cm from the surface) in The Netherlands. The data are from an area where confined animal operations (mainly swine and poultry) are located on sandy soils with a relatively low P sorption capacity and shallow water tables Breeuwsma et al., 1989). The P concentration in the uppermost layer of ground-water (about 20 cm) was very high, with total P up to 10 mg L^{-1}. Some 80% of samples had concentrations in excess of the general quality goal for surface waters

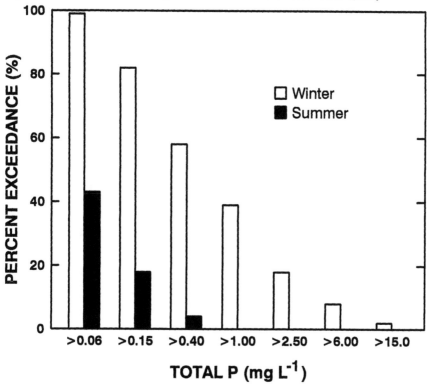

Figure 7. Percentage of observations for which a given concentration of total P in groundwater (upper 20 cm) is exceeded in the Schuitenbeek area in The Netherlands at high water levels in the winter and low levels in the summer.

in winter (0.15 mg L^{-1} of total P; Figure 7). During the summer, values were significantly lower because the groundwater was deeper. However, the general quality goal (0.15 mg L^{-1}) was still exceeded in 80% of the samples (Figure 7).

The decrease in dissolved P (DP) with soil depth should be considered when establishing quality criteria for groundwater. In The Netherlands this has been done by setting a standard for water quality at the Mean Highest Water level (MHW), which is defined as "the arithmetic mean during at least eight years of the three highest water levels per hydrological year (April 1 - March 31)." The concentration of DP at this depth should not exceed 0.10 mg L^{-1}. The soil is considered saturated with P if the concentration exceeds this value, or if model calculations indicate that this will happen in the future (Breeuwsma and Reyerink, 1992).

In a recent survey of the P concentration at MHW level, mean values were still below the critical limit, indicating the breakthrough of P is yet to come. Data of Figure 7 illustrate concentrations that can be expected after breakthrough.

III. Surface Water Quality

The main factors influencing the impact of land application of manure on surface water quality are the fate of manure N and P in surface soil and manure management. Due to adsorption of P by soil material, loss of P in surface runoff is generally of greater concern than leaching. There are two reasons for this. First, because P is adsorbed by the soil fraction most susceptible to losses in runoff (clays, oxides of Fe and Al), it is important to reduce soil erosion and, thus, transport of particulate P to streams and lakes. Second, P often accumulates in the upper few centimeters of soil, particularly under minimum tillage conditions where manures and fertilizers are not incorporated. Hence, DP levels can be quite high in the upper few cm of soil that are most interactive with surface runoff. Nitrogen can be a problem if organic matter is carried with surface runoff, e.g., surface manure applied to no-till systems, but these potential N problems are of secondary importance compared to P losses in surface runoff. Thus, this section places more emphasis on the impact of manure management on P than N loss in surface runoff.

A. Fate of Phosphorus in Soil

In the U.S. and Europe, recommendations for application rates of animal manures have been based on crop N requirements, in order to minimize NO_3 loss by leaching and potential for groundwater contamination. In most cases, this strategy has led to an increase in soil P levels in excess of crop requirements (Figure 2), due to the generally lower ratio of N:P added in manure (4:1; Table 1) than taken up by major grain and hay crops (8:1; Fertilizer Handbook, 1982). The ratio of N:P in manure can be reduced even further during stockpiling of manure because of N losses.

Buildup of P in soil is accentuated by the poor efficiency with which P added as fertilizer or manure is utilized by crops. In addition, manure application rates are often adjusted to account for only part of the N being plant available. This narrows the N:P ratio and hastens soil P buildup. For example, broiler litter has an approximate N:P ratio of 40:16.9 (Table 1); with a plant-available N value of 50%, the ratio becomes 20:16.9.

The rate of P buildup is rapid, particularly where manure is applied (Table 2). Fertilizer and manure inputs to intensive crop and livestock production areas of the Po region of Italy are 40 to 60 kg ha^{-1} yr $^{-1}$ and in central and southern Netherlands are 100 to 200 kg P ha^{-1} yr $^{-1}$ (Breeuwsma and Silva, 1992). These inputs are much greater than average annual crop removal rates of 30 to 40 kg P ha^{-1} (Table 2). Clearly, P inputs of up to 190 kg ha^{-1} greater than crop removal can quickly create environmental problems. However, considerable time is required for significant depletion of high soil test P. Johnston (1989) and McCollum (1991) found that it took about 9 years for Olsen P contents of a clay loam soil and Mehlich 3 contents of a sandy soil, respectively, to decrease to half its original value. Even so, many soils may still be above agronomic (or even environmental) critical values. Also, it

Table 1. Average P, N, and K contents (dry weight basis) of animal manures

Animal	Nitrogen	Phosphorus	Potassium
	------------------------------g kg^{-1}------------------------------		
Beef	32.5	9.6	20.8
Dairy	39.6	6.7	31.6
Poultry layers	49.0	20.8	20.8
Poultry broilers	40.0	16.9	19.0
Sheep	44.4	10.3	30.5
Swine	76.2	17.6	26.2
Turkey	59.6	16.5	19.4

(Adapted from Gilbertson et al., 1979.)

Table 2. Potential input of P in fertilized and animal manures in areas of intensive livestock production in Italy (Po Basin) and The Netherlands (Sand districts) and removal in maize and wheat

	---Potential input---		Potential removal[a]		---Balance---	
	Fertilizers	Manures	Maize	Wheat	Maize	Wheat
	-----------------------------kg P ha^{-1} yr^{-1}-----------------------------					
Italy						
Piemonte	22	16	40	30	-2	+8
Lombardia	31	30	40	30	+21	+31
Veneto	35	26	40	30	+21	+31
Emilia-Romagna	27	19	40	30	+6	+16
The Netherlands						
Salland-Twente	15	85	40	30	+60	+70
West Veluwe	15	174	40	30	+149	+159
Maijerij	15	130	40	30	+105	+115
South Peel	15	204	40	30	+179	+189

[a]Maize and wheat yields of 10 and 5 t ha^{-1} yr^{-1}, respectively.
(Adapted from Breeuwsma and Silva, 1992.)

is unlikely that depletion time will be linear; hence, the second half of depletion may be much slower.

Table 3. Phosphorus balance and efficiency of plant and animal uptake of P for the U.S. and several European countries[a]

Location	Input	Output	Surplus	Efficiency of		
				Plant uptake	Animal uptake	Total uptake
	----------kg P ha^{-1} yr^{-1}----------			---------------%---------------		
U.S.	39	13	26	56	15	33
The Netherlands	143	55	88	69	24	38
E. Germany	79	8	71	59	10	11
W. Germany	84	29	55	76	34	35

[a]Data for U.S. adapted from National Research Council (1993) and for European countries from Iserman (1990).

Efficiency of P uptake by animals from feed materials can also influence the fate of P in soil and transport in runoff. A generalized P balance and efficiency of plant and animal uptake of P for the U.S. and several European countries indicate the potential for P accumulation in agricultural systems (Table 3). Although the magnitude of P input and output varies between countries, the relative efficiency of plant and animal removal is similar. In spite of relatively efficient P utilization in crop production of 56 to 76%, P utilization by agriculture in total (animals or plants), is only 11 to 38% (Table 3). In contrast to crop production, P efficiency of animal production is only 10 to 34%. Thus, the overall efficiency of agriculture is dominated by that of animal production, as 76 to 94% of the total crop production is fed to animals (in addition to P additives) with low P utilization efficiencies. Animal-specific excretion rates, ranging from 70 to 80% for dairy cows and feeder pigs and from 80 to 90% for poultry (Iserman, 1990), substantiate the poor utilization of P by livestock. Clearly, agricultural systems including confined animal operations can determine the overall efficiency of P utilization in agriculture and, thereby, the magnitude of P surpluses or potential soil accumulations.

Once soil test P contents become very high, the potential for P loss in runoff is greater than any agronomic benefits of further P application. This is due to the dependence of P loss in runoff on soil P content (Figure 8). A wide range in soil test P contents (50 to 500 mg kg^{-1} as Mehlich 3) for a Captina silt loam in Arkansas was obtained by application of poultry litter to fescue pastures (Pote et al., 1996). The DP concentration of runoff from these soils was positively correlated to soil test P content (Figure 8; $r^2 = 0.72$; $p < 0.05$). Several other studies have also reported a dependence of runoff DP on soil test P (Olness et al., 1975; Romkens and Nelson, 1974; Schreiber, 1988).

In addition to soil test P, P added in manure can influence other forms of soil P. Total soil P increased up to 15 fold with long-term manure application (Table 4). In all studies, manure P increased inorganic P to a greater extent than organic P. Although amounts of P in all fractions increased, there was a shift in dominant inorganic P form from Al- and Fe-bound P to Ca-bound P with application of each

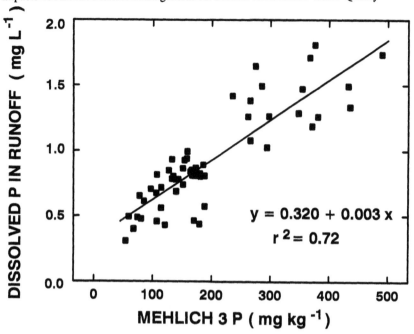

Figure 8. Relationship between the soil test P content of surface soil (0-5 cm) and dissolved P concentration of runoff from fescue in Arkansas. (Adapted from Pote et al., 1996.)

manure type except swine slurry (Table 4). The greater increase in Ca-bound P compared to other P forms in soils treated with manure may be due to the increase in soil pH and large amounts of Ca added in manure. This would favor formation of hydroxyapatite and metastable Ca-P forms (Wang et al., 1995). It is thus possible that current soil test methods, such as the acidic Bray and Mehlich extractants, may overestimate plant-available P in soil treated with manure released by dissolution of Ca-P complexes. This may also lead to an overestimation of the potential for soil P to enrich runoff with P.

B. Manure Management

The main factors affecting N and P loss in runoff from land receiving manure include the rate, method, and timing of manure application. As expected, N and P loss in runoff increases with the rate of manure application (Figure 9). The concentration of total N, ammonium (NH_4), DP, and total P increased linearly with increased poultry litter (Edwards and Daniel, 1993c) and swine manure rates (Edwards and Daniel, 1993b). The application rates used represent half, average, and twice the recommended rates for fescue in northeast Arkansas. Although

Table 4. Inorganic and organic P fractions in surface soil (0-5 cm) and treated with manure

Manure type[b]	Soil type	Study period	Soil pH	P applied	Soil total P	Inorganic P			Organic P	
						Labile	Al/Fe bound	Ca bound	Labile	Stable
		yr		kg ha⁻¹ yr⁻¹	mg kg⁻¹	%				
Beef feedlot	Pullman	–	6.8	0	353	8	26	12	11	25
	clay loam	8	7.2	273	996	15	22	29	10	17
Dairy	Spodosols	–	4.6	0	34	18	49	9	--	14
		25	6.9	d.n.a.	1680	9	8	70	--	2
Poultry litter	Kullit silt	–	5.9	0	212	6	23	13	11	20
	loam	12	6.1	130	1410	17	16	25	13	12
Swine slurry	Captina silt	–	5.0	0	273	5	24	3	5	14
	loam	9	5.9	130	566	17	28	16	4	7

[a]Labile, Al/Fe bound, and Ca bound inorganic P extracted by NaHCO$_3$, NaOH, and HCl and labile and stable organic P extracted by NaHCO$_3$, and NaOH by the method of Hedley et al. (1982).

[b]Data for beef feedlot from Sharpley et al. (1984), dairy manure from Graetz and Nair (1995), poultry litter from Sharpley et al. (1993), and swine slurry from Sharpley et al. (1991).

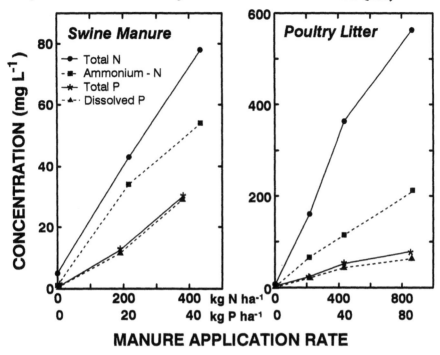

Figure 9. The concentration of N and P in runoff from a Captina silt loam in fescue as a function of the rate of surface applied manure. (Data adapted from Edwards and Daniel, 1993a,c.)

rainfall intensity and duration influence the concentration and overall loss of manure N and P in runoff, the relationship between potential loss and application rate is critical to establishing environmentally-sound manure management guidelines, as discussed in a later section.

Incorporation of manure into the soil profile either by tillage or subsurface placement reduces the potential for N and P loss in runoff. Mueller et al. (1984) showed incorporation of dairy manure by chisel plowing reduced total P loss in runoff from corn 20 fold compared to no-till areas receiving surface applications (Figure 10). However, the concentration of P in runoff did not decrease as dramatically as the mass of P lost (Figure 10). This was due to an increase in infiltration rate with manure incorporation and consequent decrease in runoff volume. In fact, runoff volume from no-till corn was greater than from conventional-till corn (Mueller et al., 1984). Thus, P loss in runoff is decreased by a lower surface soil P content and reduction in runoff with incorporation of manure.

A similar situation was observed by Pote et al. (1996) following application of poultry litter to fescue in Arkansas. Although soil test P, as altered by poultry litter, was related to the DP concentration of runoff ($r^2 = 0.72$; Figure 8), no significant relationship was obtained between the mass of DP lost (kg ha^{-1}) and soil test P (mg kg^{-1}) ($r^2 = 0.05$). Incorporation of poultry litter increased organic C content and

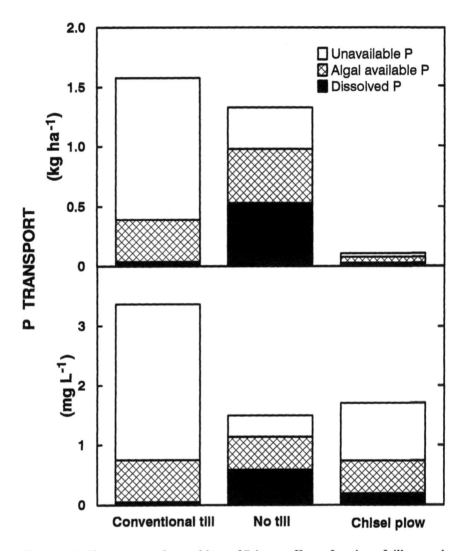

Figure 10. The concentration and loss of P in runoff as a function of tillage and method of manure application to corn in Wisconsin. (Data adapted from Mueller et al., 1984.)

decreased bulk density of the surface 5 cm of soil, with the result that runoff decreased. From the results of Pote et al. (1996), it appears that we can have soils of 59 and 381 mg kg[-1] soil test P, which both supported a DP loss of 0.18 kg ha[-1] during a 30-min runoff event, even though DP concentration was higher with elevated soil test P (0.49 and 1.26 mg L[-1], respectively). This highlights the dangers of using soil test P as the sole criteria to determine manure applications based on the potential loss of P in runoff.

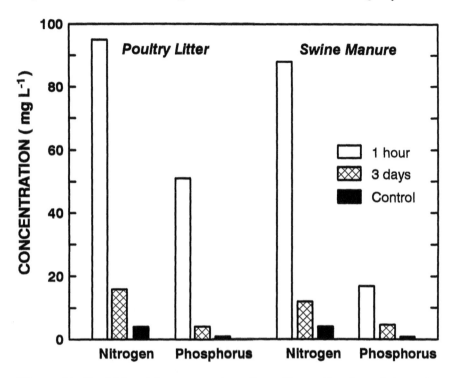

Figure 11. Total N and P concentration of runoff as a function of time after application of poultry litter and swine manure (Data adapted from Westerman and Overcash, 1980.)

Timing of manure applications relative to occurrence of rainfall influences N and P loss in runoff. The major portion of annual N and P losses in runoff generally occurs during one or two intense storms (Edwards and Owens, 1991; Smith et al., 1991b). If manure applications are made during periods of the year when intense storms are likely, then the percentage of applied N and P lost would be higher than if applications are made when runoff probabilities are lower. Burwell et al. (1975) demonstrated that runoff P loss from watersheds was greatest during the planting season, a time of intense rains, high P application, and minimum crop cover.

An increase in the length of time between applying manure and a runoff event reduces N and P transport in runoff (Figure 11). When simulated runoff was delayed from 1 hr to 3 d after poultry or swine manure application to fescue in North Carolina, Westerman and Overcash (1980) found concentrations of total N and P in runoff were reduced approximately 90%. The reduction in N and P loss was attributed to NH$_4$ volatilization and soluble N infiltration prior to runoff, and increased time for P sorption, respectively. With longer periods between swine manure application to fescue and rainfall-runoff initiation in Arkansas, Edwards and Daniel (1993b) found little effect of time on N or P loss in runoff. These studies suggest intervals of more than 3 d between manure application and runoff will not greatly affect N or P loss in runoff.

By necessity, some farmers apply manure to the surface of soils during fall and winter months, when more time is available for this operation or frozen conditions result in less physical damage to soils by application equipment. As manure is not incorporated and plant growth is minimal, sorption and plant P uptake are quite low, and potential for P loss during spring rainfalls is greatly increased. In fact, Wisconsin recommends against manure application to frozen soil of >9% slope, within 60 m of a watercourse or waterbody, or on the 10 yr or less flood plain of rivers (USDA-SCS, 1993).

The proportion of N and P added in manure that is transported in surface runoff is summarized in Table 5 for several studies. Clearly, factors controlling manure N and P loss in runoff that were just discussed are dynamic, causing a high variation in relative losses (Table 5). Although the number of runoff events and period of study varied, no one type of manure appeared to be more likely to cause N or P loss in runoff.

Less information is available on the transport of other chemicals applied in manure. Although there is a potential for accumulation of feed additives such as insecticide-based As, Cu, and Zn in surface soil (Kingery et al., 1993; 1994), quantification of their transport in runoff has not yet been reported. Even so, it is logical to assume that an accumulation of these trace elements in surface soil will increase their potential for movement from the site of application with runoff and erosion. As the amount and chemical composition of such growth stimulants and insecticides in feed become more carefully controlled, potential water quality concerns about trace elements or organics may be alleviated by changing feed composition. However, potential benefits on animal health vs. the environmental impacts of very low concentrations of metals and organics have not received a thorough risk analysis. Eliminating or reducing the amount of an economically beneficial feed additive is unlikely to be accepted by animal industries unless a clear environmental risk can be shown.

The presence of the growth hormones testosterone (0.8 to 2.9 ng L^{-1}) and estrogen (1.2 to 4.1 ng L^{-1}) has been reported in several streams of the Conestoga River Valley, Chesapeake Bay watershed (Shore et al., 1995). Levels of these hormones >0.5 ng L^{-1} are considered appreciable (Shore et al., 1995). As manure from 7 week-old broilers in the Conestoga River Valley contained 13 ng kg^{-1} testosterone and 65 ng kg^{-1} estrogen, its land application may be a potential source of these hormones (Shore et al., 1995). Five months after poultry manure application, no estrogen could be found in the surface soil but testosterone levels were at 50 ng kg^{-1}. Surface runoff from manured fields contained 215 ng L^{-1} testosterone and 19 ng L^{-1} estrogen (Shore et al., 1995). The authors concluded that runoff from land application of poultry manure may be a source of these hormones in the Chesapeake Bay watershed.

IV. Remedial Strategies

Environmentally sound management of manure and remediation of impaired water quality involves three general strategies -- improving N- and P-use efficiency in agricultural systems which include animal operations; innovative manure management programs; and targeting watershed management to environmentally-sensitive areas.

A. Improving Nitrogen-Use Efficiency

Management practices for improving N-use efficiency are vital for minimizing NO_3 losses to groundwater. Specific BMPs for N vary from region to region due to large-scale differences such as climate, geologic material, depth to water, irrigation or drainage practices, etc., and also due to small-scale differences such as soils, cropping systems, past field history, N source, etc. Therefore, BMPs for N will necessarily be quite site-specific and should be prescribed by someone who has a good knowledge of the local N cycle. In the U.S., this task has been allocated to state and federal extension, soil conservation, and research agencies. These agencies help develop tools for improved N management and do research to evaluate N management systems.

Although BMPs for N will depend on the specific hydrologic setting, field, and source of N, there are some basic N management principles which underlie these practices for minimizing N losses to groundwater or surface water. The most fundamental principle is to supply only the N needed to meet the N requirement of the next crop, and to apply it in synchrony with crop use. Steps in applying this principle include (i) estimating the N requirement of the next crop (expected yield), (ii) evaluating N available from native sources (soil N mineralization, residual soil NO_3, irrigation water, etc.), (iii) subtracting the available native N from the crop N requirement to estimate supplemental N needs, (iv) determining the source of supplemental N (manure, fertilizer, crop residues), and (v) determining the most efficient and practical management practice for the specific source of supplemental N (rate, time, and placement of the N). The remainder of this section will focus on general remedial measures for manure N, since this is the source most likely to be available in livestock based systems. Chapter 4 (Bouldin et al.) of this publication contains examples of specific applications of these principles in New York. The above procedures frequently require analysis on a monthly or weekly basis because N mineralization and crop N uptake rates change during the growing season.

1. Remedial Measures for Manure Nitrogen When Nitrogen is Deficient

Applying the above 5 steps for a manure N source is the basis for improving N efficiency and minimizing groundwater losses of NO_3 in livestock based systems.

Table 5. Proportion of N and P added in manure transported in surface runoff

Crop	Amount added		Study period	Percent loss		Reference and location
	N	P		N	P	
	----kg ha⁻¹ yr⁻¹----			-------%-------		
Dairy manure						
Corn	451	108	3 months	11.1	8.1	Klausner et al. (1976), NY
C. Bermuda grass	807	175	4 years	1.6	—	Long (1979), AL
Fescue	133	142	4 events	2.1	1.3	McLeod and Hegg (1984), SC
Corn	—	100	2 events	—	6.2	Mueller et al. (1984), WI
Fescue — dry[a]	415	104	8 events	2.8	7.9	Reese et al. (1982), AL
Fescue — slurry[a]	403	112	8 events	4.1	12.1	Reese et al. (1982), AL
Alfalfa — spring[b]	205	21	1 year	10.7	12.1	Young and Mutchler (1976), MN
Alfalfa — fall[b]	285	55	1 year	13.2	13.3	Young and Mutchler (1976), MN
Corn — spring[b]	205	21	1 year	1.0	2.4	Young and Mutchler (1976), MN
Corn — fall[b]	285	55	1 year	0.8	4.7	Young and Mutchler (1976), MN
Poultry litter						
C. Bermuda grass	1177	—	2 years	4.3	—	Dudinsky et al. (1983), GA
	699	—	5 years	4.6	—	Dudinsky et al. (1983), GA
	1397	—	5 years	10.7	—	Dudinsky et al. (1983), GA
Fescue	218	54	1 event	4.0	2.2	Edwards and Daniel (1993c), AR
	435	108	1 event	4.2	2.3	Edwards and Daniel (1993c), AR
Fescue	450	150	1 year	0.3	1.9	Heathman et al. (1995), OK
Fallow	287	165	1 event	20.0	19.0	Westerman et al. (1983), NC

Table 5. continued

Poultry manure						
Fescue	220	76	1 event	3.1	2.6	Edwards and Daniel (1992), AR
	879	304	1 event	3.3	3.2	Edwards and Daniel (1992), AR
Fescue	149	85	4 events	4.2	2.4	McLeod and Hegg (1984), SC
Fallow	428	95	1 event	5.0	12.6	Westerman et al. (1983), NC
Swine manure	217	19	1 event	2.6	7.4	Edwards and Daniel (1993a), AR
Fescue	435	38	1 event	2.9	8.4	Edwards and Daniel (1993a), AR

[a]Applied as dry manure or as a slurry.
[b]Manure applied in the spring and fall.

A step-by-step consideration of this process is useful because it emphasizes the complexity of managing N efficiently in livestock systems.

a. Crop N Requirement

Estimating crop N requirement can be divided into the tasks of determining the physiological N requirement of a crop per unit of dry matter and estimating dry matter production. For example, corn N requirement can be defined as the N concentration necessary to satisfy the physiological needs of corn, and it has been estimated as about 11 kg N t^{-1} of aboveground dry matter (Stanford 1973; Meisinger et al., 1985). Estimating dry matter production is usually done through an estimate of the most likely or the expected yield. This appraisal should be a realistic estimate of the most likely yield and should include site- and management-specific factors such as soil resources, weed control, plant population, and timeliness of field operations.

Selection of the expected yield goal is one of the most critical BMPs for N, because most fertilizer and/or manure application rates are based directly on anticipated yield. Several studies have shown that farmers, or those advising them, often have unrealistic yield expectations and that resultant overfertilization with N can be directly related to long-term increases in groundwater NO$_3$ (Hergert, 1987; Schepers et al., 1991).

The most direct way to integrate over all site-specific factors is to calculate the average yield of the specific soil-crop system over the past 3 to 5 years. One can then adjust the average yield for unusual conditions (eliminating unusually wet or dry years), for current conditions (stored soil moisture, planting date, tillage practices, etc.), or for new technologies (new varieties, new irrigation, etc.) and then calculate a final estimate of expected yield. In any case, it is important to base the estimated yield on "real world conditions," i.e., actual field yields, in order to avoid excess N applications which can lead to inefficient N use. More detailed discussion of yield estimation approaches can be found in Meisinger et al. (1992b).

b. Native N Availability

The second step is to evaluate N available from native sources, i.e., sources that are not directly manageable by the farmer. These sources include N present in the root-zone as inorganic NO$_3$, N released through organic matter decomposition (mineralized N), and N contributed through water sources (irrigation or atmospheric inputs). The literature is voluminous on estimation techniques for these sources of N, because this has been a goal of soil scientists for the past 80 years.

Reviews on crop N requirement techniques (Keeney, 1982; Stanford, 1982; Meisinger, 1984; Meisinger et al., 1992b; Dahnke and Johnson, 1990; Hergert, 1987) have generally concluded that (i) measurement of root-zone NO$_3$ just before crop establishment (preplant NO$_3$) has been conclusively shown to be important in sub-humid climates (Dahnke and Johnson, 1990; Hergert, 1987); (ii) preplant NO$_3$

can be significant in humid climates under conditions of excess N inputs the previous year, low precipitation winters, low percolation soils, and deep-rooted plants (Angle et al., 1993; Bundy and Malone, 1988; Roth and Fox, 1990; Meisinger et al., 1992a); (iii) laboratory mineralization tests to estimate N mineralization (microbial incubations, chemical extractions, total N analyses) have met with only limited success and have not been accepted as a routine "soil N test" (Keeney, 1982; Stanford, 1982; Meisinger, 1984); (iv) indirect N mineralization credits, e.g., legume N credits for previous crops of alfalfa (*Medicago sativa* L.) or soybeans (*Glycine max* Merr.) or manure N credits, are most frequently used, but these generalizations are not highly accurate for a specific site (Meisinger, 1984; Meisinger et al., 1992a); (v) N inputs from irrigation water can be directly estimated through water analysis and irrigation volumes, and estimates of atmospheric N inputs can be made from regional maps with local adjustments for NH_3 sources (e.g., Schepers and Mosier, 1991; Meisinger and Randall, 1991); and (vi) the most recent tools for including native sources of N are the pre-sidedress NO_3 test (PSNT) and the leaf chlorophyll meter (Meisinger et al., 1992b; Schepers et al., 1992a). These latest two tools will be further discussed because they offer some unique benefits for use in manured systems.

The PSNT and leaf chlorophyll tests are especially important in manured systems because they survey the N status of the soil or plant and allow adjustments in N fertilization practices based on local soil-plant conditions. The PSNT was first proposed by Magdoff et al. (1984) and measures the NO_3 concentration in the surface 30 cm of soil when the corn is 15-30 cm tall. The PSNT has been successful in many soils of humid temperate climates, especially in high N mineralizing soils, i.e., manured soils. It assumes a sidedress N management program and is based on the timely monitoring of the soil NO_3 pool (believed to result from recent spring mineralization) just before the warm-season corn crop begins its period of rapid N uptake. If soil NO_3-N concentration is above 20-25 mg N kg^{-1} soil, the site is unlikely to respond to further N inputs (Blackmer et al.,1989; Fox et al., 1989; Meisinger et al., 1992b; Magdoff et al., 1990). Further details of the principles and the practical use of the PSNT can be found in Meisinger et al. (1992a) and Magdoff (1991a,b).

The PSNT has been found to be especially useful in identifying N sufficient sites and in eliminating unneeded fertilizer N applications. This is a common situation in manured systems, because long-term manure applications are known to greatly increase the mineralization capacity of soils, as noted in the N budget work of Pratt et al. (1976a,b), which showed a large increase in soil organic matter with dry manure. Acceptance of the PSNT in the U.S. has been summarized by Bock and Kelley (1992) who reported widespread use in the Northeastern states (CT, NH, VT), in the Mid-Atlantic states (DE, MD, NJ, PA), and in two Midwestern states (IA, WI). States utilizing the PSNT have commonly found that it has saved farmers an average of 35 to 55 kg N ha^{-1} of fertilizer N (Woodward et al., 1993) with the most frequent savings occurring on manured fields.

As with all soil tests, however, the PSNT is not foolproof. Most studies have used a Cate-Nelson approach to identify a critical level and have found that the predictive accuracy of the PSNT ranges from approximately 70-85% (Fox et al.,

1992). The most frequent type of error observed has usually been a site with a low PSNT value that did not respond to additional sidedress N. In this situation both economic and environmental costs would be incurred because fertilizer N would be applied unnecessarily. Possible causes of low PSNT values at nonresponsive sites include delayed mineralization of organic N due to cool or dry spring conditions and/or leaching of NO_3 below the sampling zone but not below the crop's rooting zone. It should also be noted that there is often a very poor correlation between the PSNT value and relative yield below the critical value of 20-25 mg NO_3-N kg^{-1}. Although this means that it is difficult to make an accurate recommendation for a sidedress N rate at a low PSNT value, some states (e.g., DE, NJ, PA) have adopted "sliding scales" that decrease sidedress recommendations roughly proportional to decreases in PSNT levels.

Caution should also be noted with high PSNT values. In 1994, several Iowa farms had high PSNT values but encountered late season N deficiencies (pers. comm. Drs. Voss and Blackmer, Department of Agronomy, Iowa State Univ., Ames, IA). This problem was attributed to accelerated mineralization of N because of the warmer than normal spring.

A direct measure of crop N status (leaf N concentration or leaf chlorophyll level) can also provide very useful information for N management. However, plant N indexes are "point in time" measurements that reflect both the native soil N availability (soil NO_3 content, mineralization, etc.) and the crop ability to convert photosynthate into protein or chlorophyll (affected by water stress, sunlight, growth stage, etc.).

Leaf chlorophyll meters have been successfully developed and used as a plant measure of crop N status (Schepers et al., 1992a,b; Wood et al., 1992a,b). The meter procedure uses a local reference area in a field known to be N sufficient, to standardize factors such as cultivar, growth stage, water status, and sunlight conditions. The chlorophyll meter essentially measures the "greenness" of a specific leaf and compares it to the "greenness" of a N- sufficient reference plant leaf of the same cultivar in an adjacent strip treated the same as the test area except for extra N. The comparison of these readings can be used to indicate the need (or lack of need) for more fertilizer N (Peterson et al., 1993; Reeves et al., 1993).

Leaf chlorophyll data have been used to successfully track the dynamics of N mineralization for a diverse group of manures during the corn growing season (Schepers and Meisinger, 1994). They applied sheep, turkey, and beef feedlot manures and composted paunch manure in alternate years to meet the P needs of irrigated corn and found N immobilization by these incorporated materials frequently resulted in a crop N deficiency. The timing and duration of these deficiencies depended on the amount of residual inorganic N in soil and the climatic conditions governing microbial activity (i.e., temperature and soil water status). Soils with low levels of residual N showed a deficiency first, and materials with the widest C/N ratio encountered the deficiency over a longer period. In some cases, these short-term deficiencies reduced yields even though chlorophyll meter data indicated mineralization after mid-season was adequate to meet crop needs. Producers frequently encounter similar problems when using waste materials with a C/N ratio above 30. For this reason they may be inclined to give a lower than appropriate credit for the N contained in manures. Some producers recognize the potential for immobilization by certain types of manure and try to schedule its application so as to not result in a crop deficiency. Nonetheless, it is frequently difficult to synchronize N mineralized from manure and crop N needs.

Since N mineralization frequently continues beyond the time when N uptake by crops like corn is complete, NO_3 is likely to accumulate in soil, be denitrified, or leach below the root zone. Considering these possible N loss mechanisms, it is important for producers to use deep soil sampling for residual N, the PSNT or chlorophyll meter technology, as appropriate, to improve manure N management. The chlorophyll meter is most useful in irrigated systems where readings can be taken throughout the corn growing season and N added as needed by overhead sprinklers, e.g., in center-pivot irrigation systems. This technology is still in the research evaluation stage in most states, but there is some commercial and producer use in Nebraska.

Another, somewhat different, plant N test that has shown great promise is the late-season stalk NO_3 test developed by Binford et al. (1990). This test is based on the premise that NO_3 concentration in the basal portion of a corn stalk at maturity is an accurate indicator of the presence of excess N during the growing season. Blackmer et al. (1992) recently updated the interpretation of this test and proposed an "optimum range" of 0.7 to 2.0 g NO_3-N kg^{-1}. Maize with stalk NO_3-N values below 0.7 g kg^{-1} would likely have shown an economic response to additional sidedress N, while stalk NO_3-N values greater than 2.0 g kg^{-1} are associated with nonresponsive sites and, thus, excessive N fertilization; intermediate stalk NO_3 values were considered likely to be found near the economically optimum N rate. Sims et al. (1995) recently evaluated the stalk NO_3 test in Delaware and found that the same range successfully identified responsive and nonresponsive sites. Combining the PSNT, leaf chlorophyll meter, and stalk NO_3 tests provides farmers with rapid and inexpensive means to select the most efficient N rate for corn and to verify its accuracy in different fields and under different growing conditions.

c. Supplemental N Needed

The third step is to estimate the need for supplemental N by comparing the crop N requirement to the N available from native sources. If a preplant test is used, e.g., a deep soil NO_3 test, then this usually involves subtracting the root zone soil NO_3 content from the crop N requirement (Dahnke and Johnson, 1990), assuming that fertilizer N and soil NO_3 have equal availability. If a legume or manure N credit is used, a similar subtraction process is followed with alfalfa commonly credited for 55-150 kg N ha^{-1}, soybean 20-45 kg N ha^{-1}, and manure 2-3 kg N t^{-1} depending on the source (Meisinger, 1984). The crucial factor for N credits is knowing the N content of manures and good field calibration data.

d. Source of N

Once the need for and rate of supplemental N has been established, the source of supplemental N basically comes down to manure vs. fertilizer. Choice of source is usually dictated by factors such as field history (recent legumes), manure availability, equipment available, distance to field, and labor/time available for application. For this discussion of livestock systems we will assume that the first choice will be manure N, with fertilizer being used only as needed.

e. Management (Rate, Placement, Timing) of N Source

The above steps produce an estimate of the appropriate N rate for a realistic yield of the next crop, which is the basic principle of efficient N use. The final step is to manage the selected N source in a manner to supply N in phase with crop demand. For fertilizer N, this is a relatively easy task because it can be applied just before the period of rapid crop N uptake. For example, sidedress N applications for corn are made when it is about 30 to 45-cm tall, or at growth stage 4-5 (Feekes scale); for wheat (*Triticum aestivum* L.) when it is about 20-cm tall. The most efficient placement for fertilizer N is below the soil surface to minimize NH_3 losses, which can be especially large for urea-containing sources or for conventional N sources on high pH soils. Hence, efficient N-management is straightforward for fertilizer N - place it below the surface just before the time of rapid N uptake. In the Midwest, injecting the fertilizer (NH_3 and urea ammonium nitrate) also minimizes N immobilization by surface residues.

Manure N, however, is more difficult to manage due to uncertainties in initial composition (ration, animal age, etc.), losses during storage or handling (e.g., NH_3 volatilization or denitrification losses from liquid vs. dry systems), uncertainties in application rates (uncalibrated spreaders, uneven applications), difficulty in spreading manure to a growing crop without causing crop damage, and greater gaseous N losses with manure after application (denitrification and NH_3 volatilization, see above section on N budgets).

One step to reduce these difficulties is to analyze the manure before application. Table 6 summarizes analytical data collected through the Maryland Nutrient Management Program (V.A. Bandel, pers. comm., 1990), which offers manure analysis to farmers as part of a total farm nutrient management plan. The data clearly show a wide variation in nutrient content, even within animal type, which likely results from differences in ration, bedding material, storage system, and length of storage. These data certainly illustrate the need for manure analysis on an individual farm basis in order to efficiently estimate manure N content. Use of general "rules of thumb" for manure nutrient composition are obviously not highly accurate. Once the manure composition is known, the next step is to determine the method of application in order to estimate the importance of NH_3 loss.

Table 6. Summary of Maryland manure testing program data

Type of manure analyzed	Number of samples	--Total N (%)-- Mean	CV	-Ammonium N (%)- Mean	CV
Dairy, liquid	276	0.31	65	0.11	60
Dairy, solid	410	0.61	46	0.12	81
Poultry	894	2.88	37	0.73	39
Swine, liquid	85	0.47	142	0.18	75
Sludge	50	2.99	9	0.23	30

(Source: V.A. Bandel, Univ. MD, 1990; unpublished data.)

Ammonia losses are greatly affected by method and time of application (surface vs. incorporated), weather conditions (high ET vs. rainfall), soil characteristics (pH, CEC, texture), and manure composition (urea content, pH). Losses can range from near zero (incorporated) to 50% (surface solid application) with common values being 10 to 30% (Meisinger and Randall, 1991; Midwest Planning Serv., 1985; Gilbertson et al., 1979; Hall and Ryden, 1985; Bouldin et al., this publication). Currently the best practical method of dealing with these losses is to factor in a daily percentage loss of NH_3 for various application methods and types of manure (e.g., Klausner and Bouldin, 1983a). A more accurate estimate of NH_3 loss can be made if the NH_3 and organic N fractions are determined in the manure analysis. This approach is described in detail by Bouldin et al. (this publication).

The next important step for manure N management is to know the approximate decomposition rate of the organic N. This is generally estimated as a decay series for the particular type of manure. An example of a decay series for solid beef manure would be 40% mineralized the first year, 25% of the remaining N the next year, 6% the next year, and so on (Gilbertson et al., 1979; Pratt et al., 1973). The appropriate decay series will depend on the type of animal, the manure storage system, and the bedding material used. In the U.S., the Extension Service usually determines the decay series for common manures in their state (e.g., Klausner and Bouldin, 1983b). Determining the correct decay series for various manures is a critically important element in efficient manure N management. This is one area where much more applied research needs to be done.

The last step is to calibrate the manure spreader. Obviously it does little good to know the crop N need, the manure composition, the likely NH_3 loss, and the decomposition rate, if one cannot apply the calculated rate of manure accurately. Manure spreader calibration programs in Maryland and Pennsylvania frequently find that farmers are applying 2 to 5 times more manure than they originally estimate. Educational materials for spreader calibration (e.g., Brodie and Smith, 1985) can significantly improve manure N utilization.

Combining the above factors of manure composition, NH_3 loss estimates, decomposition rate, etc. into a final field manure application rate is given in review by Bouldin et al. (this publication), by Klausner and Bouldin (1983b), and by Vitosh et al. (1988). Manure applications are normally recommended at times coincident with crop N uptake. For grain crops, this typically means as close to planting as possible. In some areas, however, a tradition of wet spring conditions necessitates that farmers apply manure the previous fall or during the winter when soils are frozen to avoid soil compaction problems, neither of which is desirable from an environmental perspective.

For winter crops, manure applications in the fall should be discouraged because a portion of the manure N will mineralize and be at high risk for loss via leaching before winter cereals begin their period of spring N uptake. Duynisveld et al. (1988) reported NO_3-N concentrations in shallow groundwater beneath sandy soils of 50-55 mg L^{-1} after fall manuring of winter cereals and concentrations of 35-40 mg L^{-1} after spring manuring (nonmanured areas had NO_3-N levels of about 30-35 mg L^{-1}). In some cases manure can be topdressed on a growing winter cereal crop in the spring if NH_3 losses are not a major concern, e.g., topdressing winter wheat in the early spring after it breaks dormancy.

The above uncertainties in manure composition, NH_3 loss, etc., often lead farmers to discount the value of manure N. This is an area where the PSNT soil test can be especially helpful. For example, a farmer can accurately estimate his corn N need, analyze his manure, calibrate his spreader, and find he has plenty of manure N to meet his corn N needs - provided all the decay series and NH_3 loss estimates are accurate. Use of the PSNT at corn sidedress time can give such a farmer some much needed assurance that the manure has indeed supplied enough N for his corn, i.e., PSNT value above 20-25 mg N kg^{-1} soil. A low PSNT value below 20-25 mg N kg^{-1} can alert one to a N deficient situation that might have resulted from wet weather producing higher than expected denitrification losses, dry weather producing higher than expected NH_3 losses, cold soil temperatures that slowed mineralization, or N immobilization caused by incorporation of residues or waste materials with a wide C:N ratio. In any case, the PSNT is a valuable tool in an overall manure N management strategy, because it can help reduce some of the risks in relying on manure as a primary source of N.

2. Remedial Measures for Manure Nitrogen When Nitrogen Is Sufficient

It is a well known fact that manure builds soil organic matter, increases N mineralization, and increases the soil profile NO_3 content, even in humid climates (e.g., see above section on N budgets; Angle et al., 1993; Roth and Fox, 1990). This is just as true today as in the early days of agriculture when manures were used as the primary source of N. In fact, with today's large animal operations it is not uncommon to find heavier per hectare applications of manure (more animals on smaller land areas) than before the advent of commercial fertilizer. It should come as no surprise then, to find fields which are quite N sufficient, i.e., fields that can supply all the N a growing crop will need without any outside N inputs. These sites are usually associated with a long-term history of manure additions, often are fields near barns, or are fields which have an unusually high level of soil NO_3 in the root zone even after the winter leaching season (in humid climates).

Remedial measures to minimize NO_3 losses to groundwater from N-sufficient sites require careful management. The high levels of soil organic matter will continue to mineralize all year long, whenever weather conditions favor microbial activity. These sites will also normally have high levels of NO_3 throughout the soil profile due to previous mineralization and incomplete leaching from small-pore sequences. Management of these N sufficient sites is accomplished by minimizing NO_3 leaching by minimizing water movement below the root zone, and restricting the size of the soil NO_3 pool.

a. Management of Water Movement

In rain-fed dryland agriculture, not much can be done to reduce the movement of water below the root zone, other than what might be offered by the tillage system, because such movement is dependent on the timing and amount of natural rain events, on soil texture, and on soil water use. However, opportunities exist for promoting more efficient crop water (and NO_3) use through sound agronomic management, such as ameliorating subsoil acidity, extending the root zone through use of deep-rooted crops, and cover cropping to reduce water movement (see Sharpley et al., 1992 for other measures).

In irrigated agriculture there are many opportunities to reduce water movement below the root zone. Percolate can be reduced through design of more efficient irrigation systems (low volume sprinkler vs. furrows), through monitoring of soil water and weather conditions for irrigation scheduling, and through deficiency irrigation which intentionally underirrigates at noncritical growth stages. The N budget of Figures 5 and 6 illustrates that, with average irrigation management (Figure 6), leaching losses were about 30% but with careful irrigation management (Figure 5), these losses were reduced to 20%. However, it must be realized that some leaching of NO_3 is inevitable in irrigated agriculture in several states where salts must be controlled in order to maintain a viable soil-crop system.

The last area where some degree of water control is feasible is in subsurface drainage with drainage outlet management. This involves installing water level control structures in the main drainage lines or outlet ditches to change the water table height beneath a field. Drainage is usually restricted and the water table raised during the winter months in hopes of encouraging denitrification. Water table management studies in coastal North Carolina (Gilliam et al., 1979; Evans et al., 1989) have shown a reduction in the mass of NO_3-N transported from field tile lines of 7-20 kg ha^{-1} for controlled drainage vs. noncontrolled fields. Most of this N reduction, however, occurred through reductions in drainage water volume with only minor changes in NO_3 concentrations. The final impact of controlled drainage on water quality delivered to receiving waters could not be assessed in these studies, due to unstudied N transformations in drainage canals, dilution along drainage canals, etc. But it is clear that a reduction in NO_3 inputs at the head of a water course should also reduce NO_3 inputs into receiving waters, even though the magnitude of the reduction could not be quantified.

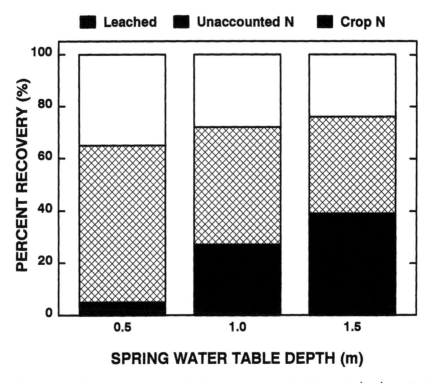

Figure 12. Nitrogen budget for liquid swine manure (190 kg N ha⁻¹ yr⁻¹) applied to a grassed sandy soil in Denmark as a function of spring water table depth. (Data adapted from Steenvoorden, 1985).

Water table management can also affect the soil N cycle by changing aeration status and promoting denitrification when water is maintained close to the soil surface. The effect of water table management on a manure N budget has been reported by Steenvoorden (1985) who summarized a 3-year lysimeter N budget for liquid swine manure (300 Mg ha⁻¹ yr⁻¹, 1900 kg N ha⁻¹ yr⁻¹) applied in November (incorporated within one day of application) to a sandy soil cropped to grass. The spring water table depths were controlled at 0.5, 1.0, or 1.5 m but summer water table depths were all allowed to fall to 1.5 to 1.7 m due to ET losses. The data (Figure 12) show that a shallow water table leads to higher crop N uptake, much lower leaching losses, and higher "unaccounted for" N losses. The "unaccounted for" losses were probably dominated by denitrification outputs because the prompt incorporation of manure would minimize NH₃ volatilization. In this example, water table management could be used to channel up to 60% of the manure N into denitrification outputs from the organic-rich liquid waste and thus minimize NO₃ losses to groundwater. However, this may contribute to air quality concerns as discussed earlier.

Unfortunately, water table management practices have specific geographic applications due to the limited areas with suitable natural resources, e.g., high natural water tables, coarse over fine-textured soils, and an infrastructure to organize drainage districts. However, where water table management is possible, it offers a large potential for NO_3 removal from manures.

b. Management of the Soil Nitrate-N Pool

Nitrate leaching into ground water can also be reduced by reducing the size of the soil NO_3 pool. One of the most obvious ways to accomplish this on N-sufficient sites is to not add N. To continue fertilizing or manuring a N sufficient site can only add to the size of the NO_3 pool and increase potential leaching. Although this seems quite straightforward, it is not always feasible due to limited land areas for manure application. In this case, the livestock enterprise should consider expanding the land base for manure application, constructing a waste treatment system such as manure lagoons for denitrification, or exporting composted manure.

The most practical method for limiting or decreasing the size of the soil NO_3 pool is to continually crop the site. Ideally, this would consist of a high N uptake crop that is completely harvested. For ruminant-based livestock systems, this can be achieved by growing a summer silage crop of corn and a winter/spring silage crop of cereal rye (*Secale cereale* L.). There are also other forage crop systems, such as high-yielding Bermuda grass [*cynodon dactylon* (L.) Pers.], orchardgrass (*Dactylis glomerata* L.), or fescue (*Festuca arundinacea* Schreber) systems, which give a high N removal over a long period of time (Wood et al., 1993).

Furrer and Stauffer (1985) summarized some long-term German lysimeter data with poultry manure and liquid swine waste. They showed a continuous grass crop virtually eliminated NO_3 leaching, compared to fallow, by accounting for about 63% of the N inputs through harvested forage; the remainder probably went into added soil organic N and denitrification losses.

For nonruminant systems one should consider the benefits of cover cropping. Cover crops can be grown between main-season grain crops, if growth and N uptake are not limited by harsh winter climates, and can serve as excellent sinks for NO_3 that could otherwise be lost to ground water. For example, Meisinger et al. (1991) reviewed the groundwater quality benefits of cover crops and concluded that grass cover crops can reduce both the NO_3 concentration and the mass of NO_3 leached by 20 to 80%, compared to no cover crop. Shipley et al. (1992) compared the ability of legume and grass cover crops to conserve fall $^{15}NO_3$ and found that cereal rye was the most efficient, recovering 45% in the aboveground dry matter, compared to hairy vetch (*Vicia villosa* Roth) or crimson clover (*Trifolium incarnatum* L.) or native weeds which recovered 10% or less. If NO_3 has accumulated deep in the profile, say greater than 1 m deep, one should consider use of deep-rooted crops such as alfalfa, sunflower (*Helianthus annuus* L.), or safflower (*Carthamus tinctorius* L.) to utilize this NO_3 and thereby reduce potential NO_3 leaching. The above examples illustrate the basic principle of minimizing leaching by reducing the soil NO_3 pool via continuous cropping and eliminating noncropped fallow periods.

Less success has been achieved with winter covers in the Northeast because (i) there is inadequate time available between harvest of the summer crop and planting of the winter cover to get good establishment; (ii) there is inadequate growth during the winter to sequester much N; (iii) there is often a need to kill the cover crop early in the spring to prevent depletion of soil moisture or allow for timely planting of next year's crop; and most important (iv) farmers see little economic advantage to adding another time-consuming operation during a busy part of the year, even if it is offered as a cost-share. Therefore, if use of cover crops is to be encouraged in short growing-season climates, they will have to be considered within the farm crop-economic system and the societal food-water quality system. Obviously, more research is needed to develop better cover crop germ plasm, to develop methods of "relay cover cropping," and to educate farmers and society of the environmental benefits of cover crops.

Another means of reducing the size of the soil NO_3 pool is to add carbonaceous residues to immobilize inorganic N. This can be crop residues, such as cover crops, or manures with high C/N ratios that are produced from high C/N bedding materials, such as straw or wood shavings. Use of imported nonfarm high C materials such as cannery, newspaper, or processing wastes may also be possible in certain areas.

Incorporation of the above materials in soil is not a substitute for analytical technologies such as deep soil NO_3 testing, the PSNT to identify N sufficient sites, and stalk NO_3 test to evaluate the effectiveness of management practices. These tests, especially annual deep NO_3 measurements, are excellent methods of monitoring the progress of management practices to reduce the size of the NO_3 pool. The PSNT can also serve to assure producers that the site is N-sufficient and thus prevent unnecessary fertilizer N or manure additions, as well as identify the time when the field may again need supplemental N.

B. Manure Management

Manure management strategies strive to increase the marketability and demand for manure as a replacement for fertilizer N and P. This may be accomplished by a combination of several factors which include (i) the development of reliable methods to determine the nutritive value of manure; (ii) basing manure applications on N or P dependent on site-characteristics; use of manure amendments that may reduce environmental hazards; (iii) establishment of mechanisms which encourage transporting manure greater distances to areas where N and P can be more effectively used; (iv) formation of cooperatives that can more efficiently compost, pelletize, or concentrate manure to increase the distance it can be economically transported from its source; and (v) expansion of education and extension programs highlighting the overall value of manure to nonmanure-producing farmers.

1. Manure Analysis

There are many variables associated with animal production systems that can affect manure quality at the time of application. These include the type and amount of bedding material used, presence of feed additives (e.g., Cu, As, alum, $FeSO_4$, pesticides, hormones), accumulation time, amount and quality of water used to flush the house, location in a storage pit from which the waste is removed, and length of storage before land application. The variability in these management factors can result in a wide range in the content of nutrients, trace elements, organic chemicals, and other feed additives in manures. It is important to note that, with the exception of N, P, and K, the bioavailability of many manure constituents has not been well-documented. This includes plant nutrients (e.g., B, Ca, Cu, Mg, S, Zn), nonessential trace elements (e.g., As) and organic additives (hormones, pesticides). Hence, even if a manure is analyzed for an element (or compound) that may be of environmental concern, there will often be little information available on transformation in soils, uptake by plants, and transport in runoff.

Farm advisors and extension agents in several states are now recommending that the N and P content of both manure and soil be determined by soil test laboratories before land application of manure. This is important because there is a tendency among landowners to underestimate the nutritive value of manure. Thus, manure analyses are a constructive educational tool showing landowners that manure represents a valuable source of N and P. Manure analyses, in combination with soil test results, can also demonstrate the positive and negative long-term effects of manure use and the time required to buildup or deplete soil nutrients. For instance, with P, they can help a farmer identify the soils in need of fertilization, those containing excess P that should not be manured, and those where moderate manure applications may be of some value.

In those states where manure analyses are conducted, total N, NH_4, and moisture content are generally determined. The use of more sophisticated analytical equipment in soil test laboratories allows multi-element analysis, total P, K, and other nutrients to also be determined and reported to the farmer upon request.

Manure application based on nutrient contents needs to be adjusted to account for nutrient availability in soil. Nitrogen availability is related to mineralization of organic N (usually 50 to 60% of the organic N fraction) and recovery of added NH_4. This availability may be adjusted further to account for the effect of storage time on N mineralization and volatilization, soil type on NH_4 fixation, and soil pH on volatilization. It is generally assumed that 75 to 80% of added total P is plant-available.

A cautionary note to basing application rates on manure analyses must be sounded because of the wide variability in nutrient contents that can be obtained. Igo et al. (1991) estimated N applied using analysis of stockpiled broiler litter on 17 farms in Delaware and compared this with the actual N loading rate based on analysis of samples collected during application to field corn. When desired application rates were applied to large field plots using commercial manure spreaders, overapplication by 10 to 20 kg N Mg^{-1} of litter commonly occurred, as did under application by 5 to 10 kg N Mg^{-1} because of inaccurate analytical data.

Therefore, the accurate application of a recommended litter rate for corn (approximately 5 Mg ha^{-1}), based on analysis of the litter, commonly resulted in the application of excess manure N approaching the total N requirement of the crop in Delaware (approximately 100 kg N ha^{-1}). Similar variabilities are associated with P, often ranging from 5 to 15 kg P Mg^{-1} manure. Thus, manure analyses should be conducted just before manure application and be used as guidelines in determining application rates. Once the correct manure rate has been determined, soil and plant tests such as the PSNT, stalk NO$_3$ test, and chlorophyll meter, can help confirm the plant availability of N; routine soil tests can monitor changes in other nutrients (e.g., P, K, Ca, Mg).

2. Nitrogen or Phosphorus Application Strategies

Manure applications to land that has been targeted as environmentally sensitive will depend to a large extent on site hydrology and topography. If the potential for NO$_3$ leaching from an application site exists, N should be a priority management consideration. However, for areas with high runoff and erosion potentials, or for those where the possibility of extensive P losses in agricultural drainage is a concern (e.g., DE, FL, NC, The Netherlands), P should be the nutrient of primary consideration in the development of environmentally sound manure application rates. A P-based management strategy for manures will almost always minimize NO$_3$ leaching losses as well. This is because crop N needs far exceed P uptake. The main disadvantage of a P-based approach is that it will eliminate land application of manure as an option for many agricultural soils with a history of continual high P inputs because of existing high or excessive soil test P values. This would force farmers to identify larger areas of land on which to utilize the generated manure, further exacerbating the problem of local land area limitations. In addition, farmers relying on manure to supply most of their crop N requirements may be forced to buy fertilizer N to supplement foregone manure N. Using a P-based strategy may resolve potential environmental issues with surface waters but would undoubtedly place additional economic burdens on farmers.

Clearly, the development of environmentally sound management systems utilizing manure is a challenge from both agronomic and economic standpoints. Several BMPs are currently being implemented or evaluated in areas north of Lake Okeechobee, Florida, where large numbers of dairies are present. Some strategies are (i) managing manure application rates based on soil test P values, (ii) controlling cattle grazing density, (iii) manipulating feed ration composition, (iv) limiting cattle access to streams or channels, and (v) creating buffer areas near streams or channels (Bottcher and Tremwell, 1995).

3. Spatial Variability

Field variability in soil chemical and physical properties and crop yield potential all affect the amount of N that must be supplied to the crop by fertilizer or manure after considering inputs from legumes, mineralization, and waterborne N. Managing the various N sources to minimize losses to the environment while maintaining productivity becomes a particular problem because most tillage, manure application, and fertilizer operations are usually performed on a whole-field basis. Yet, nutrient availability and crop needs vary throughout a field. This apparent incompatibility clearly leads to areas of nutrient excesses and deficiencies within a field.

Variability in soil nutrient availability can lead to considerable uncertainty when deciding how much manure should be applied. This decision process is further complicated by soil type and topography differences that influence leaching and runoff losses of nutrients. In the absence of environmental concerns, producers using manure as a nutrient source naturally prefer to err on the side of excess nutrient availability. The problem is that what might be considered an appropriate application of manure can lead to everything from greatly excessive to highly inadequate nutrient availability because of inherent differences in soil fertility.

The need to integrate spatial variability of soil properties into manure management decisions is receiving increased attention as global positioning systems (GPS) and geographical information systems (GIS) are developed. In particular, grid sampling activities have identified situations where the tradition of compositing soil samples prior to chemical analysis has led to inappropriate nutrient management decisions. This problem can be costly to producers in situations where one or more soil samples are taken from old farmsteads or feedlots that can no longer be identified on the landscape but where high levels of certain nutrients remain. Mixing a few very fertile subsamples with any number of subsamples with marginal nutrient availability can lead to analytical results that give producers a false sense of security. A recent example from Nebraska where scientists extensively sampled a 58-ha corn field clearly illustrates this type of situation (Hergert et al., 1994; Peterson et al., 1994). More than 2000 cores were collected on a grid basis from this field that had grown center-pivot irrigated corn for the past 17 years. Their objective was to make site-specific fertilizer N recommendations and vary the rate of fertilizer N application within the field to minimize NO_3 leaching while maintaining or improving productivity and profitability. In the process, they discovered that approximately 75% of the field would have been expected to respond to P fertilization. Yet annual soil sampling activities where subsamples were composited did not indicate the need for much if any P fertilizer. Further investigation disclosed the existence of an old farmstead that included a swine feeding operation about 20 years earlier and another area where sheep had been fed in confinement over 70 years ago. Similar situations are bound to be commonplace and will undoubtedly be the topic of many discussions related to manure management.

4. Manure Amendments

Manure amendments are being commercially developed that will alter nutrient content and solubility prior to land application. For example, slaked lime or alum can reduce NH_3 volatilization and P solubility of poultry litter by several orders of magnitude (Moore and Miller, 1994). These amendments have several beneficial effects on manure management. For example, increased weight gains by birds with reduced in-house NH_3 volatilization is of economic benefit to the farmer. Also the DP concentration of runoff from fescue treated with alum-amended litter (11 mg L^{-1}) was much lower than from fescue treated with unamended litter (83 mg L^{-1}; Shreve et al., 1995).

Perhaps the most important benefit of manure amendments (for both air and water quality), however, will be an increase in the N:P ratio of manure, via reduced N loss from manure by NH_3 volatilization. As a result of the increased N:P ratio of animal manure, it becomes closer to that of crop uptake. Thus, basing manure additions on crop N requirements would reduce the amount of P added, thereby minimizing potential soil P accumulations. Increasing the N content of manures also reduces application rates needed to meet crop N requirements, resulting in economic savings for farmers by decreasing time associated with manure handling and transport.

Enzyme additives for animal feed that will increase the efficiency of N and particularly P uptake during digestion are also being tested. Development of such enzymes that would be cost-effective in terms of animal weight gain may reduce the P content of manure. One example is the use of phytase, an enzyme that enhances the efficiency of P recovery from phytin in grains fed to poultry. It is now common to supplement poultry feed with mineral forms of P because of the low digestibility of phytin. This contributes to P enrichment of poultry manures and litters. While the phytase enzyme has been shown to decrease the need for mineral P, it is currently too expensive for use as a routine feed additive. Provision of phytase to poultry growers via cost-sharing by federal agencies would represent an innovative approach to controlling nonpoint source losses of P at the source, which would be much easier than after-the-fact control programs for land applied manures.

5. Manure Transportation

Most manures are heavy and bulky due to high water contents and/or incorporated bedding material and thus, have a lower nutrient content than mineral fertilizers (Table 1). As a result, much more manure than mineral fertilizer must be applied to achieve similar additions of N and P. The cost of transporting low-density manure more than short distances from the site of its production often exceeds its nutrient value. This has limited the acreage of land available for application of manure with most manure applied in the immediate vicinity of production. Thus, the dominant geology, soils, and topography of the local area often cannot be adequately taken into account prior to application. This inflexibility may result in application of manure to areas less suited, in terms of elevated soil N and P from previous applications and high runoff or leaching potentials, than more distant areas. Unless an infrastructure is developed that can process, market, and distribute manure to nutrient-deficient areas, the potential for transport of chemicals added in manure to surface and groundwater will remain.

Innovative measures are being used by some farmers to transport manure from the area of production. For example, following delivery of grain or feed, trucks and railcars are transporting dry manure instead of returning empty (Collins et al., 1988). Railcars bringing coal from Wyoming to power stations in southeastern Oklahoma have carried poultry litter back to ranches in Wyoming at costs comparable to the use of commercial fertilizers. In Delaware, the local poultry trade organization has established a "manure bank" network that puts farmers in need of manures in contact with small poultry growers that have inadequate land available to use all the manure generated by their operation. Cost-share monies are also made available by USDA-NRCS to subsidize the use of newer and more efficient manure application equipment.

Even so, large scale transportation of manure from producing to nonmanure producing areas is generally not occurring. Therefore, programs are needed to encourage transportation of manure from surplus to deficient areas (Bosch and Napit, 1992). In some parts of the country, farmers contract out this transportation to manure handling firms that deliver and apply manure as a supplement to mineral fertilizers.

6. Manure Handling

Recent trends in forming of cooperatives that can more cost-effectively compost and compact manure should be encouraged by cost-sharing programs. Neighboring landowners and private industry are also developing manure processing alternatives. Examples of this include centralized storage and distribution networks, regional composting facilities, and commercial pelletizing operations that can produce a value-added processed manure for distribution to other areas. Pelletization is a particularly attractive option because it results in a dried, lightweight material that can be handled, transported, and applied in much the same manner as commercial fertilizer.

The major obstacle to establishing a pelletizing operation in an agricultural area has been the reluctance of the pelletizing industry to construct a processing plant without a guaranteed, low-cost supply of manure "feedstock." Farmers, on the other hand, because of their own experience and extensive educational programs emphasizing the economic value of manures, are equally reluctant to essentially give away valuable plant nutrients to the pelletizing operation. Some form of economic stimulus (e.g., tax credits, cost-sharing) to both the industry and farmers may be required to successfully initiate a pelletizing operation even in areas with high animal densities.

By composting and compacting, the bulk density of the manure is reduced, as is the cost of transportation. If the consumer is not willing to bear a part of the financial support, then it may be necessary to require processors, producers, and landowners to take part in cooperative manure treatment programs. The level of involvement could be linked to the number of animals per farm.

Storage of manure will allow more flexibility in timing of applications. A wide range of storage methods and costs are available to landowners (Brodie and Carr,

1988). Inexpensive plastic sheeting can perform well with very low cost for some solid manures. However, all storage methods must be managed carefully to fully realize their potential in an agronomically and environmentally sound BMP.

7. Extension and Education

Several economic and institutional factors influence the structure and effectiveness of educational programs concerning confined animal operations. Farmers with limited land have turned to confined animal production as a possible source of steady income to supplement additional off-farm income. Usually, these operations are located on marginal soils due to regional factors such as slope, fertility, and erratic weather conditions. Confined animal operations have become such a popular practice, especially in the southern states, that poultry production has become the major agricultural industry. For example, in Arkansas and Delaware, approximately 47 and 65% of income from agriculture, respectively, is attributed to the sale of poultry products (National Agricultural Statistics Service, 1991). The bulk of this production occurs in the four-county area of northwest Arkansas. Similarly, one county in southern Delaware contains approximately 2600 poultry houses, located on 2450 km^2 of land, and produces over 270,000 Mg of poultry manure annually, all of which is used as a fertilizer for the 120,000 ha of grain crops in the county. In such situations, sound information regarding proper manure handling techniques needs to get in the hands of the grower to minimize any water quality impacts, because vast amounts of manure must be land-applied.

Dissemination of proper manure handling information is of particular importance to farmers operating on small acreage (< 40 ha). Often these farmers turn to confined animal operations to supplement inadequate cash returns on traditional grain and forage production due to inherent low fertility soils, erratic rainfall, and reduced crop prices. Therefore, the local need for N and P additions for crop production will be lower than in areas of intensive crop or forage production. Further exacerbating this situation, insofar as the poultry and swine industries are concerned, is the fact that many of these farms are part-time operations located on an inadequate land base. Growers often have full-time jobs in nearby businesses, several poultry or swine houses located on small areas (e.g., < 5 ha), and little expertise in (or time for) waste management. Also, in too many cases, farmers are still not fully aware of the nutritive value of applied manure. This again illustrates the need for education and extension programs as well as an infrastructure that can collect, process, and redistribute manures to areas with high local demand for the N and P.

For a variety of reasons, getting the right information from the laboratory bench to the field under these conditions differs markedly from what has been traditionally done in production agriculture. Historically, if a new corn variety is developed through industry/university testing, the mechanisms for getting this information to the grower are well proven and identifiable. These include the use of field days, extension brochures, and university/industry field personnel. At least initially, when dealing with animal manure, the state and federal agencies involved in water quality are the target audience/clientele for the information and not the producer. Before a practice can be identified as a BMP it has to be accepted by the proper agencies.

Information transfer among the agencies is complicated by evolving roles of respective agencies. Some important roles are clear while others are not. In each state, an agency is responsible for developing programs dealing with nonpoint source pollution. The issue is further complicated when federal agencies, such as the USDA-NRCS, become involved in the development of individual farm plans that have maintaining high water quality as a goal. When dealing with confined animal operations, the processor/integrator also becomes a major player, because its management program is dynamic and can have significant impacts on the amount and quality of manure produced. Somewhere in this complex process, input from the grower/farmer must also be included because they will ultimately determine which practices are implemented at the farm level. Obviously these are uncharted waters and the process can not occur in a vacuum. As information is generated, the designated state agency must inform each player and solicit their response with the ultimate goal of trying to gain consensus. Information must eventually be disseminated to the end-user or farmer. At this point the process is familiar to the Extension and National Resource Conservation Services providing the critical link between the farmers and public agencies.

Preventive and remedial strategies that minimize water quality impacts of land application of manure are often accomplished by expenditure of limited cost-shared public funds. Thus, successful strategies must obtain maximum benefits from these limited resources by systematic implementation of effective management practices in targeted locations. Extension and education programs should ensure that public funds are used where water quality improvement is likely. Otherwise, such programs will not sustain broad public support and funding. Finally, educational programs must overcome the perception that it is often much cheaper to treat the symptoms of impaired water quality rather than control the nonpoint sources.

C. Watershed Management

Watershed management strategies that minimize water quality impacts of land application of manure are dependent on reliable, yet realistic, threshold levels of N and P in surface and ground water that should not be exceeded on environmental grounds. Cost-effective remedial strategies are also dependent on identification of pollutant source areas, so that more intensive manure management may be targeted to these areas. Multi-watershed planning should also occur, because the concentrated nature of animal production in some watersheds exceeds the capacity of even the most intensive manure management programs to control nutrient losses. Coordinating manure management across several watersheds often means that local, state, and federal governments must form a sometimes uneasy alliance whereby one county is asked to help solve another's waste management problem. Nevertheless, long-term success will certainly require both individual and multi-watershed perspectives.

1. Realistic Water Quality Criteria

Water criteria for N and P have been established by the U.S. EPA (1976). Acceptable limits of NO_3-N concentration for human and livestock consumption are 10 and 100 mg L^{-1}, respectively, while NH_4-N concentrations above 0.5 and 2.5 mg L^{-1} may be harmful to humans and fish, respectively (U.S. EPA, 1976). For P, critical DP and total P concentrations of 0.01 and 0.02 mg L^{-1}, respectively, were proposed by Sawyer (1947) and Vollenweider (1968) that, if exceeded, may accelerate surface water eutrophication. These concentrations, ranging as high as 0.05 to 0.1 mg L^{-1} for DP and total P, have been modified by state agencies (U.S. EPA, 1976). For example, Florida recently identified 0.05 mg L^{-1} as the concentration of DP allowable in drainage water entering the Everglades (USA vs. South Florida Water Management District, 1994). By the year 2000, the state hopes to be able to reduce this concentration to 0.01 mg L^{-1}. Conservation practices may be successful, however, in reducing the *mass* of P lost from a field by controlling erosion, runoff, and drainage. For heavily manured soils, however, it will probably be necessary to reduce DP concentrations and soil and water losses from a field.

These water quality criteria should not be used as the sole criteria to guide manure management where N and P loss in runoff and drainage water is of concern. For example, the mean annual DP concentration of runoff from several wheat watersheds in the Southern Plains receiving mineral fertilizer P (20 kg P ha^{-1} yr^{-1}; Smith et al., 1991a) and grassed watersheds receiving various types of manure (Heathman et al., 1995; Jones et al., 1995) all exceed a "generous" critical DP value of 0.1 mg L^{-1} (Figure 13). This is also the case for unfertilized native grass watersheds (Sharpley et al., 1986). Thus, it is unlikely that any form of manure management involving large continual manure P inputs will reduce DP in runoff to below critical concentrations.

A more flexible approach considers the complex relationships between N and P loadings and physical characteristics of affected watersheds (leaching, runoff, and erosion potential) and water bodies (mean depth and hydraulic residence time) on a site-specific and recognized water-use basis. Water use will influence "desired or tolerable" nutrient loadings. For example, lakes used principally for water supply, swimming, and multi-purpose recreation will benefit from low N and P loadings. However, lakes mainly used for fish production benefit from a moderate degree of biological productivity and thus, tolerate higher N and P inputs.

Clearly, realistic water quality criteria that guide manure management within a watershed should encompass more factors than just N and P concentrations in agricultural runoff from impacted fields. Unrealistic or unattainable criteria will not be adopted unless regulated. Thus, it is essential for long-term sustainable management of manures that workable water quality criteria are initially proposed. Phasing in of environmentally managing manure should receive wider acceptance and compliance by farmers without creating severe economic hardships within rural communities. Manure applications that reduce potential N and P loss in agricultural runoff based on environmental goals have been practiced in many parts of Europe since the mid-70s. After initial resistance to adoption of these guidelines, farmers

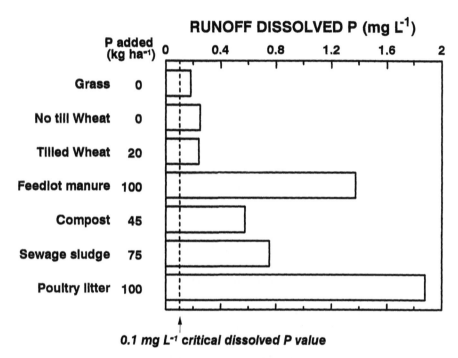

Figure 13. The mean flow-weighted concentration of dissolved P in runoff over one year as a function of watershed and manure management in the Southern Plains relative to critical values associated with accelerated eutrophication. (Data adapted from Jones et al., 1995; and Heathman et al., 1995.)

are now aware of the need for this practice and are receptive to suggestions for their implementation.

In order to meet water quality criteria, a system of buying and selling pollution credits within a given watershed, similar to that recently adopted for air quality control, has been suggested by several eastern states in the U.S. Farmers able to limit P loss below recommended levels could sell credits to a farmer unable to meet these levels. The number of credits a farmer has could be linked to the number of animals and area of the farm. As a result, P export from a watershed may be kept within predetermined limits by sharing the responsibility among farmers.

It should be noted, however, that "pollution trading" has been bitterly criticized by some environmental groups because it is perceived as allowing wealthier operations to buy the "right to pollute." Heated debate will likely precede the adoption of pollution credits for agriculture; hence, careful planning to justify their value and need will be required.

Current technology will not permit an unlimited number of animals in a region without impacting water quality. Thus, it may be necessary to redistribute animals or to limit animal numbers within an area. Several states (e.g., FL, NC, PA) now require that new animal facilities which exceed a certain size have an appropriate manure management plan. Thus, it is essential that we develop and transfer technology to implement realistic water quality criteria for environmentally-sound manure management.

Using water quality criteria to guide manure management can be likened to treating the symptoms rather than the source of the problem. Once critical NO_3 and DP concentrations in runoff are exceeded, continual manure applications must be carefully managed. Due to the relatively greater immobility in soil of P than N, soil P can rapidly accumulate with repeated manure applications. Thus, several states have attempted to identify a soil test level where management of P in fertilizer or manure must change to reduce the potential for P loss in runoff (Table 7; from Sharpley et al., 1994). In many situations, this would require reduced or no manure applications, a situation very restrictive to many farmers, and that alternative manure uses be developed. This is a clear example of the need for an integrated approach to the use of soil test P information and the expanding role that soil test laboratories will play in this process.

However, soil testing alone cannot assess the potential for soil N or P at any given site in a watershed to play a significant role in water quality impairment. Any environmental soil test must be linked to site assessment of leaching, runoff, and erosion potential and management factors affecting the vulnerability for N and P transport from a site. For example, we showed earlier (Figure 8) that adjacent fields having a similar soil test P level can support different P losses in runoff due to differing susceptibilities to runoff and erosion as a result of contrasting topography or management. Clearly, these two sites should not have similar P-based manure recommendations.

2. Targeting Remedial Strategies

Strategies to minimize N and P loss in runoff will be most effective if sensitive or vulnerable source areas within a watershed are identified, rather than implemented over a broad area. Long-term field studies that reliably evaluate N and P movement are costly, lengthy, and labor intensive. Also, models simulating the effect of agricultural management on nutrient loss in runoff often require detailed climate, soil, crop, and tillage information and computer equipment and experience to run them. These limitations often restrict the value of prediction models to relative comparisons only. Consequently, indexing approaches were developed as a field tool to readily assess site vulnerability to the loss of N by leaching and P by surface runoff. Initial site assessment involves determining if leaching or runoff dominates water loss from a specific site (Table 8). If runoff is negligible and the potential for percolation is high, the N leaching index should be used to guide manure or fertilizer applications. However, if from Table 8, surface runoff potential is high, then the P indexing system should be used.

Site vulnerability to N leaching can be ranked by considering textural and permeability characteristics limiting water movement through a soil profile (Table 9). The finest textural horizon of the soil profile is used to determine the site's textural class. Similarly, the permeability of the most limiting layer (lowest permeability) is used to determine permeability class. This simple index exemplifies one method to highlight the potential for groundwater contamination by agricultural chemicals (Kissel et al., 1982). However, such indices need to be evaluated and calibrated for each state's major soil types to determine its specific usefulness in new areas.

Lemunyon and Gilbert (1993) developed a P indexing system to identify and rank sites according to their vulnerability to P loss in runoff (Table 10). The index rates source (soil test P and fertilizer and manure management) and transport factors (runoff and erosion potential) of a site providing an overall numerical value ranking site vulnerability to P loss in runoff. Each site characteristic affecting P loss is assigned a weighting, assuming that certain characteristics have a relatively greater effect on potential P loss than others. The P loss potential is given a value (Table 10), although each user must establish a range of values for different geographic areas. An assessment of site vulnerability to P loss in runoff is made by selecting the rating value for each site characteristic from the P index (Table 10). Each rating is multiplied by the appropriate weighting factor. Weighted values of all site characteristics are summed and site vulnerability obtained from Table 11.

The index is intended for use as a tool for field personnel to easily identify agricultural areas or practices that have the greatest potential to accelerate eutrophication. It is intended that the index will identify management options available to land users that will allow them flexibility in developing remedial strategies. Based on site vulnerability to P loss in runoff using the P index, Sims (1993b) proposed management options to minimize nonpoint source pollution of surface waters by soil P (Table 12).

Buffer or vegetative filter strips can effectively reduce N and P loss in runoff from manured fields, feedlots, and dairy milkhouses (Dillaha et al., 1986; Edwards et al., 1983; Magette, 1988; Schwer and Clausen, 1989; Yang et al., 1980). Chaubey et al. (1993) found the loss of N and P in runoff from a source area treated with poultry litter (5 Mg ha^{-1}) was reduced an average of 40% after passage through a 3.1-m fescue filter strip and 80% for a 21.4-m length (Figure 14). Effectiveness of the filters did not significantly increase beyond 9.2 m for total N and P and 15.2 m for NH_4 and DP (Chaubey et al., 1993). Total N and P loss in feedlot runoff in Minnesota was reduced 88% by a 25.4 m orchard grass filter strip, while total coliform, fecal coliform, and fecal streptococci were reduced about 70% (Young et al., 1980). However, no significant reduction in N, P, or bacteria concentration in barnyard runoff by a 23-m mixed grass filter strip in Vermont was observed by Schellinger and Clausen (1992), due to an excessive hydraulic loading rate. This emphasizes, as the authors pointed out, the need for careful design, targeting, and maintenance of vegetative filter strips to ensure their success as a remedial measure.

In addition to targeting appropriate measures to environmentally sensitive areas, manure applications may also be confined to times of the year with lower rainfall-runoff probabilities to maximize crop N and P utilization and minimize water

A. Sharpley et al.

Table 7. Soil P interpretations and management guidelines

State	Critical value	Management recommendation	Rationale[a]
Arkansas	150 mg kg^{-1} Mehlich 3 P	At or above 150 mg kg^{-1} soil P: 1. Apply no P from any source. 2. Provide buffers next to streams. 3. Overseed pastures with legumes to aid P removal. 4. Provide constant soil cover to minimize erosion.	CV: Ohio sewage sludge data. MR: reduce soil P and minimize movement of P from field.
Delaware	120 mg kg^{-1} Mehlich 1 P	Above 120 mg kg^{-1} soil P: Apply no P from any source until soil P is significantly reduced.	CV: greater P loss potential from high P soils. MG: protect water quality by minimizing further soil P accumulations.
Ohio	150 mg kg^{-1} Bray P1	Above 150 mg kg^{-1} soil P: 1. Institute practices to reduce erosion. 2. Reduce or eliminate P additions.	CV: greater P loss potential from high P soils as well as role of high soil P in zinc deficiency. MR: protect water quality by minimizing further soil P accumulations.
Oklahoma	130 mg kg^{-1} Mehlich 3 P	30 to 130 mg kg^{-1} soil P: Half P rate on >8% slopes. 130 to 200 mg kg^{-1} soil P: Half P rate on all soils and institute practices to reduce runoff and erosion. Above 200 mg kg^{-1} soil P: P rate not to exceed crop removal.	CV: greater P loss potential from high P soils. MR: protect water quality, minimize further soil P accumulation, and maintain economic viability.

Table 7. continued

Michigan	75 mg kg⁻¹ Bray P1	Above 75 mg kg⁻¹ soil P: P application must not exceed crop removal. Above 150 mg kg⁻¹ soil P: Apply no P from any source.	CV: minimize P loss by erosion or leaching in sandy soils. MR: protect water quality and encourage wider distribution of manures.
Texas	200 mg kg⁻¹ Bray P1 or Texas A&M P	Above 200 mg kg⁻¹ soil P: P addition not to exceed crop removal.	CV: greater P loss potential from high P soils. MR: Protect water quality by minimizing further soil P accumulations.
Wisconsin	75 mg kg⁻¹ Bray P1	Above 75 mg kg⁻¹ soil P: 1. Rotate to P demanding crops. 2. Reduce manure application rates. Above 150 mg kg⁻¹ soil P: Discontinue manure applications.	CV: at that level, soils will remain nonresponsive to applied P for 2-3 years. MR: Minimize further soil P accumulations.

[a]CV represents critical value rationale and MR, management recommendation rationale. (Adapted from Sharpley et al., 1994 and Gartley and Sims, 1994.)

Table 8. Runoff index to assess surface runoff potential

Curve number	Annual precipitation (cm)				
	< 30	31-70	71-110	111-150	> 150
< 65	Low	Low	Low	Medium	High
65-75	Low	Low	Low	Medium	High
76-82	Low	Medium	Medium	High	Very high
> 83	Low	Medium	High	Very high	Very high

Table 9. Nitrogen leaching index

Textural class	Permeability class (cm hr^{-1})			
	50-15	14-5	4-1.5	< 1.5
Sand to coarse sandy loam	High	Medium	Low	Very low
Loamy fine sand to sandy loam	Medium	Medium	Low	Very low
Loam to fine sandy loam	Low	Low	Low	Very low
Clay loam to loam	Very low	Very low	Very low	Very low

(Adapted from Kissel et al., 1982.)

quality impacts. A modeling approach (Erosion-Productivity Impact Calculator, Sharpley and Williams, 1990) was used by Edwards et al. (1992) to estimate optimal timing windows for poultry litter application in Arkansas (Table 13). Three criteria were used based on water quality impacts of annual N and P loss in runoff, annual crop production, and a combination of both (Edwards et al., 1992). Optimal timing windows varied between three locations receiving litter in Arkansas (Table 13). Nutrient loss in runoff was obviously related to rainfall and runoff probability and warm, moist soil conditions favoring N mineralization (Edwards et al., 1992). Bermuda grass, a warm-season crop requiring nutrients during spring and summer months, was used in the simulation. Other crop types may influence the timing windows.

As Edwards et al. (1992) suggest, it may be possible to meet the most restrictive criterion; both goals are not mutually exclusive. However, the appropriate criteria for optimization must be carefully selected. For example, Edwards et al. (1992) assumed that N and P were of equal concern. In some areas, the primary concern may be P, while in others, NO_3 movement to groundwater may be of concern.

Table 10. The P indexing system to rate the potential P loss in runoff from site characteristics

Characteristic (Weight)	Phosphorus loss potential (Value)				
	None (0)	Low (1)	Medium (2)	High (4)	Very high (8)
Transport factors					
Soil erosion (1.5)[a]	Negligible	< 10	10-20	20-30	> 30
Runoff class (0.5)	Negligible	Very low or low	Medium	High	Very high
Source factors					
Soil test P (1.0)	Negligible	Low	Medium	High	Excessive
Fertilizer P application rate (0.75)[b]	None applied	1-15	16-45	46-75	> 76
Fertilizer P application method (0.5)	None applied	Placed with planter deeper than 5 cm	Incorporated immediately before crop	Incorporated > 3 months before crop or surface applied < 3 months before crop	Surface applied > 3 months before crop
Organic P application rate (0.5)[b]	None applied	1-15	16-30	30-45	> 45
Organic P source application method (1.0)	None	Injected deeper than 5 cm	Incorporated immediately before crop	Incorporated > 3 months before crop or surface applied < 3 months before crop	Surface applied > 3 months before crop

[a]Units for soil erosion are Mg ha^{-1}.
[b]Units for P application are kg P ha^{-1}.
(Adapted from Lemunyon and Gilbert, 1993.)

Table 11. Site vulnerability to P loss as a function of total weighted rating values from the index matrix

Site vulnerability	Total index rating value
Low	< 10
Medium	10 - 18
High	19 - 36
Very high	> 36

Table 12. Soil management options based on the P index

P index	Management options to minimize nonpoint source pollution of surface waters by soil phosphorus
< 10 (Low)	Soil testing: Have soils tested for P at least every three years to monitor buildup or decline in soil P. *Soil conservation:* Follow good soil conservation practices. Consider effects of changes in tillage practices or land use on potential for increased transport of P from site. *Nutrient management:* Consider effects of any major changes in agricultural practices on P losses before implementing them on the farm. Examples include increasing the number of animal units on a farm or changing to crops with a high demand for fertilizer P.
10 - 18 (Medium)	*Soil testing:* For areas with low P index values, have soils tested for P at least every three years to monitor buildup or decline in soil P. Conduct a more comprehensive soil testing program in areas that have been identified by the P ndex as being most sensitive to P loss by erosion, runoff, or drainage. *Soil conservation:* Implement practices that control P losses via erosion, runoff, or drainage in the most sensitive fields. Examples include reduced tillage, wider field border strips, grassed waterways, and improved irrigation and drainage management. *Nutrient management:* Any changes in agricultural practices may affect P loss. Carefully consider the sensitivity of fields to P loss before implementing any activity that will increase soil P. Avoid broadcast applications of P fertilizers and apply manures only to fields with low P index values.
19 - 36 (High)	Soil testing: A comprehensive soil testing program should be conducted on the entire farm to determine fields that are most suitable for further additions of P. For fields that are excessive in P, estimates of the time required to deplete soil P to optimum levels should be made for use in long-range planning. *continued next page--*

Table 12. continued

19-36 (High)	*Soil conservation:* Implement practices to control P losses via erosion, runoff, or drainage. Examples are reduced tillage, wider field border strips, grassed waterways, and improved irrigation and drainage management. Consider using crops with high P removal capacities in fields with high P index values. *Nutrient management:* In most situations fertilizer P, other than a small amount used in starter fertilizers, will not be needed. Manure may be in excess on the farm and should only be applied to fields with low P index values. A long-term P management plan should be considered.
> 36 (Very high)	*Soil testing:* For fields that are excessive in P, estimates of the time required to deplete soil P to optimum levels should be made for use in long-range planning. Consider the use of new soil testing methods that may provide more information on environmental impact of soil P. *Soil conservation:* Implement practices that control P losses via erosion, runoff, or drainage. Examples include reduced tillage, wider field border strips, grassed waterways, and improved irrigation or drainage management. Consider using crops with high P removal in fields with high P index values. *Nutrient management:* Fertilizer and manure P will not be required for at least three years and perhaps longer. A comprehensive, long-term P management plan must be developed and implemented.

(Adapted from Sims, 1993b.)

D. What Can We Learn from Europe?

Preventive and remedial measures in Europe generally follow the same lines as described above. Incorporation of manure into the soil by tillage or subsurface placement (injection) is now commonly accepted and required by legislation. The same holds for the timing of manure application, which is restricted to early spring and the growing season. Applications in fall and winter are generally prohibited to reduce the transport of N and P to ground and surface waters.

Recommended or legally imposed application rates of manure are either N-based, as in Denmark, or P based, as in The Netherlands. In Denmark, N is considered to cause more environmental problems than P (Danish Ministry of Environment, 1991). High concentrations of P in groundwater seldom occur, and N is not only a threat to the quality of drinking water, but it is also the major factor in eutrophication of marine waters in the region. In both countries, P is considered to regulate eutrophication in many freshwater lakes. In addition, the problem of P-saturated soils had a large impact on Dutch environmental policy.

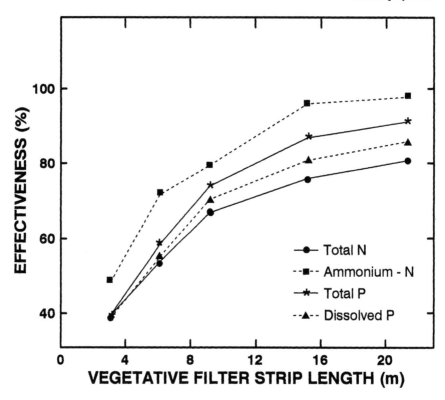

Figure 14. Effectiveness of a fescue filter strip in reducing N and P loss in runoff from land receiving poultry litter. (Data adapted from Chaubey et al., 1993.)

Table 13. Optimal timing windows for poultry litter application in Arkansas

Management criterion	Locations within Arkansas		
	Texarkana	Stuttgart	Fayetteville
Minimize N and P loss in runoff	June to December	July to April	August to May
Maximize grass yield	November to April	December to April	December to April
Minimize runoff N and P and maximize grass yield	November	December to March	November to April

(Adapted from Edwards et al., 1992.)

However, the most decisive factor for P-based legislation in The Netherlands was the fact that P losses from soil were considered easier to identify and were relatively small compared to N losses. In other words, it would be easier to establish acceptable application rates for manure in terms of P than of N. Moreover, P-based application rates also proved appropriate in meeting the quality standards for N in groundwater.

Legislation does not yet take into account soil P status. Because of the high percentage of soils with a very high soil P level, such regulations would require a significant transport of manure outside the country. The current manure policy, developed by the Ministries of Agriculture and Environment, aims at reducing application rates stepwise to "equilibrium rates" in the year 2000. Such rates should meet crop requirements plus allow losses that are environmentally acceptable.

In addition to adoption of recommended application rates and methods, remedial measures involve manuring practices, manure amendments, feed additives, storage and handling at the farm, transport to other farms, and processing in large-scale operations (L' Hermite et al., 1993). Remedial measures not only address water pollution but also air pollution (NH_3 and odor emissions). In The Netherlands, a special program, "Manure and Ammonia Research," was set up in 1986 to study these remedial measures. Major results include the development of methods to reduce NH_3 loss from buildings, storage facilities, and land spreading; reduce losses of N and P by leaching and runoff; improve the uptake of P from animal feeds; and process manure in large-scale operations.

The importance of transporting manure from surplus to deficient, nonproducing areas to solve the manure surplus problem in The Netherlands is demonstrated by Table 14. The transport of manure has increased significantly, partly through the national distribution network. Recent estimates indicate that in 1995, 37% of the national manure surplus can be distributed within the surplus areas and 32% transported to deficient areas. Based on data for 1993, manure export is expected to be at least 6% and the processing of manure 5%. This would still leave a gap of about 20% of the total surplus that has to be managed.

Mandatory transport of manure from surplus areas to nearby farms where the nutrients are needed faces several significant obstacles. First, it must be shown that the current location is unsuitable, based on soil properties, crop nutrient requirements, hydrology, etc. From European experiences this may be difficult to justify scientifically due to the large temporal and spatial variability in the factors controlling N and P mobility in soils and transport to ground or surface waters. Second, in many areas there is no clearly defined legal basis for requiring farmers in one physiographic area to perform management practices that are not required on neighboring farms. The greatest success with redistribution of manure nutrients is likely to occur when the general goals set by a national (or state) government are supported by local governments, the farm community, and the animal industry involved. This may initially require cost-sharing to facilitate an overall assessment of the situation and the subsequent transport of manures from one area to another.

Table 14. Estimated production and fate of P in manure in The Netherlands in 1995

	kg P (million)	% of surplus
Production[a]	95	
Use on the farm[a]	57	
Surplus[a]	38	
Transport within surplus areas[a]		37
Transport to deficit areas[a]		32
Export[b]		6
Processed in large-scale operations[b]		5
Remaining surplus		21

[a]National Environment Survey 3, 1993-2015, Table 3.6.1 from RIVM (1993).
[b]Data for 1992, from National Manure Board (September 1993).

V. Conclusions

Environmentally sound management of animal manures is likely to be the most problematic issue facing agriculture for the next decade. The concentrated nature of modern animal production operations makes it virtually inevitable that excesses of N and P will occur in geographic areas with an inadequate land base to successfully use these nutrients. Since animal-based agriculture is frequently in close proximity to shallow groundwater aquifers or sensitive surface waters, the leachability of NO_3 and ease of P loss in erosion, runoff, or drainage will further contribute to degradation of water quality unless more intensive manure management practices are developed and implemented. Other issues likely to arise in the near future are the fate and environmental impact of manure constituents such as trace elements, pesticides, antibiotics, and growth-hormones.

Modern society seeks a sustainable agriculture, one that provides adequate food with little or no adverse effect on the quality of our environment. Scientific research has identified many of the approaches needed to ensure that animal-based agriculture can sustain crop and animal production, build soil tilth and fertility, and maintain water quality. New soil and plant N tests (e.g., the PSNT, leaf chlorophyll meter, stalk NO_3) have evolved from our research on soil N budgets and N cycling. Means to identify (and control) areas where the fate and transport of biologically available particulate and dissolved P in watersheds is environmentally significant are now being developed. Information is slowly emerging on the effects of trace elements and organics in manures. Together, these and other efforts provide a solid research base for future actions.

Unfortunately, many recent scientific advances have not been widely implemented at the field, farm, or watershed scale. Economic and labor constraints, the lack of technical expertise in manure management on the part of many farmers, and the inherent difficulty in controlling nonpoint losses of N and P have all combined to present a formidable challenge to an agricultural community facing more and more questions about the environmental impact of animal manures.

Future BMPs for animal manures must be comprehensive and should be conceived and implemented at several scales. Individually, farmers cannot resolve all environmental problems posed by animal manures. A collective effort that integrates resources and expertise of the animal production industry as a whole, the advisory and regulatory agencies of local, state and national governments, and the research base available from universities and federal research agencies will be necessary. Planning should be done at regional, watershed, farm, and field scales for maximum effectiveness. Innovative approaches to cost-sharing and other economic stimuli should be considered by legislators and policy-makers. Finally, a rigorous risk assessment evaluation should be conducted to determine the most pressing environmental issues.

A number of important questions must be addressed to best prioritize the distribution of funding between research, technical assistance, educational, and regulatory programs for animal manure management. When do we focus on NO_3 leaching to groundwaters rather than P losses to sensitive lakes and bays? How important are the pesticides, trace metals, and antibiotics or hormones in animal manures? Who is best positioned to educate consumers and producers of animal products on environmental issues and need for improved management practices? Success will ultimately depend on the extent of cooperation between a number of diverse groups, each with their own agenda. Without this we are likely to see a difficult transition period between the slow and inconsistent implementation of improved management practices and the gradual imposition of regulations on the use of animal manures.

References

Alley, M.M. 1991. Environmental implications for phosphorus management in the Mid-Atlantic United States. p. 63-71. In: Proc. Western Phosphate and Sulfur Workshop. Ft. Collins, CO.

Amdur, M.O., J. Dull, and E.D. Klassen (eds.). 1991. *Casarett and Doull's Toxicology*. Fourth Ed., Pergamon Press, New York, NY.

Angle, J.S., C.M. Gross, R.L. Hill, and M.S. McIntosh. 1993. Soil nitrate concentrations under corn as affected by tillage, manure, and fertilizer applications. *J. Environ. Qual.* 22:141-147.

Binford, G.D., A.M. Blackmer, and N.M. El-Hout. 1990. Tissue test for excess nitrogen during corn production. *Agron. J.* 82:124-129.

Blackmer, A.M., T.F. Morris, and G.D. Binford. 1992. Predicting N fertilizer needs for corn in humid climates: Advances in Iowa. p. 57-72. In: B.R. Bock and K.R. Kelley (eds.), *Predicting N fertilizer needs for corn in humid regions*. DNC. SSSA Symp., Oct. 28, 1991, Denver, CO, Tenn. Valley Authority, Bull. Y-226, TVA, Muscle Shoals, AL.

Blackmer, A.M., D. Pottker, M.E. Cerrato, and J. Webb. 1989. Correlations between soil nitrate concentrations in late spring and corn yields in Iowa. *J. Prod. Agric.* 2:103-109.

Bock, B.R. and K.R. Kelley. 1992. Predicting N fertilizer needs for corn in humid regions. Proc. SSSA Sym., Oct. 28, 1991, Denver, CO, Tenn. Valley Auth. Bull. Y-226, TVA, Muscle Shoals, AL.

Bosch, D.J. and K.B. Napit. 1992. Economics of transporting poultry litter to achieve more effective use as a fertilizer. *J. Soil Water Conserv.* 47:342-346.

Bottcher, A.B. and T. Tremwell. 1995. Best Management Practices for water quality improvement. *Ecol. Eng.* (in press).

Breeuwsma A. and J.G.A. Reyerink. 1992. Phosphate saturated soils: A new environmental issue. In: G.R.B. ter Meulen, W.M. Strigliani, W. Salomons, E.M. Bridges, and A.C. Imerson (eds.), Chemical Time Bombs, Proceedings European State-of-the-Art Conference, September 1992, Foundation for Ecodevelopment, Hoofddorp, The Netherlands.

Breeuwsma, A. and S. Silva. 1992. *Phosphorus fertilization and environmental effects in The Netherlands and the Po region (Italy)*. Rep. 57. Agric. Res. Dep. The Winand Staring Centre for Integrated Land, Soil and Water Research. Wageningen, The Netherlands.

Breeuwsma A., J.G.A. Reyerink, D.J. Brus, H. van het Loo, and O.F. Schoumans. 1989. Phosphate loading of soils, groundwater and surface waters in the catchment area of the Schuitenbeek (in Dutch), DLO Winand Staring Centre, report 10, Wageningen, The Netherlands.

Brodie, H.L. and L.E. Carr. 1988. Storage of poultry manure in solid form. p. 115-119. In: E.C. Naber (ed.), Proc. Natl. Poultry Waste Management Symp., Columbus, OH. Ohio State Univ. Press, Columbus, OH.

Brodie H.L. and G.L. Smith. 1985. Calibrating manure spreaders. Univ. Maryland Coop. Ext. Serv. Fact Sheet No. 419, College Park, MD.

Bundy, L.G. and E.S. Malone. 1988. Effect of residual profile nitrate on corn response to applied nitrogen. *Soil Sci. Soc. Am. J.* 52:1377-1383.

Burwell, R.E., D.R. Timmons, and R.F. Holt. 1975. Nutrient transport in surface runoff as influenced by soil cover and seasonal periods. *Soil Sci. Soc. Am. Proc.* 39:523-528.

Carton, O.T., J.J. Lenehan, W.L. Magette, and A.J. Douglas. 1993. TEAGASC - Agriculture and Food Development Authority - slurry research programme. Environment and Land Use Depart., Johnstown Castle, Wexford, Ireland. 33 pp.

Chaubey, I., D.R. Edwards, T.C. Daniel, and D.J. Nichols. 1993. Effectiveness of vegetative filter strips on controlling losses of surface applied poultry litter constituents. Am. Soc. Agric. Eng., Paper 932011. 17 pp. St. Joseph, MI.

Collins Jr., E.R., J.M. Halstead, H.V. Roller, W.D. Weaver Jr., and F.B. Givens. 1988. Application of poultry manure -- Logistics and economics. p. 125-132. E.C. Naber (ed.), Proc. Natl. Poultry Waste Mangt. Symp., April 1988, Columbus, OH.

Dahnke, W.C. and G.V. Johnson. 1990. Testing soils for available nitrogen. p. 127-140. In: R.L. Westerman (ed.), *Soil Testing and Plant Analysis* (3rd ed.). Soil Sci. Soc. Am., Madison, WI.

Danish Ministry of Environment. 1991. *Environmental Impacts of Nutrient Emissions in Denmark*, Copenhagen, Denmark.

Dillaha, T.A., S.D. Lee, V.O. Shanholtz, S. Mostaghimi, and W.L. Magette. 1986. Use of vegetative filter strips to minimize sediment and phosphorus losses from feedlots: Phase I, experimental plot studies. VPI-VWRRRC-Bull. 151. Virginia Polytechnical Inst., Blacksburg, VA.

Dudinsky, M.L., S.R. Wilkinson, R.N. Dawson, and A.P. Barnett. 1983. Fate of nitrogen from NH_4NO_3 and broiler litter applied to Coastal Bermudagrass. p. 373-388. In: R. Lowrance, R. Todd, L. Asmussen, and R. Leonard (eds.), *Nutrient Cycling in Agricultural Ecosystems*. Univ. Georgia Agric. Exp. Sta. Spec. Bull. 23, Athens, GA.

Duynisveld, W.H.M., O. Strebel, and J. Bottcher. 1988. Are nitrate leaching from arable land and nitrate pollution of groundwater unavoidable? *Ecological Bull.* 39:116-125.

Edwards, D.R. and T.C. Daniel. 1992. Potential runoff quality effects of poultry manure slurry applied to fescue plots. *Trans. Am. Soc. Agric. Eng.* 35:1827-1832.

Edwards, D.R. and T.C. Daniel. 1993a. Runoff quality impacts of swine manure applied to fescue plots. *Trans. Am. Soc. Agric. Eng.* 36:81-80.

Edwards, D.R. and T.C. Daniel. 1993b. Drying interval effects on runoff from fescue plots receiving swine manure. *Trans. Am. Soc. Agric. Eng.* 36:1673-1678.

Edwards, D.R. and T.C. Daniel. 1993c. Effects of poultry litter application rate and rainfall intensity on quality of runoff from fescuegrass plots. *J. Environ. Qual.* 23:361-365.

Edwards, W.M. and L.B. Owens. 1991. Large storm effects on total soil erosion. *J. Soil Water Conserv.* 46:75-77.

Edwards, D.R., T.C. Daniel, and O. Marbun. 1992. Determination of best timing for poultry waste disposal: A modeling approach. *Water Res. Bull.* 28:487-494.

Edwards, W.M., L.B. Owens, and R.K. White. 1983. Managing runoff from a small, paved beef feedlot. *J. Environ. Qual.* 12:281-286.

Evans, R.O., J.W. Gilliam, and R.W. Skaggs. 1989. *Effects of agricultural water table management on drainage water quality*. Univ. N. Carolina Water Resour. Res. Inst. Rpt. No. 237, N. Carolina St. Univ., Raleigh, NC.

Fertilizer Handbook. 1982. The fertilizer handbook. In: W.C. White and D.N. Collins (eds.), *The Fertilizer Institute*, Washington, D.C. p. 274.

Fox, R.H., J.J. Meisinger, and J.T. Sims. 1992. Predicting N fertilizer needs for corn in humid regions: Advances in the Mid-Atlantic States. p. 43-56. In: *Predicting N fertilizer needs for corn in humid regions*. National Fert. and Environ. Res. Center, Tenn. Valley Authority, Muscle Shoals, AL.

Fox, R.H., G.W. Roth, K.V. Iverson, and W.P. Piekielek. 1989. Soil and tissue nitrate tests compared for predicting soil nitrogen availability to corn. *Agron. J.* 81:971-974.

Francis, D.D., J.S. Schepers, and M.F. Vigil. 1993. Post-anthesis nitrogen loss from corn. *Agron. J.* 85:659-663.

Furrer, O.J. and W. Stauffer. 1985. Influence of sewage sludge and slurry application on nutrient leaching losses. p. 108-115. In: A. Dam Kofoed et al. (eds.), *Efficient Land Use of Sludge and Manure*. Proc. Round-table seminar on efficient land use of sludge and manure, June 25-27, 1985, Commission European Communities, Environ. Res. Prog., Brorup-Askov, Denmark. Elsevier Applied Sci. Pub., New York, NY.

Gartley, K.L. and J.T. Sims. 1994. Phosphorus soil testing: Environmental uses and implications. *Commun. Soil Sci. Plant Anal.* 25:1565-1582.

Gilbertson, C.B., F.A. Norstadt, A.C. Mathers, R.F. Holt, A.P. Barrett, T.M. McCalla, C.A. Onstad, and R.A. Young. 1979. Animal waste utilization on cropland and pastureland--A manual for evaluating agronomic and environmental effects. U.S. Environ. Prot. Agency and U.S. Dept. of Agriculture, U.S. EPA Rep. No. EPA 600/2-79-059 and USDA Rep. No. URR 6. U.S. Government Printing Office, Washington, D.C.

Gillham, R.W. 1989. Nitrate contamination of groundwater in southern Ontario: A hydrogeologic perspective. p. 69-83. In: J.A. Stone and L.L. Logan (eds.), *Agriculture Chemicals and Water Quality in Ontario*. Proc. of Workshop, Nov. 17-18, 1988, Ontario Water Management Res. and Serv. Comm., Kitchener, Ontario.

Gilliam, J.W., R.W. Skaggs, and S.B. Weed. 1979. Drainage control to diminish nitrate loss from agricultural fields. *J. Environ. Qual.* 8:137-142.

Graetz, D.A. and V.D. Nair. 1995. Fate of phosphorus in Florida spodosols contaminated with cattle manure. *Ecol. Eng.* (in press).

Hall, J.E. and J.C. Ryden. 1985. Current UK research into ammonia losses from sludges and slurries. p. 180-192. In: A. Dam Kofoed et al. (eds.), *Efficient Land Use of Sludge and Manure*. Proc. Round-table seminar on efficient land use of sludge and manure, June 25-27, 1985, Commission European Communities, Environ. Res. Prog., Brorup-Askov, Denmark. Elsevier Applied Sci. Pub., New York, NY.

Hamilton, P.A. and R.J. Shedlock. 1992. Are fertilizers and pesticides in the ground water? A case study of the Delmarva peninsula. U.S. Geol. Surv. Circ. 1080. Denver, CO. 15 pp.

Heathman, G.C., A.N. Sharpley, S.J. Smith, and J.S. Robinson. 1995. Poultry litter application and water quality in Oklahoma. *Fert. Res.* 37:40:165-173.

Hedley, M.J., J.W.B. Stewart, and B.S. Chauhan. 1982. Changes in inorganic and organic soil phosphorus fractions induced by cultivation practices and by laboratory incubations. *Soil Sci. Soc. Am. J.* 46:970-976.

Hergert, G.W. 1987. Status of residual nitrate-nitrogen soil tests in the United States of America. p. 73-88. In: J.R. Brown et al. (eds.), *Soil Testing: Sampling, Correlation, Calibration, and Interpretation*. Soil Sci. Soc. Am. Spec. Pub. No. 21, Soil Sci. Soc. Am., Madison, WI.

Hergert, G.W., R.B. Ferguson, C.A. Cotway, and T.A. Peterson. 1994. Developing accurate nitrogen rate maps for variable rate application. *Agron. Abstr.* 1994, p. 399. Am. Soc. Agron., Madison, WI.

Holtan-Hartwig, L. and O.C. Beckman. 1994. Ammonia exchange between crops and air. *Norwegian J. Agric. Sci.*, Suppl. No. 14, 1-41.

Igo, E.C., J.T. Sims, and G.W. Malone. 1991. Advantages and disadvantages of manure analysis for nutrient management purposes. *Agron. Abstr.* p. 154.

Iserman, K. 1990. Share of agriculture in nitrogen and phosphorus emissions into the surface waters of Western Europe against the background of their eutrophication. *Fert. Res.* 26:253-269.

Johnston, A.E. 1989. Phosphorus cycling in intensive arable agriculture. p. 123-136. In: H. Tiessen (ed.), *Phosphorus Cycles in Terrestrial and Aquatic Systems.* Proceedings of a SCOPE and UNEP Workshop, May 1 to May 6, Czerniejewo, Poland.

Jones, O.R., W.M. Willis, S.J. Smith, and B.A. Stewart. 1995. Nutrient cycling of cattle feedlot manure and composted manure applied to Southern High Plains drylands. p. 265-272. In: K. Steele (Ed.), *Impact of Animal Manure and the Land-Water Interface.* Lewis Pub., CRC Press, Boca Raton, FL.

Keeney, D.R. 1982. Nitrogen-availability indices. In: A.L. Page et al. (ed.), Methods of soil analysis, Pt. 2 (2nd ed.). *Agronomy* 9:711-733.

Kingery, W.L., C.W. Wood, D.P. Delaney, J.C. Williams, and G.L. Mullins. 1994. Impact of long-term land application of broiler litter on environmentally related soil properties. *J. Environ. Qual.* 23:139-147.

Kingery, W.L., C.W. Wood, D.P. Delaney, J.C. Williams, G.L. Mullins, and E. van Santen. 1993. Implications of long-term land application of poultry litter on tall fescue pastures. *J. Prod. Agric.* 6:390-395.

Kissel, D.E., O.W. Bidwell, and J.F. Kientz. 1982. Leaching classes of Kansas soils. Kansas Agric. Exp. Sta. Bull. 641:3-11. Manhattan, KS.

Klausner S. and D.R. Bouldin. 1983a. Managing animal manure as a resource. Part I. Basic principles. New York State Coop. Ext. Serv., Ithaca, NY.

Klausner S. and D.R. Bouldin. 1983b. Managing animal manure as a resource. Part II. Field management. New York State Coop. Ext. Serv., Ithaca, NY.

Klausner, S.D., P.J. Zwerman, and D.F. Ellis. 1976. Nitrogen and phosphorus losses from winter disposal of dairy manure. *J. Environ. Qual.* 5:47-49.

Kolle, W., O. Strebel, and J. Bottcher. 1985. Formation of sulfate by microbial denitrification in a reducing aquifer. *Water Supply* (Berlin, Ser. B) 3:35-40.

Kotak, B.G., E.E. Prepas, and S.E. Hrudey. 1994. Blue green algal toxins in drinking water supplies: Research in Alberta. *Lake Line* 14:37-40.

Kotak, B.G., S.L. Kenefick, D.L. Fritz, C.G. Rousseaux, E.E. Prepas, and S.E. Hrudey. 1993. Occurrence and toxicological evaluation of cyanobacterial toxins in Alberta lakes and farm dugouts. *Water Res.* 27:495-506.

Lawton, L.A. and G.A. Codd. 1991. Cyanobacterial (blue-green algae) toxins and their significance in UK and European waters. *J. Inst. Water Environ. Managt.* 5:460-465.

Lemunyon, J.L. and R.G. Gilbert. 1993. Concept and need for a phosphorus assessment tool. *J. Prod. Agric.* 6: 483-486.

L' Hermite, P., P. Sequi, and J.H. Voorgburg. 1993. Scientific bases for environmentally safe and efficient management of livestock farming, European Conference, Chamber of Commerce, Manatova, Italy.

Long, F.L. 1979. Runoff water quality as affected by surface-applied dairy cattle manure. *J. Environ. Qual.* 8:215-218.

Madison, R.J. and J.O. Brunett. 1985. Overview of the occurrence of nitrate in ground water of the United States. p. 93-105. In: National water summary 1984 - hydrologic events, selected trends and ground water resources. Water Supply Paper 2275, U.S. Geologic Survey, U.S. Gov. Print. Off., Washington D.C.

Magdoff, F.R. 1991a. Understanding the Magdoff pre-sidedress nitrate test for corn. *J. Prod. Agric.* 4:297-305.

Magdoff, F.R. 1991b. Field nitrogen dynamics: Implications for assessing N availability. *Comm. Soil Sci. Plant Anal.* 22:1507-1517.

Magdoff, F.R., D. Ross, and J. Amadon. 1984. A soil test for nitrogen availability for corn. *Soil Sci. Soc. Am. J.* 48:1301-1304.

Magdoff, F.R., W.E. Jokela, R.H. Fox, and G.F. Griffin. 1990. A soil test for nitrogen availability in the Northeastern U.S. *Comm. Soil Sci. Plant Anal.* 21:1103-1115.

Magette, W.L. 1988. Runoff potential from poultry manure applications. p. 102-106. In: E.C. Naber (ed.), *Proc. Natl. Poultry Waste Management Symp.*, Columbus, OH, 1988. Ohio State Univ. Press, Columbus, OH.

Marriott, L.F. and H.D. Bartlett. 1975. Animal waste contributions to nitrate nitrogen in soil. p. 296-298. In: *Managing Livestock Wastes*. Proc. 3rd Int. Symp. on Livestock Wastes - 1975, April 21-24, 1975, Univ. of IL, Urbana-Champaign, IL. Am. Soc. Agric. Eng., St. Joseph, MI.

Martin, A. and G.D. Cooke. 1994. Health risks in eutrophic water supplies. *Lake Line* 14:24-26.

Mathers, A.C. and B.A. Stewart. 1971. Crop production and soil analyses as affected by applications of cattle feedlot waste. p. 229-234. In: *Livestock Waste Management and Pollution Abatement*. Proc. Int. Symp. on Livestock Wastes, April 19-22, 1971, Ohio St. Univ., Columbus, OH. Am. Soc. Agric. Eng., St. Joseph, MI.

McCollum, R.E. 1991. Buildup and decline in soil phosphorus: 30-year trends on a Typic Umprabuult. *Agron. J.* 83:77-85.

McLeod, R.V. and R.O. Hegg. 1984. Pasture runoff water quality from application of inorganic and organic nitrogen sources. *J. Environ. Qual.* 13:122-126.

Meisinger, J.J. 1984. Evaluating plant-available nitrogen in soil-crop systems. p. 391-416. In: R.D. Hauck et al. (eds.), *Nitrogen in Crop Production*. Am. Soc. Agron., Madison, WI.

Meisinger, J.J. and G.W. Randall. 1991. Estimating nitrogen budgets for soil-crop systems. p. 85-124. In: R.F. Follett et al. (eds.), *Managing Nitrogen for Groundwater Quality and Farm Profitability*. Proc. Symp. Am. Soc. Agron., Nov. 30, 1988, Anaheim, CA, Soil Sci. Soc. Am., Madison, WI.

Meisinger, J.J., F.R. Magdoff, and J.S. Schepers. 1992a. Predicting N fertilizer needs for corn in humid regions: Underlying principles. p. 7-27. In: B.R. Bock and K.R. Kelley (Eds.), *Predicting N Fertilizer Needs for Corn in Humid Regions*. Proc. SSSA Sym., Oct. 28, 1991, Denver, CO. Tenn. Valley Auth. Bull. Y-226. TVA, Muscle Shoals, AL.

Meisinger, J.J., V.A. Bandel, J.S. Angle, B.E. O'Keefe, and C.M. Reynolds. 1992b. Presidedress soil nitrate test evaluation in Maryland. *Soil Sci. Soc. Am. J.* 56:1527-1532.

Meisinger, J.J., V.A. Bandel, G. Stanford, and J.O. Legg. 1985. Nitrogen utilization of corn under minimal tillage and moldboard plow tillage. I. Four-year results using labeled N fertilizer on an Atlantic Coastal Plain Soil. *Agron. J.* 77:602-611.

Meisinger, J.J., W.L. Hargrove, R.B. Mikkelsen, J.R. Williams, and V.W. Benson. 1991. Effect of cover crops on groundwater quality. p. 57-68. In: W.L. Hargrove (ed.), *Cover Crops for Clean Water*. Proc. Int. Conf., April 9-11, 1991, Jackson, TN, Soil and Water Conserv. Soc. Am., Ankeny, IA.

Midwest Planning Service--Livestock Waste Subcommittee. 1985. Livestock waste facilities handbook. Midwest Planning Serv. Rep. MWPS-18 (2nd ed.), Iowa State Univ., Ames, IA.

Monsanto Agricultural Co. 1990. *The National Alachlor Well Water Survey, Data Summary*. The Monsanto Co., St. Louis, MO.

Moore, P.A., Jr. and D.M. Miller. 1994. Decreasing phosphorus solubility in poultry litter with aluminum, calcium and iron amendments. *J. Environ. Qual.* 23:325-330.

Mueller, D.H., R.C. Wendt, and T.C. Daniel. 1984. Phosphorus losses as affected by tillage and manure application. *Soil Sci. Soc. Am. J.* 48:901-905.

National Agricultural Statistics Service. 1991. Poultry production and value. Bull. POU 3-1 (April). USDA, Washington, D.C.

National Manure Board, 1993. Mest in Balans 2, Nijkerk, The Netherlands.

National Research Council. 1993. *Soil and Water Quality: an Agenda for Agriculture*. National Academy Press, Washington, D.C.

Olness, A.E., S.J. Smith, E.D. Rhoades, and R.G. Menzel. 1975. Nutrient and sediment discharge from agricultural watersheds in Oklahoma. *J. Environ. Qual.* 4:331-336.

Palmstrom, N.S., R.E. Carlson, and G.D. Cooke. 1988. Potential links between eutrophication and formation of carcinogens in drinking water. *Lake Reserv. Managt.* 4:1-15.

Peterson, T.A., T.M. Blackmer, D.D. Francis, and J.S. Schepers. 1993. Using a chlorophyll meter to improve N management. Univ. Neb. Coop. Ext. Serv. Neb-Guide No. G93-1171-A. Lincoln, NE.

Peterson, T.A., J.S. Schepers, Changhe Chen, C.A. Cotway, R.B. Ferguson, and G.W. Hergert. 1994. Interpreting yield and soil parameter maps in the evaluation of variable rate nitrogen applications. *Agron. Abstr.* 1997, p. 397. Am. Soc. Agron., Madison, WI.

Potash and Phosphate Institute. 1990. Soil test summaries: Phosphorus, potassium, and pH. *Better Crops* 74:16-19.

Pote, D., T.C. Daniel, A.N. Sharpley, P.A. Moore, Jr., D.R. Edwards, and D.J. Nichols. 1996. Relating extractable phosphorus in a silt loam to phosphorus losses in runoff. *Soil Sci. Soc. Am. J.* (in press).

Pratt, P.F., F.E. Broadbent, and J.P. Martin. 1973. Using organic wastes as nitrogen fertilizers. *Calif. Agric.* 27:10-13.

Pratt, P.F., S. Davis, and R.G. Sharpless. 1976a. A four year field trial with animal manures. I. Nitrogen balances and yield. *Hilgardia* 44:99-112.

Pratt, P.F., S. Davis, and R.G. Sharpless. 1976b. A four year field trial with animal manures. II. Mineralization of nitrogen. *Hilgardia* 44:113-125.

Reese, L.E., R.O. Hegg, and R.E. Gantt. 1982. Runoff water quality from dairy pastures in the Piedmont region. *Trans. Am. Soc. Agric. Eng.* 25:697-701.

Reeves, D.W., P.L. Mask, C.W. Wood, and D.P. Delaney. 1993. Determination of wheat nitrogen status with a hand-held chlorophyll meter. *J. Plant Nutr.* 16:781-796.

Ritter, W.F. and A.E.M. Chirnside. 1987. Influence of agricultural practices on nitrates in the water table aquifer. *Biol. Wastes* 19:165-178.

RIVM. 1993. National Environment Survey 3, 1993-2015 (in Dutch) National Institute of Public Health and Environment Protection, Bilthoven, The Netherlands.

Robertson, F.N. 1977. *The Quality and Potential Problem of Ground Water in Sussex County, DE*. Water Res. Ctr., Univ. of Delaware, Newark, DE.

Rolston, D.E. and F.E. Broadbent. 1977. Field measurement of denitrification. U.S. EPA Res. Rep. Ser. EPA-600/2-77-23, Ada, OK.

Rolston, D.E., F.E. Broadbent, and D.A. Goldhamer. 1979. Field measurement of denitrification: II. Mass balance and sampling uncertainty. *Soil Sci. Soc. Am. J.* 43:703-708.

Rolston, D.E., D.L. Hoffman, and D.W. Toy. 1978. Field measurement of denitrification: I. flux of N_2 and N_2O. *Soil Sci. Soc. Am. J.* 42:863-869.

Romkens, M.J.M. and D.W. Nelson. 1974. Phosphorus relationships in runoff from fertilized soil. *J. Environ. Qual.* 3:10-13.

Roth, G.W. and R.H. Fox. 1990. Soil nitrate accumulations following N-fertilized corn in Pennsylvania. *J. Environ. Qual.* 19:243-248.

Safley, L.M. Jr., J.C. Barker, and P.W. Westerman. 1992. Loss of nitrogen during sprinkler irrigation of swine lagoon liquid. *Bioscience Technol.* 40:7-15.

Sandstedt, C.A. 1990. *Nitrates: Sources and Their Effects upon Humans and Livestock*. The American Univ., Washington, D.C.

Sawyer, C.N. 1947. Fertilization of lakes by agricultural and urban drainage. *J. New England Water Works Assoc.* 61:109-127.

Schellinger, G.R. and J.C. Clausen. 1992. Vegetative filter treatment of dairy barnyard runoff in cold regions. *J. Environ. Qual.* 21:40-45.

Schepers, J.S. and J.J. Meisinger. 1994. Field indicators of nitrogen mineralization. p. 31-47. In: J.L. Havlin and J.S. Jacobsen (eds.), *Soil testing: Prospects for Improving Nutrient Recommendations*. Proc. SSSA Sym., Nov. 8, 1993, Cincinnati, OH. Soil Sci. Soc. Am. Spec. Pub. #40, Madison, WI.

Schepers, J.S. and A.R. Mosier. 1991. Accounting for nitrogen in nonequilibrium soil-crop systems. p. 125-138. In: R.F. Follett et al. (eds.), *Managing Nitrogen for Groundwater Quality and Farm Profitability*. Proc. Sym. Am. Soc. Agron., Nov. 30, 1988, Anaheim, CA, Soil Sci. Soc. Am., Madison, WI.

Schepers, J.S., T.M. Blackmer, and D.D. Francis. 1992a. Predicting N fertilizer needs for corn in humid regions: Using chlorophyll meters. p. 105-114. In: B.R. Bock and K.R. Kelley (eds.), *Predicting N Fertilizer Needs for Corn in Humid Regions*. Proc. SSSA Symp., Oct. 28, 1991, Denver, CO. Tenn. Valley Auth. Bull. Y-226. TVA, Muscle Shoals, AL.

Schepers, J.S., D.D. Francis, M. Vigil, and F.E. Below. 1992b. Comparison of corn leaf nitrogen and chlorophyll meter readings. *Comm. Soil Sci. Plant Anal.* 23:2173-2187.

Schepers, J.S., M.G. Moravek, E.E. Alberts, and K.D. Frank. 1991. Maize production impacts on groundwater quality. *J. Environ. Qual.* 20:12-16.

Schreiber, J.D. 1988. Estimating soluble phosphorus (PO_4-P) in agricultural runoff. *J. Miss. Acad. Sci.* 33:1-15.

Schwer, C.B. and J.C. Clausen. 1989. Vegetative filter treatments of dairy milkhouse wastewater. *J. Environ. Qual.* 18:446-451.

Sharpley, A.N. and J.R. Williams (eds.). 1990. EPIC-Erosion/Productivity Impact Calculator. 1. Model documentation. USDA Technical Bull. 1768. 235 pp. U.S. Govt. Print. Office, Washington, D.C.

Sharpley, A.N., J.T. Sims, and G.M. Pierzynski. 1994. Innovative soil phosphorus availability indices: Assessing inorganic phosphorus. p. 113-140. In: J. Havlin, J. Jacobson, P. Fixen, and G. Hergert (eds.), *Soil Testing Prospects for Improving Nutrient Recommendations*. Soil Sci. Soc. Am. Spec. Publ. 40, Madison, WI.

Sharpley, A.N., S.J. Smith, and W.R. Bain. 1993. Nitrogen and phosphorus fate from long-term poultry litter applications to Oklahoma soils. *Soil Sci. Soc. Am. J.* 57:1131-1137.

Sharpley, A.N., S.J. Smith, and R.G. Menzel. 1986. Phosphorus criteria and water quality management for agricultural watersheds. *Lake Reserv. Managt.* 2:177-182.

Sharpley, A.N., J.J. Meisinger, J.F. Power, and D.L. Suarez. 1992. Root extraction of nutrients associated with long-term soil management. *Adv. Soil Sci.* 19:151-217.

Sharpley, A.N., S.J. Smith, B.A. Stewart, and A.C. Mathers. 1984. Forms of phosphorus in soil receiving cattle feedlot waste. *J. Environ. Qual.* 13:211-215.

Sharpley, A.N., B.J. Carter, B.J. Wagner, S.J. Smith, E.L. Cole, and G.A. Sample. 1991. Impact of long-term swine and poultry manure applications on soil and water resources in eastern Oklahoma. Okla. State Univ., Stillwater, OK, Tech. Bull. T169, 51 pp.

Sherwood, M. 1985. Nitrate leaching following application of slurry and urine to field plots. p. 150-157. In: A. Dam Kofoed et al. (eds.), *Efficient Land Use of Sludge and Manure*. Proc. Round-table seminar on efficient land use of sludge and manure, June 25-27, 1985, Commission European Communities, Environ. Res. Prog., Brorup-Askov, Denmark. Elsevier Applied Sci. Pub., New York, NY.

Shipley, P.R., J.J. Meisinger, and A.M. Decker. 1992. Conserving residual corn fertilizer nitrogen with winter cover crops. *Agron. J.* 84:869-876.

Shore, L.S., D. Correll, and P.K. Chakraborty. 1995. Sources and distribution of testosterone and estrogen in the Chesapeake Bay Watershed. p. 155-162. In: K. Steele (ed.), *Impact of animal manure and the land-water interface.* Lewis Pub., CRC Press, Boca Raton, FL

Shreve, B.R., P.A. Moore, Jr., T.C. Daniel, and D.R. Edwards. 1995. Reduction of phosphorus in runoff from field-applied poultry litter using chemical amendments. *J. Environ. Qual.* 23:106-111.

Sims, J.T. 1992. Environmental management of phosphorus in agriculture and municipal wastes. p. 59-64. In: F.J. Sikora (ed.), *Future Directions for Agricultural Phosphorus Research.* Nat. Fert. Environ. Res. Cent., TVA, Muscle Shoals, AL.

Sims, J.T. 1993a. Environmental soil testing for phosphorus. *J. Prod. Agric.* 6:501-507.

Sims, J.T. 1993b. *The phosphorus index: A phosphorus management strategy for Delaware's agricultural soils.* Dept. Plant and Soil Sci., Univ. Delaware, Newark, DE. 4 pp.

Sims, J.T., B.L. Vasilas, K.L. Gartley, B. Milliken, and V. Green. 1995. Evaluation of soil and plant nitrogen tests for maize on manured soils of the Atlantic Coastal Plain. *Agron. J.* 87:213-222.

Smith, R.A., R.B. Alexander, and M.G. Wolman. 1987a. Analysis and interpretation of water quality trends in major U.S. rivers. 1974-81. Water Supply Paper 2307, U.S. Geologic Survey, U.S. Gov. Print. Off., Washington D.C.

Smith, R.A., R.B. Alexander, and M.G. Wolman. 1987b. Water quality trends in the nations rivers. *Science* 235:1607-1615.

Smith, S.J., A.N. Sharpley, J.W. Naney, W.A. Berg, and O.R. Jones. 1991a. Water quality impacts associated with wheat culture in the Southern Plains. *J. Environ. Qual.* 20:244-249.

Smith, S.J., A.N. Sharpley, J.R. Williams, W.A. Berg, and G.A. Coleman. 1991b. Sediment-nutrient transport during severe storms. p. 48-55. In: S.S. Fan and Y.H. Kuo (eds.), *Fifth Interagency Sedimentation Conf.*, March 1991, Las Vegas, NV. Federal Energy Regulatory Commission, Washington, D.C.

Stanford, G. 1973. Rationale for optimum nitrogen fertilization in corn production. *J. Environ. Qual.* 2:159-166.

Stanford, G. 1982. Assessment of soil nitrogen availability. In: F.J. Stevenson (ed.), Nitrogen in agricultural soils. *Agronomy* 22:651-688. Am. Soc. Agron., Madison, WI.

Steenvoorden, J.H.A.M. 1985. Nutrient leaching losses following application of farm slurry and water quality considerations in the Netherlands. p. 168-176. In: A. Dam Kofoed et al. (eds.), *Efficient Land Use of Sludge and Manure.* Proc. Round-table seminar on efficient land use of sludge and manure, June 25-27, 1985, Commission European Communities, Environ. Res. Prog., Brorup-Askov, Denmark. Elsevier Applied Sci. Pub., New York, NY.

Svendsen, L.M. and B. Kronvang. 1991. Phosphorus in the Nordic countries: Methods, bioavailability, effects, and measures. A report by the National Environmental Research Institute, Denmark. 201 pp.

USA vs. South Florida Water Management District. 1994. U.S. District Court/Southern District, Case number 88-1880-CIV.

U.S. Dept. of Agriculture. 1991. Nitrate occurrence in U.S. waters (and related questions): A reference summary of published sources from an agricultural perspective. U.S. Dept. of Agric. Working Group on Water Quality, U.S. Gov. Print. Off., Washington, D.C.

U.S. Dept. of Agriculture - Soil Conservation Service. 1993. Wisconsin Field Office Technical Guide, Spec 590. SCS, Madison, WI. 8 pp.

U.S. Environmental Protection Agency. 1976. Quality criteria for water. U.S. Govt. Print. Office, Washington, D.C.

U.S. Environmental Protection Agency. 1990a. Nonpoint sources agenda for the future. WH-556. Office of Water, U.S. EPA, Washington, D.C. 3 pp.

U.S. Environmental Protection Agency. 1990b. National pesticide survey - nitrate. U.S. Environ. Prot. Agency, Washington, D.C.

U.S. Environmental Protection Agency. 1994. National water quality inventory. 1992 Report to Congress. U.S. EPA 841-R-94-001. Published March 1994. Office of Water. U.S. Govt. Printing Office, Washington, D.C.

U.S. Geologic Survey. 1988. State summaries of ground water quality. p. 135-546. In: National Water Quality Summary 1986. Water Supply Paper 2325, U.S. Geologic Survey, U.S. Gov. Print. Off., Washington D.C.

van Beek, C.G.E.M. and F.A.M. Hettinga. 1989. The effects of manure spreading and acid deposition upon groundwater quality at Vierlingsbeek, The Netherlands. p. 155-162. In: L.M Abriola (ed.), *Groundwater Contamination*. Proc. Int. Assoc. Hydrological Sci. Sym., May 1989, Baltimore, MD. Int. Assoc. Hydrological Sci. Pub. No. 185, Wallingford, Oxfordshire, UK.

Vitosh, M.L., H.L. Person, and E.D. Purkhiser. 1988. Livestock manure management for efficient crop production and water quality preservation. Mich. St. Coop. Ext. Serv. Bull. WQ12, East Lansing, MI.

Vollenweider, R.A. 1968. Scientific fundamentals of the eutrophication of lakes and flowing waters with particular reference to nitrogen and phosphorus. OECD Rep. DAS/CSI/68.27. 182 pp., Paris, France.

Wang, H.D., W.G. Harris, and K.R. Reddy. 1995. Stability of phosphorus forms in dairy impacted soils under column leaching by synthetic rain. *Ecological Eng.* (in press).

Weil, R.R., R.A. Weismiller, and R.S. Turner. 1990. Nitrate contamination of groundwater under irrigated coastal plain soils. *J. Environ. Qual.* 19:441-448.

Westerman, P.W. and M.R. Overcash. 1980. Short-term attenuation of runoff pollution potential for land-applied swine and poultry manure. In: *Livestock Waste - A Renewable Resource*. Proc. 4th Int. Symp. on Livestock Wastes. Am. Soc. Agric. Eng., St. Joseph, MI.

Westerman, P.W., T.L. Donnely, and M.R. Overcash. 1983. Erosion of soil and poultry manure - a laboratory study. *Trans. ASAE* 26:1070-1078, 1084.

Wood, C.W., H.A. Torbert, and D.P. Delaney. 1993. Poultry litter as a fertilizer for bermudagrass: Effects on yield and quality. *J. Sust. Agric.* 3:21-36.

Wood C.W., D.W. Reeves, R.R. Duffield, and K.L. Edmisten. 1992a. Field chlorophyll measurements for evaluation of corn nitrogen status. *J. Plant Nutri.* 15:487-500.

Wood C.W., P.W. Tracy, D.W. Reeves, and K.L. Edmisten. 1992b. Determination of cotton nitrogen status with a hand-held chlorophyll meter. *J. Plant Nutri.* 15:1435-1448.

Woodward, M.D., V.A. Bandel, and B.R. Bock. 1993. Summary of soil nitrate tests for corn, 1993 state surveys. Tenn. Valley Auth. Misc. Pub., Muscle Shoals, AL.

Yang, S., J.H. Jones, F.J. Olsen, and J.J. Paterson. 1980. Soil as a medium for dairy liquid waste disposal. *J. Environ. Qual.* 9:370-372.

Young, R.A. and C.K. Mutchler. 1976. Pollution potential of manure spread on frozen ground. *J. Environ. Qual.* 5:174-179.

Young, R.A., T. Huntrods, and W. Anderson. 1980. Effectiveness of vegetated buffer strips in controlling pollution from feedlot runoff. *J. Environ. Qual.* 9:483-487.

Processing Manure: Physical, Chemical and Biological Treatment

D.L. Day and T.L. Funk

I. Introduction

Treatment of manure, an organic material, usually falls into three major categories: physical, biological, and chemical. The manure-land cycle is a large complex biological process (Figure 1), whereas liquid-solids separation is a physical process and odor control may involve a chemical process. A combination of physical, biological, and chemical methods are commonly used in complete treatment systems. Considerable material for this paper is from Day (1988) and Day and Arogo (1993).

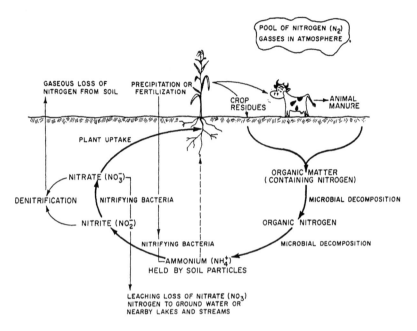

Figure 1. Animal, plant and soil biological interactions. (From Day and Arogo, 1993.)

Livestock and poultry products (meat, milk, and eggs) provide a large portion of the protein needs of the American people. Raising the large number of livestock and poultry to provide these food needs generates huge volumes of manure commonly called wastes. Manure properly collected, stored, and treated, however, can provide fertilizer, energy, and feed resources. Historically, manure has been spread onto croplands for its fertilizer and soil amendment value. While the practice will continue, alternatives are needed, especially where sufficient cropland is not readily available or when the manure is more important as a source of feed or fuel than it is as a fertilizer.

II. Physical Treatment

Physical treatment of livestock manure usually is accomplished with solid-liquid separation by sedimentation or various methods of screening or centrifuging. Other physical treatments include drying and incineration but increasing fuel costs have diminished interest in these methods. Solar energy, however, may help make drying economically viable. A discussion of physical treatment of animal manure, including constructed wetlands, is given by Moore (1993).

A. Solid-Liquid Separation

Separation of solids from the insoluble fraction of animal wastewaters is becoming more important as new methods of processing manures are developed. Advantages of solid-liquid separation schemes include: volume reduction, concentration of solids for separate treatment or reuse, reduction of settleable solids (such as pretreatment of liquid manure going to a lagoon), reduction of the pollution potential of wastewater, prevention of clogging and ease of hydraulic handling, moderation of odors, etc. Major disadvantages are high energy and investment costs of mechanical devices as well as labor requirements for maintenance and operation (Blaha, 1977).

Solid-liquid separation may be achieved by several methods: sedimentation, evaporation, leaching of liquid from stockpiled manure, or by mechanical solids removal. There are a variety of mechanical devices that have become of interest in recent years. White (1978) and Hepherd and Douglas (1973) reviewed some of the devices including: stationary screens, vibrating screens, centrifuges, pressure rollers, screw presses, and perforated belts and rollers. Fiber obtained from separation may be utilized as a soil conditioner, for compost, livestock bedding, or even processed into building materials. The liquid may be used as plant nutrients, for flushing to remove manure, and as a substrate for biogas production (Rorick et al., 1981).

There are two fundamental means of solid-liquid separation. The first uses the difference in density between the particulate matter and the liquid (sedimentation and centrifuging), and the second uses the shape and size of the particles to effect separation (screening and filtering).

Four liquid-solid separation devices were tested by Shutt et al. (1975) — a stationary screen, a vibrating screen, a liquid cyclone, and a settling chamber. The first three are depicted in Figures 2, 3, and 4. The devices were evaluated on the basis of separation efficiency, concentration of solids fraction, and flow capacity. The wastewater tested was untreated swine manure flushed out of finishing barns by discharging large quantities of water down the gutters. The best separation efficiency was obtained by the stationary screen, 35% of the Total Solids (TS) volume as compared to 22% and 26% for the vibrating screen and cyclone, respectively. The vibrating screen gave the highest concentration of solids. On the basis of removal of TS and Chemical Oxygen Demand (COD), the settling chamber gave the best performance.

A centrisieve, decanter centrifuge and vacuum filter were tested by Glerum et al. (1971), Figures 5, 6 and 7, as well as a vibroscreen and sedimentation silo, Figures 8 and 9. Results of testing swine manure are given in Table 1. For separating liquid and solid parts of the swine manure, the centrisieve was best based on results, capacity, and initial expense. Approximately 30 to 40% of the dry matter was removed at a rate of 8 to 9 cubic meters per hour. The separated material had a dry matter content of 14 to 19%. The sedimentation silo gave a lower dry matter in the separated solids than the centrisieve, but a Biochemical Oxygen Demand (BOD) reduction of 50% was obtained, indicating that a larger

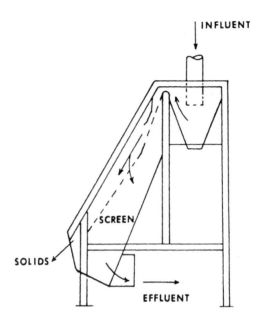

Figure 2. Stationary, rundown screen. (From Shutt et al., 1975.)

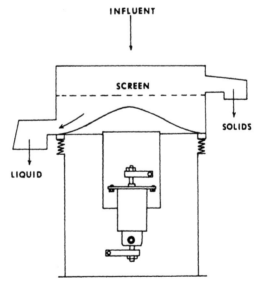

Figure 3. Vibrating screen. (From Shutt et al., 1975.)

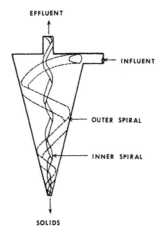

Figure 4. Liquid cyclone. (Shutt et al., 1975.)

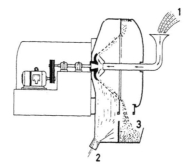

Figure 5. Centrisieve (1 slurry, 2 liquids, 3 solids). (From Glerum et al., 1971.)

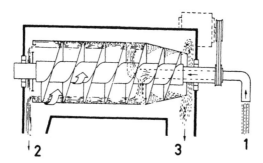

Figure 6. Decanter centrifuge (1 slurry, 2 liquids, 3 solids). (From Glerum et al., 1971.)

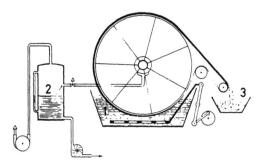

Figure 7. Rotary vacuum filter (1 slurry, 2 liquids, 3 solids). (From Glerum et al., 1971).

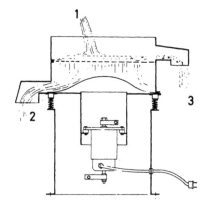

Figure 8. Vibroscreen (1 slurry, 2 liquids, 3 solids). (From Glerum et al., 1971.)

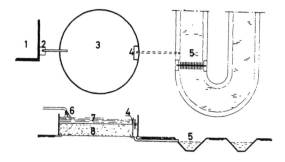

Figure 9. Sedimentation silo (1 piggery, 2 pump, 3 silo, 4 overflow, 5 oxidation ditch, 6 slurry, 7 liquids, 8 solids). (From Glerum et al., 1971.)

Table 1. Dry matter reduction and removed quantities of material by tests made to remove solid parts from pig slurry

Type	Installation Mesh-size screen, micron	Necessary hp	Price, dollars	Dry matter content in percent Slurry	Filtrate	Reduction, percent	Separated material Percent of slurry	Dry matter content, percent	Capacity, $m^3 \, hr^{-1}$
Centrisieve	31	15	2100	2.48	1.47	41	51	3.45	>10
				3.09	2.56	17	14	6.47	9
				3.59	3.80	32	12	18.7	
				6.28	3.97	37	18	16.7	
				7.7	4.7	39	31	14.5	—
				12.0	5.4	55	47	19.4	9
Decanter centrifuge									
No. 1		25	12500	1.92	1.43	25	3	17.9	1.1
No. 2			8350	7.58	2.61	66	14	37.4	0.6
Vacuum-filter	290	4	8350	7.54	3.68	51	22	21.5	0.25
Vibroscreen	85	0.25	700						0.2
Sedimentation silo, cap. 175 m^3			1100	11.29	1.54	86	69	15.6	
				8.29	1.60	81	45	16.5	

(From Glerum et al., 1971.)

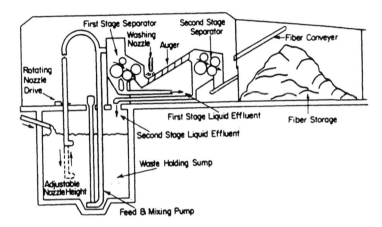

Figure 10. Schematic diagram of pressure roller liquid-solid separator. (From Rorick et al., 1981.)

proportion of the fine settleable solids was removed by the silo than by the centrisieve.

Chiumenti et al. (1987) reported tests of five separators, four of them commercially available and one pilot model, using cattle manure. The equipment included a horizontal decanter, vertical decanter, centrifugal separator, rotorpress, and an experimental pressure filter. Highest separation efficiency in terms of fraction of TS removed, 81%, was obtained by the horizontal decanter, and the centrifugal separator performed the least satisfactorily in that regard, at only 32%. From an energy standpoint, the rotorpress was the most efficient, consuming only one-tenth as much input energy per volume of manure processed as the horizontal decanter.

Performance of a perforated pressure roller liquid-solid separator for dairy manure was tested by Rorick et al. (1981), Figure 10. Some performance data are given in Table 2.

Conventional gravity settling consists of detaining a fairly dilute wastewater in a quiescent state so that the particles in the waste can settle to the bottom or float to the top (MWPS-19, 1975, p. 12). Equipment is then needed to scrape solids off the bottom and skim the scum off the top as in municipal waste treatment plants (Figure 11). Simpler systems, however, have been used for farm settling tanks wherein solids are pumped from a collection sump in the bottom and the scum or crust is left in place to help control odors until the tank is emptied, at which time the crust is mixed into the liquor.

Table 2. Constituents of slurry and products

Constituent	Slurry, TS, %		
	4.5	6.7	9.9
Slurry			
VS, % of TS	72	71	68
SS, % of TS	72	80	79
COD, mg/L	56.7	66.4	84.4
TN, mg/L	6160	3687	5245
First liquid			
TS, %	4.3	5.6	7.8
VS, % of TS	71	66	64
SS, % of TS	68	72	74
COD, mg/L	52.3	58.5	74.7
TN, mg/L	4714	3992	6384
Second liquid			
TS, %	3.8	4.5	6.6
VS, % of TS	74	67	68
SS, % of TS	64	63	68
COD, g/L	34.8	44.8	67.0
TN, mg/L	5460	3624	6019
Fiber			
TS, %	26	27	30
VS, % of TS	93	92	81

Data represent mean values for the wash and no-wash treatments.
(From Rorich et al., 1981.)

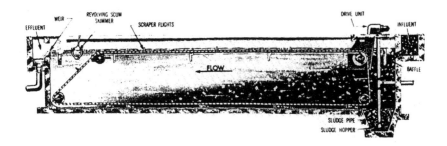

Figure 11. Rectangular primary settling basin with combination chain and scraper surface skimmer and sludge collector (courtesy Rexnord Envirex, Inc.). (From Loehr, 1974.)

Table 3. Heat of vaporization for water

Temperature (°C)	Latent heat (kcal/kg)
0	596.5
4.5	593.8
15.5	588.2
26.5	581.0
38.0	574.4
49.0	568.3
71.0	555.0
93.0	542.2
100	537.8

(From Esmay, 1977.)

B. Drying

Drying can be accomplished with unheated air, or with heated air at a higher rate. Mechanical driers operate at about 50% or less thermal efficiency (Esmay, 1977). Theoretical energy requirements to evaporate 1 kg of water range from 538 to 600 kcal, Table 3. Figure 12 gives a graphical means of calculating drier heat energy requirements for various water evaporation capacities. The bottom curve labeled 555 kcal/kg is for a theoretical 100% thermal efficient drier. The 1100 kcal/kg curve may be used for a typical mechanical manure drier (Esmay et al., 1975). For example, a drier with an evaporative capacity of 3000 kg of water per hour would require about 3.3 million kcal of energy per hour. Afterburners for driers, if required for control of air pollution from exhaust gases, may double fuel requirements per kg of water evaporated, particularly for small driers of 100 kg/hr of water or less. The upper two curves of Figure 12 may be used for making such estimates.

Assuming the rate of water evaporation capability of a manure drier remains fairly constant during operation, the output of dried manure is then inversely related to the moisture content of the wet manure going into the drier. Figure 13 gives a graphical means of estimating the production of dried product for various sizes of driers with known energy inputs (fuel requirements) and an assumed thermal efficiency of 50% (1100 kcal/kg of water evaporated). For example, for a 4 million kcal/hr energy input drier, the dry matter outputs can be obtained on the vertical scale for the input of wet manure at various moisture levels. It can be noted that the output of dried manure goes from about 900 kg/hr with an input of moisture content manure to about 4500 kg/hr with 50% moisture content manure. This is an increase of dry matter output to approximately five times.

The conventional and most reliable type of manure drier is the rotary drum type that is charged with wet manure internally and fired internally. Drum-type driers

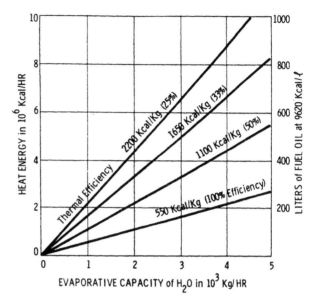

Figure 12. Heat energy requirements for manure driers with various rates of efficiencies. (From Esmay, 1977.)

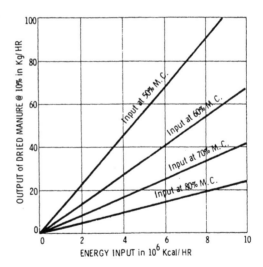

Figure 13. Production of 10% moisture content dried animal or poultry waste with driers having different energy inputs and with input of wet manure at various levels of moisture content. Drier performance of 1100 kcal/kg of water evaporated is assumed. (From Esmay, 1977.)

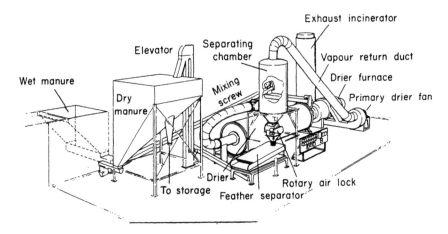

Figure 14. Drum drier. (From Esmay, 1977.)

are traditionally used for dehydration of many different biological products. Drum driers can be designed and constructed for a wide range of capacities. The rotating drum provides effective mixing of the manure and thus good exposure to the combustion energy for evaporation of water. The drum requires a minimum of power for the rotation movement and also has a minimum of moving parts. A typical drum drier is illustrated in Figure 14.

There are other types of driers manufactured. Increasing fuel costs, however, have decreased interest in drying manure with heated air driers. Solar driers are gaining in popularity.

C. Incineration

Incineration is a process in which the volume and weight of organic matter is reduced by burning. The combustible fractions of the waste are burned, and the mineral matter is left as an ash (MWPS, 1975, p. 78). Materials having a low moisture content will support combustion, but high moisture materials will require a supplemental fuel supply. Livestock manures are presently being incinerated only on a very limited scale. It is possible to incinerate with production of a minimum amount of odor due to the high temperatures used. The process has some application in dense populations and where land is not available for spreading manure.

Some laboratory-scale studies of the incineration of livestock manure were done by Davis et al. (1972). As much as 90% weight reduction and 85% volume reductions were obtained. Both fluidized bed and rotary kilns were used. Wet manure could be fed to the rotary kiln, but only dry manure could be handled by the fluidized bed. Heats of combustion of oven-dried manures ranged from 5.85 MJ/kg (2520 Btu/lb.) to 18.2 MJ/kg (7810 Btu/lb). The lowest value was attributed to contamination by sand. Wet manure incinerated in the rotary kiln at 650-750°C (1000-1380°F) required natural gas at 0.16 m³/kg (2.5 ft³/lb) wet feed to sustain combustion. The ash remaining after combustion was pelleted and tested chemically for its fertilizer properties; it ranged from 13-16% P_2O_5 and 8-14% K_2O. The nitrogen was discharged as NH_3 in the flue gas, although the flue gas could have been processed to recover NH_3.

Annamalia et al. (1985) built and tested a pilot scale fluidized bed combustion unit for incineration of dried beef cattle manure. The study determined useable moisture contents of manure for self-sustaining combustion, and the amount of excess air for maximum oxidation efficiency and CO emissions.

Incinerating equipment is designed for either batch-loading or continuous-flow operations. Batch loading requires a large amount of labor. It is also inefficient because the incinerator cools each time it is charged. Continuous-feed types of incinerating equipment are more expensive.

Air pollution can be generated by incineration equipment. Smoke from the incinerator can carry odors from the burning organic matter. Afterburners are used on some incinerators to remove the odor from the smoke before it is discharged into the air. Other incinerators incorporate water-spray systems, mechanical fly-ash collectors, or electrostatic precipitators to control air pollution.

D. Pyrolysis

Organic material may be pyrolyzed by holding it at 250-1000°C (480-1830°F) in an oxygen-deficient atmosphere (MWPS 1975, p. 79). The products are gases, oils, and ash. The gases given off include hydrogen, water, methane, carbon monoxide, and ethylene (Shuster, 1970). Shuster's work showed that pyrolysis in a low-oxygen atmosphere tended to produce more hydrocarbons. The yield of low-boiling tar-like compounds became less as the reaction temperature was increased. The substrates used in this work were paper, dried sewage sludge, and dried leaves. Because the study was concerned with identifying the compounds yielded by the reaction, few data were presented relating to quantitative yields or energy yields. White and Taiganides (1971) pyrolyzed various livestock manures and newspaper. They classified the gaseous products as CO_2, CO, H_2, illuminants, combustibles, O_2, and N_2. The pyrolysis was at atmospheric pressure, and about 1 g (dry basis) of material was used for each test. Dairy feces produced the most gas per unit of dry solids, followed by chicken, beef, and swine feces. Some 50%-60% of the product gas was combustible. The heating values of the gases produced ranged from 3.3 MJ/kg (1400 Btu/lb) of dry swine feces to 4.4 MJ/kg (1990 Btu/lb) of dry feces. The heat value of the carbon char remaining after pyrolysis was 4 MJ/kg (1700 Btu/lb) of dry swine feces, but 8.1 MJ/kg (3500 Btu/lb) for dairy. White and Taiganides also discussed the energy required to dry fresh feces compared with the energy available from the pyrolyzed products. It was noted that pyrolysis of swine feces would not be self-sustaining.

Organic wastes also may be converted to low-sulfur oils by exposing them to CO, and Na_2CO_3 as a catalyst, at high pressures and temperatures. Appell et al. (1971) were able to obtain 40% yields of oil from newsprint by holding it at 12.7 MPa (1840 lb/in^2) at 250°C (480°F) for 1 hr. One of the attractions of this oil production process is that some water is beneficial, playing a definite part in the reactions. For example, beef manure containing 45.5% water was processed at 380°C (710°F) to yield 47% of its dry weight as oil. No catalyst was needed because the calcium, potassium, and sodium salts present in manure performed that function. A pilot plant using sucrose as a substrate had been run on a continuous basis yielding 33% of the sucrose dry weight as oil. This complex oil had a heating value of 35.3 MJ/kg (15,200 Btu/lb).

Although pyrolysis and oil production are scientifically possible, the equipment for such processes has not been developed for processing livestock manure, and this will tend to limit adoption of the processes to very large operations. As with anaerobic digestion for methane production, pyrolysis and oil production will be best suited to livestock operations using climate-controlled housing, because manure collection and transport already will be part of the system.

III. Biological Treatment

Biological degradation of manure is a natural process that has occurred since the beginning of time, as manure is a good substrate for microorganisms. The biodegradable components of manure solids are shown in Figure 15. Controlled and uncontrolled biological systems are commonly in use. The systems may treat liquids or solids; may be aerobic, anaerobic, or facultative; and may be in a structure or unconfined on the land. Examples of biological treatment processes include the land/crop cycle as well as anaerobic digestors, septic tanks, oxidation ponds, aerated lagoons, oxidation ditches, and composting. Biological systems may have advantages over chemical systems for purposes of recycling, since biological systems do not involve the addition of compounds that may cause toxicity problems at later stages in the cycle.

In organic degradation, manure and other sources of organic matter are utilized as a source of energy by a succession of living microbial organisms. A series of biochemical reactions are set in motion and eventually the waste materials are decomposed and returned to nature. This decomposition of organic matter is brought about by a highly complex bacterial metabolism. The bacteria break down the complex organic substances such as carbohydrates, proteins, and fats into simple organic substances. Under aerobic conditions, the bacteria carry on respiratory metabolism to reduce the organic compounds to carbon dioxide and water without giving off offensive odors. Under anaerobic conditions, bacteria break down the complex organic substances into simple organic acids and then ferment those acids to ultimately form methane and carbon dioxide. Uncontrolled anaerobic treatment

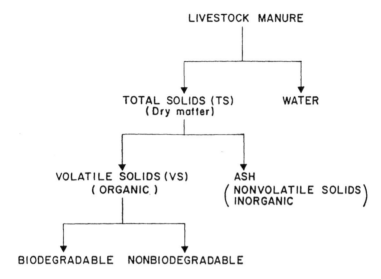

Figure 15. Solids and water in livestock manure. (From Day, 1988.)

can produce foul smelling odors as well as other gases. The presence or absence of free molecular oxygen determines which process predominates. Both have inherent advantages and disadvantages. Facultative bacterial processes can occur with or without oxygen, but are usually unstable.

Manure production rates and characteristics as excreted are given in Table 4 (ASAE, 1994). The potential for biological processes is reflected by the Volatile Solids content (VS), which is also proportional to the Biochemical Oxygen Demand (BOD) and the Chemical Oxygen Demand (COD). The ratio of VS as a percent of Total Solids (TS) is normally about 75 to 80% for fresh manure. As this value decreases, due to natural volatilization with time and weather conditions, the potential value for biological treatment also decreases. Storage and handling methods usually cause a corresponding decrease in fertilizer value due to loss of nutrients. Loading rates for lagoons, oxidation ponds, digestors and other biological treatment processes are usually expressed as the amount of VS input per day per volume of the treatment process. Other factors affecting biological treatment must be kept in proper operating ranges. Some of these factors are: pH, temperature, carbon to nitrogen ratio, and concentrations of toxic materials. Temperature ranges can be categorized as: psychrophilic, -5 to 35°C (23 to 95°F); mesophilic, 18 to 45°C (64 to 113°F); and thermophilic, 45 to 85°C (113 to 185°F). Biological processes, as well as chemical processes, occur at faster rates with increasing temperatures until sterilization temperature is reached.

Table 4. Manure production and characteristics as produced

Parameter	Units[b]		Animal type[a]										
			Dairy	Beef	Veal	Swine	Sheep	Goat	Horse	Layer	Broiler	Turkey	Duck
Total manure[c]	kg	Mean[d]	86	58	62	84	40	41	51	64	85	47	110
		S.D.	17	17	24	24	11	8.6	7.2	19	13	13	**
Urine	kg	Mean	26	18	**	39	15	**	10	**	**	**	**
		S.D.	4.3	4.2	**	4.8	3.6	**	0.74	**	**	**	**
Density	kg/m³	Mean	990	1000	1000	990	1000	1000	1000	970	1000	1000	**
		S.D.	63	75	**	24	64	**	93	39	**	**	**
Total solids	kg	Mean	12	8.5	5.2	11	11	13	15	16	22	12	31
		S.D.	2.7	2.6	2.1	6.3	3.5	1.0	4.4	4.3	1.4	3.4	15
Volatile solids	kg	Mean	10	7.2	2.3	8.5	9.2	**	10	12	17	9.1	19
		S.D.	0.79	0.57	**	0.66	0.31	**	3.7	0.84	1.2	1.3	**
Biochemical oxygen demand, 5-day	kg	Mean	1.6	1.6	1.7	3.1	1.2	**	1.7	3.3	**	2.1	4.5
		S.D.	0.48	0.75	**	0.72	0.47	**	0.23	0.91	**	0.46	**
Chemical oxygen demand	kg	Mean	11	7.8	5.3	8.4	11	**	**	11	16	9.3	27
		S.D.	2.4	2.7	**	3.7	2.5	**	**	2.7	1.8	1.2	**
pH	kg	Mean	7.0	7.0	8.1	7.5	**	**	7.2	6.9	**	**	**
		S.D.	0.45	0.34	**	0.57	**	**	**	0.56	**	**	**

Table 4. continued

Total Kjeldahl nitrogen[e]	kg	mean	0.45	0.34	0.27	0.052	0.42	0.45	0.30	0.84	1.1	0.62	1.5
		S.D.	0.096	0.073	0.045	0.21	0.11	0.12	0.063	0.22	0.24	0.13	0.54
Ammonia nitrogen	kg	mean	0.079	0.086	0.12	0.29	**	**	**	0.21	**	0.080	**
		S.D.	0.083	0.052	0.016	0.10	**	**	**	0.18	**	0.018	**
Total phosphorus	kg	mean	0.094	0.092	0.066	0.18	0.087	0.11	0.071	0.30	0.30	0.23	0.54
		S.D.	0.024	0.027	0.011	0.10	0.030	0.016	0.026	0.081	0.053	0.093	0.21
Orthophosphorus	kg	mean	0.061	0.030	**	0.12	0.032	**	0.019	0.092	**	**	0.25
		S.D.	0.0058	**	**	**	0.014	**	0.0071	0.016	**	**	**

[a] All values wet basis.

[b] Differences within species according to usage exist, but sufficient fresh manure data to list these differences were not found. Typical live animal masses for which manure values represent are: dairy, 640 kg; beef, 360 kg; veal, 91 kg; swine, 61 kg; sheep, 27 kg; goat, 64 kg; horse, 450 kg; layer, 1.8 kg; broiler, 0.9 kg; turkey, 6.8 kg; and duck, 1.4 kg.

[c] Feces and urine are voided.

[d] Parameter means within each animal species are comprised of varying populations of data. Maximum numbers of data points for each species are: dairy, 85; beef, 50; veal, 5; swine, 58; sheep, 39; goat, 3; layer, 74; turkey, 18; duck, 6.

[e] All nutrients and metals values are given in elemental form.

(Modified from ASAE, 1994.)

A. Anaerobic Treatment

Anaerobic processes take place without free oxygen. A principal end product of controlled, enclosed anaerobic digestion is methane gas, which can be collected and used as a fuel. However, manure in an uncontrolled, open, anaerobic state, such as pits and gutters, produces objectionable gases and odors such as hydrogen sulfide and ammonia (Figure 16). Production and release rates of ammonia from swine manure pits have been studied by Zhang et al. (1994) and a prediction program developed to aid better design of ventilation systems. The anaerobic systems of interest in livestock waste treatment are anaerobic lagoons (as opposed to earthen storages), anaerobic digestors, and septic tanks.

1. Anaerobic Lagoons

Anaerobic lagoons developed from their distant relative, the municipal oxidation pond or aerobic waste stabilization pond. However, due to the large surface area required for oxidation ponds, there is little similarity in terms of processes involved between an anaerobic lagoon and an aerobic oxidation pond. Aerobic oxidation ponds are designed shallow to achieve maximum oxygenation, while anaerobic lagoons should be as deep as practical to achieve maximum temperature stability and to minimize the escape of odors from the water surface.

Anaerobic lagoons have found widespread application in the treatment of animal wastes because of their low initial cost, ease of operation, and convenience of loading by gravity flow from the livestock buildings. The main disadvantage is the release of odors from the surface of the lagoon, especially during spring warmup. Mosquitoes can also be a problem.

Volume requirements for one-cell and two-cell lagoons are shown in Figure 17. Loading rates, kg VS/day per m³, are shown in Figure 18 (SCS, 1992). Discharge to a stream is prohibited by water quality standards so there should not be an outlet for regular discharge. Freeboard should be maintained at all times to contain the design 25-year, 24-hour storm event (SCS, 1992).

2. Anaerobic Digestors

Anaerobic digestors decompose manure in air-tight chambers while producing on-site fuel (biogas). Biogas is about 60% methane and 40% carbon dioxide with trace amounts of hydrogen sulfide. The biogas can be used for heating, in boilers to produce hot water, in engine-generators to produce electricity, and in absorption coolers to produce refrigeration. Biogas is difficult to liquefy, so it is best suited for stationary uses. The disadvantages are the initial cost of the digestor and the operational management required. Also, manure must be collected, transported, and fed into the digestor; and the accumulating sludge must be disposed of routinely.

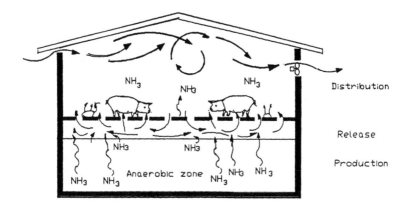

Figure 16. An under-floor manure pit in a swine building and gases released to the atmosphere. (From Christianson et al., 1993.)

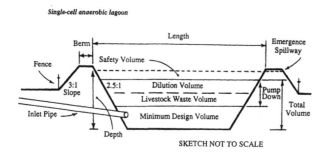

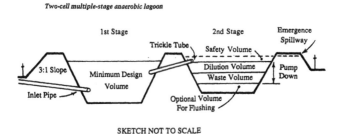

Figure 17. Volume requirements for one-cell and two-cell lagoons. (From Funk et al., 1993.)

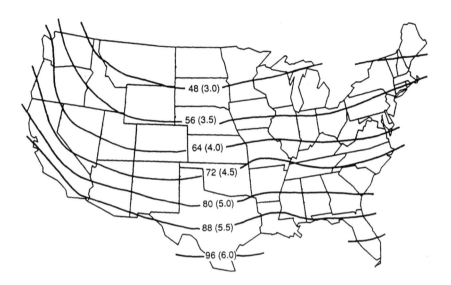

Figure 18. Anaerobic lagoon loading rates, g VS/day-m^3(lb VS/day-ft^3). (Modified from SCS, 1992.)

The sludge, however, has low odor and retains the nutrients that came into the digestor, as no nutrients are in the biogas.

A diagram of a basic digestor is given in Figure 19. There are numerous variations including heating, mixing, and multiple stages. Anaerobic digestion involves acid-forming bacteria and methane-forming bacteria with various intermediate pathways, as shown in Figure 20 (Hashimoto, 1981). Recommended digestor sizes and biogas yields are given in Table 5 (Smith, 1981). Practical designs are given in NRAES (1984).

3. Septic Tanks

Septic tanks have been considered as a treatment method for livestock wastes, with the effluent going to a soil absorption field or to a lagoon. In practice, septic tanks for livestock wastes appear to operate mainly as sedimentation tanks; absorption fields have been inadequate except for very small livestock facilities. Septic tanks for home sewage are sized on daily loading rates of approximately 0.03 to 0.05 pounds daily volatile solids per cubic foot of tank capacity. It is unlikely that absorption fields will be adequate for disposing of the effluent. Therefore, the effluent must be held in a lagoon or other such facility. Periodic removal of sludge

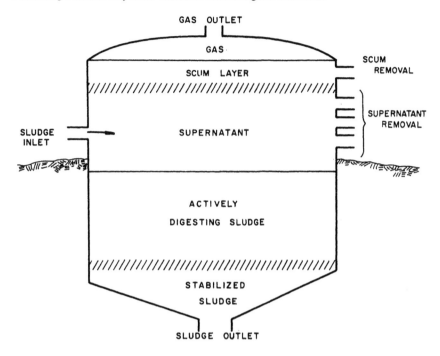

Figure 19. Schematic diagram of an anaerobic digester. (From Day, 1972.)

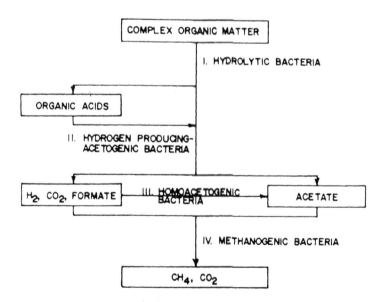

Figure 20. The four bacterial groups involved in the complete anaerobic degradation of organic matter. (From Hashimoto et al., 1981.)

Table 5. Suggested anaerobic digester operationg variables for farm-scale, mesophilic (35 °C) operation[a]

Species	Defected TS, %	Influent TS, %	Detention time, day	Loading rate, kg VS/m³ d⁻¹	Digester volume[b], m³/1000kg	Functional VS reduction	Methane[c,d] production, m³/m³ d⁻¹	Influent slurry temp., °C	Slurry heating[e], m³/m3 d⁻¹	NH₃-N mg/L
Dairy	12.7	12.7	15	5.7	1.5	0.40	0.9	0	0.23	2600
Beef	11.6	5.8	10	5	1.18	0.45	0.93	0	0.41	2000
Pigs	9.2	5.6	15	3	1.96	0.55	0.72	12	0.18	3000
Poultry	25.2	13.9	40	2.4	3.86	0.55	0.62	12	0.067	5900

[a] The values in this table have been obtained by examining values in the literature and adjusting these values to conform to average manure characteristics found in J.R. Miner and R.J. Smith (eds.), 1975. Livestock waste management with pollution control. (North Central Regional Research Publication 222). Midwest Plan Service Handbook MWPS-19. Midwest Plan Service, Iowa State University, Ames.

[b] The value in this column is the digester volume per 1000 kg animal live weight.

[c] All gas volumes are at 20°C and 1 atm.

[d] Gas production is estimated by using the stoichiometric relation 1 kg COD = 0.25 kg CH$_4$ = 0.37 m³ at 20° C, 1 atm. The daily gas production was then estimated from: Loading rate (kg VS/M³ d⁻¹) × VS reduction × (COD/VS) × 0.37 = gas production (m³/m³ d⁻¹).

[e] The values in this column are expressed in daily volumes of gas per volume of digesting liquor, taking a value of 37.3 MJ/m³ as the heating value of CH$_4$. No correction has been made for heating-equipment efficiency. (From Smith, 1981.)

from the tank will be required more often than for home septic tanks because of livestock feed particles that settle and biodegrade very slowly.

B. Aerobic Treatment

Aerobic treatment is a natural process of digesting organic matter while suppressing odors and diseases. If organic matter is represented by the elements C, O, H, N, and S, the order of abundance of the elements in biological systems, then the following overall reactions occur during aerobic biological oxidation (Simpson, 1960):

Microbial cells + COHNS + O_2 → more cells + CO_2 + H_2O + NH_3

and

Nitrifying cells + NH_3 + O_2 → (via NO_2^-) → NO_3^- + H_2O + more nitrifying cells

Nitrate nitrogen, required by plants, is an end product of aerobic treatment.

Although aerobic treatment is an odorless process, the method is not widely accepted by livestock producers because of the operating costs. Nevertheless, some form of aerobic treatment will likely be a desirable alternative to intolerable odors, especially for the producer with a high risk of odor nuisance liability. Also, the resulting aerobic bacterial cells are high in protein and have been used as a feed supplement. The major uses of aerobic treatment for livestock wastes are: oxidation ponds (naturally aerated lagoons), aerated lagoons (mechanically aerated), and oxidation ditches.

1. Oxidation Ponds

An oxidation pond is usually a shallow basin 3 to 5 feet deep for the purpose of treating sewage under climatic conditions (warmth, light, and wind) that promote the introduction of atmospheric oxygen, and that favor the growth of algae to produce oxygen (Figure 21). Bacterial decomposition of the wastes releases carbon dioxide, which promotes heavy growth of algae. Ammonia and other plant growth substances are used up by the algae, and dissolved oxygen is kept at a high level. The driving force in this type of self-purification is photosynthesis, supported by a symbiosis between saprophytic bacterial and algae. Loading rates, kg BOD/day/hectare, are shown in Figure 22 (SCS, 1992).

Because of the large surface area required, oxidation ponds have not found favor with livestock producers. Their use has been essentially limited to receiving effluent from anaerobic lagoons or other treatment units. However, they have been considered for receiving dairy milking-plant wastes.

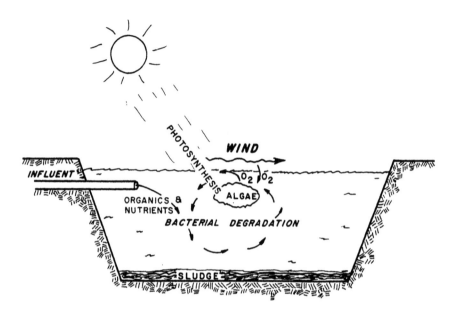

Figure 21. Schematic diagram of an oxidation pond; a naterually aerated, waste-treatment pond, showing the zones of algal synthesis, aerobic bacterial action, and anaerobic decomposition in the bottom sludge layer. (From Day, 1972.)

2. Mechanically Aerated Lagoons

A mechanically aerated lagoon is one that has a device that beats or blows air into the water with a portion of the oxygen being dissolved. Because the lagoon is not dependent on wind or algae for the oxygen supply, the surface area can be made smaller than that of an oxidation pond and the depth can be greater, which reduces temperature fluctuations.

There are numerous methods for aerating lagoons. Floating aerators (Figure 23) appear to be satisfactory, but other schemes such as compressed air entering through diffusers (perforated pipes), rotating aerators, and rotary blowers may also work satisfactorily.

Satisfactory aerobic treatment of livestock wastes has been obtained in aerated lagoons that have a volume of approximately 50 times the daily manure production. However, if the aerated lagoon is considered as a long-time storage facility of the waste residues, a much larger size is needed (SCS, 1992). If one intends to de-sludge the lagoon yearly or more often, the size may be reduced. Otherwise, a detention time of two to three years is recommended and the volumes approach those recommended for anaerobic lagoons.

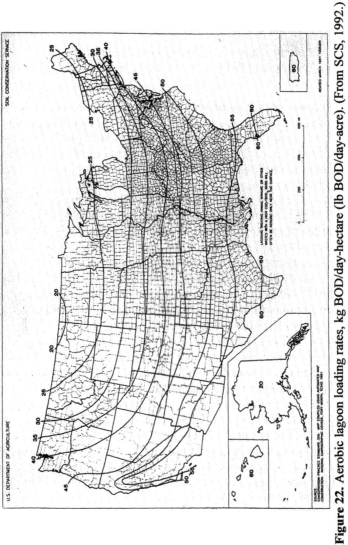

Figure 22. Aerobic lagoon loading rates, kg BOD/day-hectare (lb BOD/day-acre). (From SCS, 1992.)

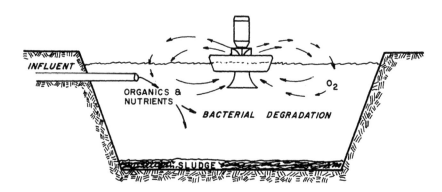

Figure 23. Schematic diagram of a floating surface aerator for mechanical aeration of a lagoon. (From Day, 1972.)

A mechanically aerated lagoon should be aerated continuously. A large storage of dissolved oxygen in water is impossible, since the oxygen saturation range is only about 6 to 9 milligrams of oxygen per liter of water (ppm). After saturation, additional oxygen is not held by the solution; further aeration is of little use and would add unnecessary expense. The ideal system then is one in which oxygen is being supplied at a rate equal to oxygen demand.

The rate of decomposition is slowed as the temperature decreases. Laboratory studies have shown that below 4.5°C, bacterial decomposition is greatly reduced. On this basis, it appears that little decomposition is accomplished by operating exposed aerators in extremely cold weather. However, the aerator should be started as soon as the temperature begins to warm up in the spring, so that aerobic bacterial action can be reestablished. Some objectionable odors can be expected during the startup period. Although the rate of decomposition is greatly reduced after about 30 days, decomposition does continue, and it is believed that in 1-1/2 to 2 years, the volatile solids may be reduced by as much as 60 to 70%.

Even with good degradation, solids (sludge) will eventually build up in the lagoon until removal is necessary. Late fall appears to be a good time to remove sludge from lagoons. The solids are most stabilized at that time, and the odors are low if the lagoon has been well aerated during the previous seven to eight months. When excess water must be disposed of, irrigating mixed liquor (supernatant plus suspended solids) from an aerobic lagoon has worked satisfactorily. Sludge buildup is retarded because suspended solids are removed by the irrigation unit.

Some hydrolysis and anaerobic decomposition may take place in the bottom of the lagoon, thus reducing some of the solids. The products of anaerobic decomposition are then further degraded in the upper aerobic levels of the lagoon. Oxidation in the upper portion reduces the release of odors.

3. Oxidation Ditches

The oxidation ditch was developed during the 1950s at the Research Institute for Public Health Engineering (TNO) in the Netherlands as a low-cost method for treating sewage emanating from small communities and industries. It has been referred to as the "Pasveer ditch" (Pasveer, 1963), Figure 24. The oxidation ditch is a modified form of the activated-sludge extended aeration process. The activated-sludge process has the characteristic that, if aeration and mixing are stopped for 30 to 60 minutes, the bacterial floc and other solids will settle, leaving clarified water on top. This principle is utilized to separate solids from the final effluent under quiescent conditions in settling tanks. The oxidation ditch has two principal parts -- a continuous open-channel ditch shaped like an oval racetrack and an aeration rotor that supplies oxygen and circulates the ditch contents to keep the solids in suspension.

The in-the-building oxidation ditch for livestock wastes is a completely mixed aerobic method having a long detention time (approximately 50 days). It is a modified form of the odorless "Pasveer" oxidation-ditch treatment plant used for municipal wastes. But it differs from the "Pasveer" plant in that (a) the ditch is located beneath self-cleaning slotted floors in a confinement building (this approaches continuous loading of the ditch; (b) the liquid volume is about 1.87 m³ per kg of daily BOD, or less than half that of municipal ditches, and the loading is in much more concentrated form (30,000 to 50,000 milligrams per liter instead of 300 to 500 milligrams per liter); (c) the liquid depth is shallow (usually less than 0.6 m) so there is sufficient velocity to keep solids suspended; and (d) for the operator's convenience, a constant rotor height is maintained. The liquid depth is also kept constant by using an overflow for the mixed liquor. The slotted-floor confinement building with an oxidation ditch is shown in Figure 25. Design recommendations are given by Jones et al. (1971).

C. Composting

1. Conventional Composting Methods

Composting generally refers to the decomposition of organic matter by aerobic thermophilic organisms to produce a stable humus-like material. Composting of manure or crop residue may be used to:
  •reduce the pollution potential, especially the odor,
  •reduce the weight and volume,
  •produce a stable material permitting handling and storage with no further decomposition, and
  •produce a material useful as a potting soil and soil amendment.

The composting process differs from conventional aerobic waste treatment because it is achieved at a much lower water content. This allows the development

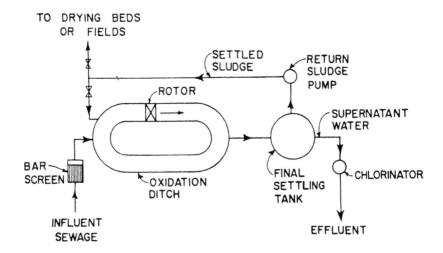

Figure 24. A flow diagram of an oxidation ditch treatment plant for municipal wastes. (From Jones et al., 1971.)

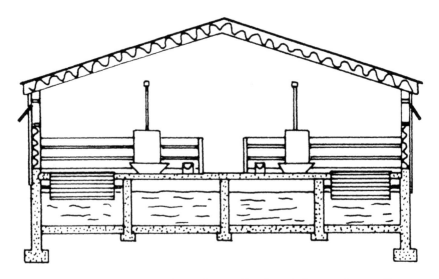

Figure 25. Vertical cross-section of a totally slotted floor swine-confinement building with an oxidation ditch beneath the self-cleaning slotted floors. (From Jones et al., 1971.)

of a loose matrix of material that can be aerated with less mixing than required by a liquid system. The biological activity in good compost produces sufficient heat energy to drive the temperature into the thermophilic range (50 to 70°C) without external heat supplies. The aerobic, thermophilic conditions are inhibitory to most pathogenic organisms and because the process is aerobic, it is largely free of offensive odors. The product is a relatively stable material reduced in weight and volume from the original waste, mainly due to moisture loss. It does not attract flies, and can be used as a soil amendment to improve structure, cation exchange capacity, fertility, and other characteristics. Most weed seeds and insects are killed by the high composting temperatures.

Composting is usually accomplished by stirring or turning the manure to get oxygen from the air. Most manures are too high in moisture content to get proper aeration. Bedded manure, manure produced in arid regions where moisture evaporates rapidly, and some poultry manures are better suited for composting (SCS, 1992).

The composting process is brought about by naturally occurring micro-organisms as they metabolize protein and carbohydrates. The principal by-products are carbon dioxide, water, ammonia, and microbial cells as shown in Figure 26 (NRAES, 1992). While composted material is stable and of low pollution potential, it is also low in nitrogen content because a considerable amount of the ammonia nitrogen is volatilized away. Recommended conditions for rapid composting are also given in Table 2.1 in NRAES (1992).

There are two major methods of composting: natural and mechanically induced. Natural methods of composting are variations of the traditional schemes originally developed by Sir Albert Howard in India. The method developed by Howard (1935) involved the formation of a layered pile about 1.5 m high using a variety of materials such as manure, garbage, sewage sludge, and leaves. The process was initially anaerobic and required up to six months to compost. The process was later modified by turning the pile over at least twice, which reduced the composting time to about three months.

A common method of mechanical composting is the windrow process. The materials are stacked into windrows which are shaped to shed rain water and to preserve temperature and humidity. Periodically the windrows are turned and mixed with specially designed mechanical equipment. Windrows are commonly 3 m wide by 1 m deep with mixing being done every 4 to 7 days. Windrow composting takes about 1 to 3 months and is considered a batch process.

A high-rate method of mechanical composting is accomplished in a trench or bin in a continuous flow operation. This system uses a mechanical mixer and may have forced aeration. High-rate composting can be accomplished in 1 to 3 weeks and will stabilize the readily degradable portions of manure, but extended storage of 2 to 6 months is required to degrade the cellulosic portion.

In recent years, composting of dead poultry has become a popular method of dead bird disposal (SCS, 1992). The method is also being developed for dead pigs (Fulhage, 1993).

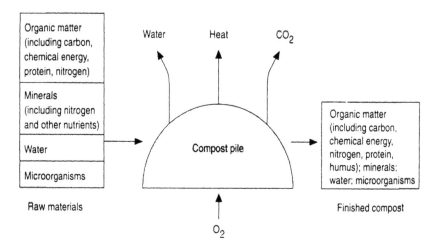

The carbon, chemical energy, protein, and water in the finished compost is less than that in the raw materials. The finished compost has more humus. The volume of the finished compost is 50% or less of the volume of raw material.

Figure 26. The composting process. (From NRAES, 1992.)

Another form of composting being used in Europe is deep litter (Novarotto and Bonazzi, 1990 and Voermans, 1992). In the deep litter method, pigs are housed on a layer of wood shavings initially about 85 cm (33 in) deep. Periodically the litter is turned and mixed but it remains in use up to about 22 months, at which time it is removed and a new layer installed. It is claimed that the major advantages of this system are the elimination of liquid manure and odor, as well as the improvement of animal welfare.

2. Thermophilic Aerobic Treatment

Another development is thermophilic aerobic treatment of liquid manure. In the usual aerobic treatment of liquids, the microbes are limited in activity by rate of oxygen supply from aerators, and by the substrates available. The metabolic heat generated is not large and it is quickly dissipated by the large surface area of tanks and lagoons, through the air by mechanical aerators, and other such methods. Therefore aerobic liquid systems do not run at much, if anything, above ambient temperature. However, if the liquid can be confined in an insulated tank, and if

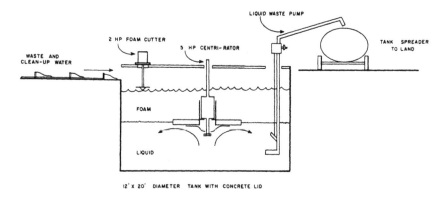

Figure 27. Schematic of a liquid composting system for swine wastes. (From Terwilleger and Crauer, 1975.)

sufficient oxygen can be supplied in a way that does not cause much cooling, then the aerated slurry will heat up from metabolic heat. The increased temperature can cause increased metabolic activity of the microbes and still more heat will be generated.

Theoretically, increasing metabolic activity of microbes can raise the habitat temperature sufficiently high to kill the organisms, but in practice, heat losses and heat production will balance out at some suitable temperature for microbial growth and the system will stabilize. The design and use of such high-temperature aeration systems is in the development stage and is often referred to as liquid composting. The "Licom" process is a thermophilic one operating with waste of 5 to 10% TS in a continuous-flow system with a detention time of 7 to 14 days. The aeration tank is insulated and the aerator pulls air down into the liquid rather than spraying liquid into the air, thus reducing heat loss and reducing foaming problems. A schematic diagram of a tank and aerator is shown in Figure 27 (Terwilleger and Crauer, 1975).

3. Vermicomposting

Vermicomposting is a process of using earthworms to produce a peat-like material suitable for plant growth and an earthworm biomass protein which can be used as food for fish, poultry, and swine (Edwards et al., 1985 and Pederson, 1993). *Eisenia Foetido* is an earthworm species commonly used.

IV. Chemical Treatment

A. Overview of Chemical Treatments

Chemicals are used in waste treatment for various reasons such as: for precipitation of certain particulates and colloidal matter, for pH control, for odor control, and for enhancement of biological treatment. Precipitation of solids is usually for reducing the oxygen demand of the remaining wastewater. Soluble organic compounds are not easily removed by this process. Soluble inorganic compounds, such as phosphates, can be removed if insoluble precipitates can be formed. The coagulants commonly used to cause precipitation in wastewaters are alum, ferric salts such as ferric sulfate or ferric chloride, and lime.

Anionic, cationic, and nonionic organic polyelectrolytes can also be used to precipitate colloidal matter in wastewater either separately or in combination with organic coagulants (Chiumenti et al., 1987; Hanna et al., 1985). Adding polymers to sludge has helped in belt liquid/solid separators.

Acid or alkaline treatment has been used to break down cellulose and lignified fibers to make them more amenable to biological digestion. Sodium hydroxide and sodium peroxide have been used.

Chemicals have been successful in reducing the buildup of struvite (magnesium ammonium phosphate) in pipes and pumps that are used in pumping flush water from secondary lagoons. Weak acid solutions such as acetic, sulfuric, and hydrochloric have been used. Disposal of the acids can be a problem.

B. Manure Additives for Odor Control

Probably the most common use of chemicals in livestock waste management is for odor control and as disinfectants. The sensation of odor is a personal response. Not all people are equally sensitive to odors, nor do they always agree regarding the severity of a particular odor. Since there is no measuring instrument for odors that is nearly as sensitive as the human nose, it is not possible to document precisely the strength and offensiveness of odors at a particular site. Tests for the effectiveness of manure treatment compounds have generally involved the use of the human nose to document results.

There are a few methods of odor control that form the basis for all the odor control chemical and biochemical treatments being offered today. Each method can be more or less effective in certain situations. Costs and duration of effectiveness vary widely.

One or more of the following interactions occur from mixing of two or more odorants:

- •odor magnification: the intensity of the mixture is stronger than that of any one component,
- •odor cancellation: the intensity or offensive qualities of one or both odorants are reduced,

•odor masking: one odor masks the other(s) so that its odor dominates, and
•odor synergism: one odor is made stronger in intensity or quality by interaction
   with others.

There are several types of chemical and biochemical odor control agents, which can be classified by their application:

treatment of atmospheric odorants

•masking agents: usually mixtures of aromatic oils that cover but do not reduce
   odor,
•counteractants: neutralize the odor with aromatic oils, leaving no overriding
   odor, similar to the effects of odor cancellation,
•deodorants: mixture of compounds that chemically destroy odors, and
•digestive deodorants: combinations of digestive enzymes and aerobic and
   anaerobic bacteria that eliminate odors through biochemical digestive
   processes.

treatment of manure to reduce odorant production based on chemicals or biochemical agents which:

•cause more orderly decomposition,
•reduce or inhibit decomposition,
•oxidize or otherwise alter the chemical composition of the volatiles,
•retain the volatiles in the manure, or
•absorb moisture required for anaerobic activity (in the case of a manure pack).

   Several references listed contain details of odor-control testing that has been done over more than two decades (Kreis, 1978; Warburton et al., 1979; ASAE, 1993). These tests range from laboratory studies done on barrels of manure, to field tests on liquid manure pits and solid manure packs on farms, to open feedlot studies.
   Of the atmospheric treatments, counteractants and masking agents were found to be somewhat effective, with counteractants more effective than masking agents. Masking agents were limited in that some people found the chemicals to smell worse than the manure odors being masked. Ozone was used in one study as a deodorant for destruction of methylamine, ammonia, and hydrogen sulfide. However, the reaction rate was slower than desired.
   Atmospheric treatments may be considered emergency measures, since they tend to cost too much to be used long-term.
   Chlorine, lime, and paraformaldehyde as biological inhibitors; hydrated lime and sodium hydroxide for pH control; and potassium permanganate, hydrogen peroxide, and paraformaldehyde as oxidizing agents have all been tried in various scenarios for treating liquid manure. Sodium hypochlorite, chlorine dioxide and potassium permanganate, all oxidants; and activated carbon, an absorbant, were tested for short-term effectiveness on dairy manure and found to be unsuccessful. Dried bacteria, orthodichlorobenzene, formaldehyde, and sodium nitrate were ineffective

as long-term treatments also. Hydrogen peroxide, sodium hypochlorite, chlorine dioxide, and potassium permanganate at 500 ppm greatly reduced sulfide and odor levels in liquid swine manure, for a short time. Sodium nitrate changed the odor, reduced sulfide levels, and caused suspended solids to float on liquid manure storages.

pH adjustments with lime treatments reduced hydrogen sulfide but increased ammonia. Spraying enzymatic materials containing amylolytic, eclylolytic, proteolytic, and lypolytic enzymes on the surface of lagoons was unsuccessful in reducing odors.

Cattle feedlots were treated with several chemicals for odor control evaluation. Potassium permanganate, potassium nitrate, paraformaldehyde, a proprietary formulation of ortho-chlorobenzene, hydrogen peroxide and another proprietary formulation were sprayed on the feedlots. Potassium permanganate was the most effective, at a rate of 22 kg/hectare (20 lb/acre) in a one percent solution in water. That chemical was also shown in the same series of tests to be effective in reducing odors in liquid beef manure at the rate of 28 kg/tonne (56 lb/ton) of slurry.

Another feedlot test showed some proprietary materials to be effective for five days after application, but after ten days the effect was gone. The cost of the products (1975 report) ranged from $740/hectare to $1481/hectare ($300/acre to $600/acre) for the odor production season.

A laboratory test of 22 commercial products and some generic chemicals showed disappointing results. Although chlorine and a couple of lab-grade chemical formulations performed well in reducing odors, their costs were very high.

Chemical additives and treatments for reduction of odors in manure should be considered a last resort, after manure management and general housekeeping have been optimized for the farmstead. Although there are many chemical and biochemical treatments that will work in odor control, no single treatment seems to work for everyone. A product needs to be tested in the specific situation, using small quantities of manure and treatment product mixed at the manufacturer's recommended rate. Some products may only work for a short time. The ration being fed can be expected to have an effect on manure odors and the usefulness of a given treatment product.

Chemicals can be extremely hazardous to keep and use in large quantities and they are equally hazardous to mix with wastes. For example, hyperchlorites give off chlorine, which causes respiratory damage. Some chemicals may be toxic when spread in manure on cropland, and need to be evaluated in that light as well as for their effectiveness in controlling odors.

V. Treating Wastes and By-Products for Feed

Various physical, chemical, and biological treatments can be used to convert manures into feed ingredients and improve properties of materials handling. Processing animal excreta can recover nutrients, destroy pathogens, control odors, and improve palatability. Major methods of processing include: drying (with both natural and heated air), chemical treatment, ensiling, liquid-solids separation, aerobic liquid treatment, and composting. These methods are briefly described by Day (1977). Arndt et al. (1979) gave comparisons of these methods; an extensive review of feeding processed manure is given in CAST (1978). Chemical treatments include mixing a biodegradable bactericide with raw manure to give a short-term treatment and use of solvents to extract protein. Biological treatments include ensiling that preserves the nutrients and microbial fermentation by aerobic and anaerobic methods that upgrade nonprotein nitrogen to single-cell quality protein.

Ruminants play a significant role in the use of recycled manure in feeds because of their unique ability to use nonprotein nitrogen and cellulose. Other animals can also use manure but to a lesser extent. A considerable amount of research has been conducted on feeding dried poultry manure to poultry (Surbrook et al., 1971; Tanabe et al., 1985; McCaskey et al., 1990; Naber et al., 1990; Stephenson et al., 1990). Also, the nitrogen in manure can be upgraded to protein quality through an aerobic fermentation process in an oxidation ditch. Such material has been successfully fed to nonruminants -- swine and poultry (Harmon and Day, 1975; Martin et al., 1976). Energy requirements for drying and anaerobic treatment, however, make these methods less competitive than the less expensive methods available for refeeding manure to ruminants. Also, heat drying of the manure results in increased nitrogen losses (Fontenot et al., 1971).

Numerous other methods of producing feed from manure have been tested, including: growing algae, *spirulina* and *chlorella vulgaris*; growing yeasts; growing fungi; growing flies; and vermicomposting with earthworms. Hydroponic growing of various plants has also been tried, as well as electrochemical processes (Day, 1988).

VI. Summary and Discussion

Biological and chemical methods of treating livestock wastes are discussed. In most cases water quality criteria will not allow a discharge (effluent) into a stream. Recommended loading rates and volumes are given for various facilities. These recommendations assume that all the excrement goes to the facility at a uniform rate and is distributed uniformly throughout the facility. The recommended volumes should be adjusted to larger values for nonuniform loading, but rates can be adjusted to lower values if some of the manure is removed before reaching the facility. Also, if combination systems are used in series, the loading rates to subsequent components can be reduced by the treatment accomplished in preceding components.

Anaerobic lagoons have found widespread use because of ease of construction, maintenance, and attractiveness of discharging liquid wastes into them by gravity. However, lagoons can produce objectionable odors at times, especially during spring warmup. Location, construction details, sealing the bottom, inlet design, fencing, mosquito control, and surplus water and sludge removal are important design considerations, but are not discussed in this paper. Liberation of ammonia from lagoon surfaces is now being studied, as it can contribute to acid rain.

Anaerobic digestors to produce biogas have several potential advantages for degrading livestock manure and producing methane, but initial cost and operating skill have discouraged their acceptance. Septic tanks have operated primarily as settling tanks, and absorption fields are usually inadequate, necessitating a holding lagoon or regular emptying of the tanks.

The cost of aerobic treatment has discouraged its use, but some form of aerobic treatment is likely to be used in livestock waste-management schemes because of the low level of odors associated with this method of treatment. Oxidation ponds, aerated lagoons, oxidation ditches, and composting are discussed.

Surplus liquids and residues from treatment schemes should be recycled onto land or as feed ingredients, as treatment sufficient to meet stream quality criteria does not appear feasible. The main purpose for reviewing these waste treatment methods is to provide ideas for making decisions about waste management alternatives. A variety of components is usually needed for a complete waste management system. In the future, disposal will give way to recycling and utilization, contributing to better environmental stewardship.

VII. Research Needs

There are a number of research needs in the area of manure treatment. Some of the more immediate needs follow:

Chemical treatment

Investigate methods for incorporating energy-efficient dewatering of swine and dairy slurry to make the pyrolysis system energy-positive. Products of manure pyrolysis are usable as plant nutrients and fuels. However, existing processes require manure with moisture content of no more than about 65%.

Biological treatment

Build and demonstrate farm-scale anaerobic digestors that are low-maintenance and less capital-intensive than have been produced to date. Anaerobic digestor sludge is stabilized, thus having much less odor than fresh manure, yet retains all the plant nutrients. Digestor gas is approximately 60% methane, and has been shown to be a viable energy source for on-farm heating and electricity generation.

Build and demonstrate farm-scale composting facilities that integrate livestock manure treatment with disposal of carbonaceous materials such as crop residue. Aerobic composting shows promise as a method for greatly reducing or eliminating obnoxious odors in manure. The composted material can be used as a fertilizer, soil amendment, or landscaping material. However, the livestock producer needs a system that can receive and compost large volumes of high-moisture manure in a labor-efficient manner.

Decision support

As new manure treatment methods are developed to the point of economic viability and environmental acceptability, decision-support tools for livestock and crop producers must also be built so that the technology will be used appropriately.

References

ASAE. 1993. Control of manure odors. Engineering Practice ASAE EP379.1. ASAE Standards 1993. ASAE, St. Joseph, MI.

ASAE, 1994. Manure production and characteristics. ASAE D384.1, ASAE Standards, 1994. St. Joseph, MI.

Annamalia, K., A.M. Madan, J.M. Sweeten, and D. Chi. 1985. Fluidized bed combustion of manure: experiment and theory. p. 37-45. In: Proceedings of the Fifth International Symposium on Agricultural Wastes, Chicago, IL, American Society of Agricultural Engineers.

Appell, H.R., Y.C. Fu, S. Friedman, P.M. Yavorsky, and I. Wender. 1971. Converting organic wastes to oil: A replenishable energy source. U.S. Dept. of Interior, Bureau of Mines Report of Investigation 7560.

Arndt, D.L., D.L. Day, and E.E. Hatfield. 1979. Processing and handling of animal excreta for refeeding. *J. Anim. Sci.* 48(1):157-162.

Blaha, K. 1977. Solids separation and dewatering. p. 183-195. In: E.P. Taiganides (ed.), *Animal Wastes,* Ch. 16, Applied Sci. Publ.

CAST. 1978. Feeding animal waste. Summary of CAST, Council for Agr. Sci. and Tech. Report No. 75.5(4):30-33.

Chiumenti, R., L. Donantoni, and S. Guercini. 1987. Liquid/solid separation tests on beef cattle liquid manure. p. 34-44. In: Seminar of the 2nd Technical Section of the C.I.G.R. on Latest Developments in Livestock Housing, University of Illinois at Urbana-Champaign, American Society of Agricultural Engineers.

Christianson, L.L., R.H. Zhang, D.L. Day, and G.L. Riskowski. 1993. Effects of building design, climate control, housing system, animal behavior and manure management at farm levels on N-losses to the air. Proc. First International Symposium on Nitrogen Flow in Pig Production and Environmental Consequences. EAAP Publication 69. Pudoc Scientific Publishers, Wageningen, The Netherlands.

Davis, E.G., I.L. Feld, and J.H. Brown. 1972. Combustion disposal of manure wastes and utilization of the residues. U.S. Dept. of the Interior. Bureau of Mines Solid Waste Research Program Technical Progress Report 46.

Day, D.L. 1972. Aerobic and anaerobic treatment for livestock wastes. Proc. Livestock Wate Management Conference, Mar. 1-2, 1972, Champaign, IL. Dept. Agric. Engr., Univ. Ill. at Urbana-Champaign.

Day, D.L. 1977. Utilization of livestock wastes as feed and other dietary products. In: E.P. Taiganides (ed.), *Animal Wastes,* London: Applied Sci. Publ., Ch. 23:295-314.

Day, D.L. 1988. *Livestock Manure Management*, text and reference book. Agricultural Engineering Dept., Univ. of IL at Urbana-Champaign.

Day, D.L. and J. Arogo. 1993. Biological and chemical treatment of animal manure. p. 53-67. In: Proc., Meeting the Environmental Challenge, NPPC Symposium held Nov. 17-18 at Minneapolis, MN.

Edwards, C.A., I. Burrows, K.E. Fletcher, and B.A. Jones. 1985. The use of earthworms for composting farm wastes. p. 229-242. In: J.K.R. Gasser (ed.), *Composting of Agriculture and Other Wastes*, Elsevier Applied Sci. Pub., London.

Esmay, M.L. 1977. Dehydration systems for feedlot wastes. Chapter 17. In: E.P. Taiganides (ed.), *Animal Wastes*, Applied Science Publishers LTD, London.

Esmay, J.L., C.J. Flegal, J.B. Gerrish, J.E. Dixon, C.C. Sheppard, H.C. Zindel, and T.S. Chang. 1975. Proc. Third International Symposium on Livestock Wastes, Amer. Soc. Agr. Engr. Proc. 275:468-472.

Fontenot, J.P., B.W. Harmon, R.E. Tucker, and W.E.C. Moore. 1971. Studies of processing, nutritional value and palatability of broiler litter for ruminants. ASAE Proc. 271, Amer. Soc. Agr. Engr., St. Joseph, MI. p. 301-305.

Fulhage, C. 1993. Composting Dead Swine. p. 15-22. In: Proceedings of the Livestock Waste Management Conference. Champaign, IL.

Funk, T., G. Bartgis, and J. Treaqust. 1993. Designing and managing livestock waste lagoons in Illinois. Agriculture Cooperative Extension Service, University of Illinois. Circular 1326.

Glerum, J.C., G. Klomp, and H.R. Poelma. 1971. The separation of solid and liquid parts of pig slurry. Proc. Second International Symposium on Livestock Wastes. ASAE Proc. 271:345-347.

Hanna, M., D.M. Sievers, and J.R. Fischer. 1985. Chemical coagulation of methane producing solids from flushing wastewaters. p. 632-637. In: Proceedings of the Fifth International Symposium on Agricultural Wastes, Chicago, IL, American Society of Agricultural Engineers.

Harmon, B.G. and D.L. Day. 1975. Nutrient availability from oxidation ditches. ASAE Proc. 275, Amer. Soc. Agr. Engr., St. Joseph, MI. p. 199-202.

Hashimoto, A.G., Y.R. Chen, and V.H. Varel. 1981. Theoretical aspects of methane production: state-of-the-art. ASAE Proc., Amer. Soc. Agr. Engr., St. Joseph, MI. p. 86-91, 95.

Hepherd, R.Q. and J.C. Douglas. 1973. Equipment and methods for the solid-liquid separation of slurries. *The Agricultural Engineer* 28(2):77-83.

Howard, A. 1935. The manufacture of humus by the indoor process. *J.R. Soc. Arts* 84:25.

Jones, D.D., A.C. Dale, and D.L. Day. 1971. Aerobic treatment of livestock wastes. Illinois Agricultural Experiment Station Bulletin 737, Univ. of IL at Urbana-Champaign.

Kreis, R.D. 1978. Control of animal production odors: the state-of-the-art. Environmental Protection Technology Series, United States Environmental Protection Agency.

Loehr, R.C. 1974. Agricultural Waste Management, Problems, Processes, and Approaches. Academic Press, N.Y. 444 pp.

Martin, J.H., D.F. Sherman, and R.C. Loehr. 1976. Refeeding aerated poultry wastes to hens. ASAE Paper No. 76-4513.

McCaskey, T.A., A.H. Stephenson, and B.G. Ruffin. 1990. Factors that influence the marketability and use of broiler litter as an alternative feed ingredient. p. 197-203 In: Proceedings of the Sixth International Symposium on Agricultural and Food Processing Wastes, Chicago, IL, American Society of Agricultural Engineers.

Moore, J.A. 1993. Physical treatment of animal waste. Proc., Meeting the Environmental Challenge, NPPC Symposium held Nov. 17-18 at Minneapolis, MN. p. 31-40.

MWPS. 1975. Livestock waste management with pollution control. North Central Regional Research Publication 222. MWPS-19. Ames, IA.

MWPS. 1985. Livestock waste facilities handbook. MWPS-18. MidWest Plan Service, Ames, IA.

Naber, E.C., J.A. deGraft-Hanson, J.F. Stephens, and O.C. Thompson. 1990. The feeding value for broiler chicks of cage layer manure rapidly composted in a closed system with various carbon sources. p. 204-211. In: Proceedings of the Sixth International Symposium on Agricultural and Food Processing Wastes, Chicago, IL, American Society of Agricultural Engineers.

Novarotto, P.L. and G. Bonazzi. 1990. A new kind of litter to eliminate slurry production in piggeries. p. 183-190. In: Recent Developments in Animal Waste Utilization, FAO REVR Tech. Series 17, Reggio Emilia, Italy.

NRAES. 1984. On-Farm Biogas Production. Northeast Regional Agr. Engr. Service, Coop. Ext. Service, NRAES-20, Ithaca, NY.

NRAES. 1992. On-farm composting handbook. Northeast Regional Agr. Engr. Services, Coop. Ext. Service, NRAES-54, Ithaca, NY.

Pasveer, A. 1963. Developments in activated sludge treatment in The Netherlands. p. 291-297. In: W.W. Eckenfelder, Jr. and B.J. McCabe (eds.), *Advances in Biological Waste Treatment*, Macmillan Co., New York.

Pederson, M. 1993. Worms take recycler full circle. *Agr. Engr.* 74(5):24-25.

Rorick, M.B., D.J. Warburton, S.L. Spahr, and D.L. Day. 1981. Performance of a perforated pressure roller solid/liquid separator on dairy manure. Proc. 4th International Symposium on Livestock Wastes, Amarillo, TX. ASAE 281:426-429.

SCS. 1992. Agricultural Waste Management Field Handbook. USDA, SCS. Washington, DC.

Shuster, W.W. 1970. Partial Oxidation of Solid Organic Wastes. U.S. Public Health Service Publication 2133.

Shutt, J.W., R.K. White, E.P. Taiganides, and C.R. Mote. 1975. Evaluation of solids separation devices. Proc. 3rd International Symposium on Livestock Wastes. ASAE Proc. 275:463-467.

Simpson, J.R. 1960. Some aspects of the biochemistry of aerobic organic waste treatment. p. 1-30. In: P.C.G. Isaac (ed.), *Waste Treatment*, Proc. of the Second Symposium on the Treatment of Waste Waters, Pergamon Press, England.

Smith, R.J. 1981. Practicality of methane production from livestock wastes: state-of-the art. In: Proc. 4th International Symposium on Livestock Wastes/1980. *ASAE. Proc.* 2-81:109-114.

Stephenson, A.H., T.A. McCaskey, and B.G. Ruffin. 1990. Management practices that affect the value of poultry litter as a feed ingredient. p. 219-226. In: Proceedings of the Sixth International Symposium on Agricultural and Food Processing Wastes, Chicago, IL, American Society of Agricultural Engineers.

Surbrook, T.C., C.C. Sheppard, J.S. Boyd, H.C. Zindel, and C.J. Flega. 1971. Drying poultry waste. ASAE Proc. 271, Amer. Soc. Agr. Engr., St. Joseph, MI. p. 192-194.

Tanabe, Y., K. Tanaka, and Z. Kato. 1985. Feeding poultry waste to chickens, swine, and cattle. p. 138-145. In: Proceedings of the Fifth International Symposium on Agricultural Wastes, Chicago, IL, American Society of Agricultural Engineers.

Terwilleger, A.R. and L.S. Crauer. 1975. Liquid composting applied to agricultural wastes. ASAE Proc. 275. Amer. Soc. Agr. Engr., St. Joseph, MI. p. 501-505.

Voermans, J.A.M. (ed.). 1992. Proc. Workshop deep litter systems for pig farming. Res. Inst. for Pig Husbandry, Rosmalen, Netherlands.

Warburton, D.J., J.N. Scarborough, D.L. Day, A.J. Muehling, A.H. Jensen, and S.E. Curtis. 1979. A study of commercial products for odor control and solids liquification of swine manure. Agricultural Engineering Research Report. Department of Agricultural Engineering, Agricultural Experiment Station, College of Agriculture, University of Illinois at Urbana-Champaign.

White, R.K. 1978. The role of liquid-solid separation in today's livestock waste management systems. Paper presented at the 70th Annual Meeting of the Amer. Soc. Anim. Sci. at Mich. State Univ., July 12.

White, R.K. and E.P. Taiganides. 1971. Pyrolysis of livestock wastes. In: Proc. Second International Symposium on Livestock Wastes. ASAE Proc. 271:190, 191, 194.

Zhang, R.H., D.L. Day, L.L. Christianson, and W.P. Jepson. 1994. A computer model for predicting ammonia release rates from swine manure pits. *J. Agr. Engr. Res.*, UK (58)223-229.

A Systems Engineering Approach for Utilizing Animal Manure

D.L. Karlen, J.R. Russell, and A.P. Mallarino

I. Introduction

Animal manure and crop residues are undoubtedly the oldest resources applied to the soil to cycle plant nutrients. Prior to World War II, the use of crop rotations and application of animal manure were the primary soil management practices used to maintain soil fertility, improve soil structure, and sustain crop yields (Karlen et al., 1994b). Change is never easy and replacement of animal manure as the primary nutrient resource by manufactured products was not always popular. The conflict is illustrated by the verse from Smith (1952), which was previously quoted by Schulte and Kroeker (1976):

I remember the time when the stable would yield,
Whatsoever was needed to fatten a field;
But chemistry now into tillage we lugs,
And we drenches the earth with a parcel of drugs;
All we poison, I hope, is the slugs.
(Published in *Punch*, 1846)

Why then has animal manure changed from being a "resource" to being considered a "waste" in the past 40 years? Schulte and Kroeker (1976) stated that livestock producers have faced a severe profit squeeze that has resulted in fewer but larger operations. Smaller operators experienced difficulty in obtaining adequate labor due largely to lack of profit. There were few, if any, government programs directed at improving methods of manure management; and unlike corporate business, the farmer could not transfer the cost of pollution control to the consumer. Unfortunately, the trend toward larger, more consolidated, and specialized operations has continued through the early 1990s.

Schulte and Kroeker (1976) suggested systems analysis as a tool for developing meaningful solutions to agricultural waste management problems. They suggested that by working together, practitioners and researchers could solve producer problems more effectively. This recommendation is consistent with a systems engineering philosophy (Wymore, 1993), which states that to be successful, the problem to be solved must be clearly defined. This includes identifying and seriously considering all stakeholder concerns that any proposed solution may affect.

Manure management problems for swine, poultry, and beef enterprises (Figure 1) have emerged largely because animal and crop production enterprises have been separated. Separation of these enterprises also increased the potential for decreased soil quality, because without periodic manure applications, input of carbon (C) into the soil is often decreased. Lower C input may subsequently decrease the water stability of soil aggregates, microbial activity, earthworm populations, and water retention, and create a less favorable biological, chemical, and physical environment for crop growth (Karlen and Doran, 1993).

In this chapter, we demonstrate how the principles of systems engineering, as outlined by Sage (1992) and Wymore (1993), can be used to develop management strategies for using animal manure. This approach requires an evaluation of the environmental, economic, and social factors that often determine whether animal manure will be considered a resource or a waste. Our approach will be to: (1) define the principles of systems engineering; (2) discuss how those principles were used with a small group of USDA-ARS research scientists to establish research priorities for addressing the utilization of beef feedlot manure, and (3) show how systems engineering was used to design a comprehensive, integrated soil-crop-animal management research program.

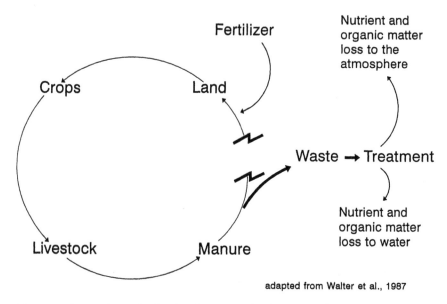

adapted from Walter et al., 1987

Figure 1. Separation of animal and crop production enterprises and the resultant animal waste and soil quality problems.

II. Principles of Systems Research

Sage (1992) defines systems engineering as the design, production, and maintenance of trustworthy systems within cost and time constraints. Wymore (1993) defines it as an intellectual, academic, and professional discipline, the principal concern of which is the responsibility to ensure that all requirements for a bioware/hardware/software system are satisfied throughout the life cycle of the system. For both definitions, a system is any situation where there are inputs (materials, requests, components, knowledge, etc.) that are manipulated or processed in some manner to produce an output (product, solution or plan, etc.).

A systems approach toward research, business, or large-scale problem solving is not new. However, without considering the entire system, it will often be very difficult if not impossible to focus on the most critical issues facing an enterprise or to effectively solve real problems. Systems research requires a comprehensive plan that clearly defines the problem to be solved. It must also provide a mechanism for transferring the information and strategies that emerge from solving that problem to the stakeholders, and accommodate feedback from those persons for subsequent adjustment or refinement of the system. Systems engineering is the process through which the boundaries or limits of the system are established using input from all clients, customers, and stakeholders who might be involved in or affected by any solution.

Table 1. Life-cycle phases associated with systems engineering

Wymore (1993) 7-Phase Description	Sage (1992) 3-Phase Description[a]
Phase 1 Requirements development	Phase 1 System definition
Phase 2 Concept development	
Phase 3 Full-scale engineering development	Phase 2 System design and development
Phase 4 System development, production, manufacturing, and deployment	
Phase 5 System test and integration	
Phase 6 Operations, support, and modification	Phase 3 System implementa- tion and maintenance
Phase 7 Retirement and replacement	

[a]The 22-phase description by Sage (1992) includes perception of need, requirements definition, draft proposal (RFP), comments on RFP, final RFP, proposal develop-ment, and source selection in phase 1; conceptual development, subsystem partitioning, subsystem specification development, component development, subsystem integration, overall system integration, and user training and develop-ment into phase 2; and implementation, final acceptance testing, operational test and evaluation, final acceptance, identification of change requirements, negotiation for system changes or support, system maintenance change development, and maintenance testing by support contractor in phase 3.

Systems research requires a holistic approach that is supported by mechanistic and component research that has been traditionally supplied by reductionist scientists in both public and private sector research organizations (Karlen et al., 1994a). To accommodate numerous and often simultaneous goals, systems engineering uses tools and techniques associated with systems science. The latter is another engineering philosophy with its foundations in constructing both conceptual and mathematical models (Bird et al., 1990).

For successful systems engineering, both Sage (1992) and Wymore (1993) stress the importance of examining the entire life-cycle associated with the system to be designed. Wymore (1993) describes the life-cycle of a system in 7 phases, while Sage (1992) offers a simplified 3 phase description, or a more detailed, 22 phase process (Table 1). All three descriptions stress the importance of using an iterative process to clearly define the real problem that needs to be solved, to allow for feedback and refinement, and to ensure understanding and communication among all stakeholders at all times.

With respect to using animal manure as a resource, rather than considering it a waste, we will briefly examine potential problems that might be encountered while developing the requirements for addressing problems focusing on the utilization of swine, poultry, or beef feedlot manures. Our goal is to demonstrate a systems planning process that may help determine the best use for these animal manures. We do not presume to be providing a definitive answer for solving the problems associated with any of these materials.

III. Magnitude of the Animal Manure Problem

Chapters within this book have addressed many of the critical issues that must be addressed by manure management programs, but several key concerns associated with swine, poultry, and beef feedlot manure are summarized here to help define the problem to be solved.

A. Swine Manure

Swine production systems in the 1990s have become larger, more specialized, and more dependent on purchased feed supplies than in the past. Production of pork is a major agricultural enterprise in the Midwest and is rapidly increasing in North Carolina, Colorado, and Utah (Hatfield et al., 1994). Estimates are that swine account for 12 to 15% of the total livestock manure produced each year in the U.S. (VanDyne and Gilbertson, 1978). Based on a 1988 inventory, production of manure from all hogs exceeds 14.1 million Mg (15.5 million tons). This material has a nitrogen (N) content of 0.66 million Mg (0.73 million tons), phosphorus (P) content of 0.42 million Mg (0.46 million tons), and potassium (K) content of 0.66 million Mg (0.73 million tons). Environmentally safe and socially acceptable handling of this manure and utilization of these nutrients as a resource have become major issues as the size of production units has steadily shifted to a scale of more than 1000 head. Furthermore, many of the most recent units accommodate more than 10,000 head, which means that the problem is becoming increasingly significant.

Large livestock production enterprises have largely been separated from crop production enterprises. As a result, retention of nutrients in the manure is often not an important goal for large swine production units, and the amount of land available for manure application is often limited. Consequently, loss of nutrients to the environment may be encouraged to reduce the quantity of manure that must be handled. For example, large swine production units generally use an anaerobic lagoon to digest manure solids and to allow the manure to be handled as a liquid. The N is converted to ammonia in the lagoons and most is lost to the atmosphere. With current technology, anaerobic lagoons can volatilize 70 to 90% of the N in the manure. With this manure management practice, the producer is required to have only 10% of the land required for application of slurry manure.

The digestibility of nutrients in animal feeds has been quantified, but less is known regarding nutrient retention by the animals. As a result, the effects of diet on the composition of manure are also unknown. Furthermore, most studies that have evaluated manure nutrient composition have focused on the material after a period of storage rather than before.

There are several environmental concerns related to both water and air quality associated with swine manure. Water quality concerns involve nitrate and phosphorus in both surface water and groundwater resources. Air quality questions focus on effects of ammonia and methane, not only for their social impact, but also as they relate to the overall production of volatile gases that affect global warming (Hatfield et al., 1994). Odors and their impact on the surrounding communities is also a major air quality concern.

B. Poultry Manure

Integrated poultry production in the U.S. is concentrated in the midsouth region, with over 40% of the nation's cash receipts derived from the sale of poultry products originating in Arkansas, Georgia, North Carolina, and Alabama (Moore et al., 1994). Increased demand for low-cholesterol meat products led to the rapid and concentrated growth of the poultry industry in these areas. This expansion has resulted in increased concern regarding the utilization or disposal of poultry wastes and its potential impact as a nonpoint source of agricultural pollution.

Litter from broiler production, manure generated by laying operations (hens and pullets), and dead birds are the three primary wastes creating the environmental concern (Edwards and Daniel, 1992). In 1990, approximately 13 million Mg (14.3 million tons) of poultry litter and manure were produced on U.S. farms, most of which (68%) was broiler litter. With the exception of a small amount that is used in animal feed, the major portion (90%) is applied to agricultural land.

Poultry litter is generally considered the most valuable animal manure for use as a fertilizer because of its low water content. There have been reports of reduced corn germination and growth, because excessive poultry litter application has increased soil salinity (Moore et al., 1994). However, poultry litter has also been recognized as a way to ameliorate salt-affected soils and used to promote plant growth on brine-contaminated soils in Arkansas (Hileman, 1973).

The average fertilizer equivalent value of poultry litter was 3-3-2 (N-P_2O_5-K_2O) when measured "as spread" (Stephenson et al., 1990). It also contained substantial quantities of B, Ca, Cu, Fe, Mg, Mn, S, and Zn. When applied using best management practices such as buffer zones between treated areas and waterways, soil testing, and correct timing and placement of manure, adverse impacts from land application of poultry litter can generally be prevented (Moore et al., 1994). Research needs associated with poultry manure include developing practices that decrease: (1) ammonia volatilization, (2) leaching of nitrate, (3) P runoff, and (4) pathogen release. Information is also needed to determine whether loading rates should be based on providing crops with their N or P requirement.

C. Beef Cattle Feedlot Manure

Approximately two-thirds of the beef cattle fed in U.S. feedlots can be found in Nebraska, Texas, Kansas, Iowa, and Colorado (Eghball and Power, 1994). More than 80% of these cattle are in feedlots containing more than 1000 head. At any one time, there are at least 10 million head of beef cattle on feed, and each animal excretes approximately 145 g (0.32 lb) of N in fresh manure daily. Approximately 125 g (0.28 lb) of N can be collected in the manure, indicating that approximately 457,000 Mg (505,000 tons) of N could be recovered in beef feedlot manure each year. Based on similar calculations, 157,000 Mg (173,000 tons) of P and 482,000 Mg (531,000 tons) of K could also be collected from this manure source. If purchased as fertilizer, the value of the N, P, and K in this manure would be approximately 111, 180, and 170 million dollars, respectively (Eghball and Power, 1994).

Factors that can affect the mineral composition of animal manure include the animal size and species, housing and rearing facilities, ration, manure storage practices, and climate. Transport of the manure to the field is an important part of any management system. Fortunately, in the area of the U.S. where most beef cattle are fed, most land is under cultivation, so there is seldom a shortage of farmland available for manure application. However, the system as a whole may have problems because the feedlot operators often do not control much of this cropland.

Problems associated with feedlot manure management that need additional research include: (1) the management of abandoned or understocked feedlots; (2) the need to develop technology for using manure with no-till crop production practices so that crop residues can be maintained on the soil surface for erosion control; (3) the need to prevent salt accumulation in the drier regions; and (4) the need to reduce the atmospheric loss of ammonia and other gases.

IV. Critical Issues Facing Animal Manure Systems

The information on swine, poultry, and beef feedlot manures identifies several factors (Table 2) that must be considered when using a systems engineering approach to design strategies for using animal manure as a resource. The next step in the systems engineering process is to conceptualize potential solutions that would be acceptable to all stakeholders. This can often be accomplished by using a Delphi or consensus building approach (Sage, 1992).

After the problem to be solved is defined, the quality of an acceptable solution (performance standards) must be agreed upon by all stakeholders (Table 3). During this process, technologies that might be useful or even required for solving the problem are identified (Table 4), and specific requirements or restrictions on this technology are agreed to by all stakeholders.

With a clear focus on: (1) the problem to be solved, (2) the standards that any solution must meet, and (3) any restrictions on the technology that can or cannot be used, we can now move forward with the systems engineering process to design systems that would be capable of solving the problem. Throughout the concept development phase, all the criteria and constraints specified by the performance standards and technologies that can or cannot be used must serve as guidelines.

Table 2. Critical issues that must be addressed by animal waste management systems

Issue	Possible causes
Odors and contribution to greenhouse gases	Management practices associated with feedlots, confinement buildings, lagoons, or manured fields
Nutrient use efficiency	Excessive manure application per unit of land, or an inability to accurately measure and determine the amount and availability of nutrients that will be available to subsequent crops from the manure
Groundwater contamination by NO_3-N	Excessive land application because of insufficient land resources, or improper design of manure storage facilities
Surface water contamination (Eutrophication and/or fish kills)	Movement of pathogenic microorganisms, P, or N into the waters through surface runoff as a result of improper application time or location

Table 3. Performance standards that must be addressed by animal waste management systems

Standard	Issue to be resolved
Soil-test P level	What are acceptable P levels in soils receiving animal manure? How does manure-supplied P affect micronutrient availability?
Nitrogen use efficiency	What fraction of manure-supplied N is lost by denitrification and/or volatilization? How fast does manure-supplied N become available to plants?
Crop residue management	What amount of surface crop residue is lost during manure injection? How uniform can animal manure be applied?
Soil quality impact	How severe is the issue of compaction? What amount of salt accumulates per Mg of manure? How fast will manure application affect aggregate stability?

Table 4. Technologies that may be useful for the design of animal waste management systems

Issue	Technological question
Manure additives	What role should manure additives have in an animal waste management system?
Manure enrichment	Is enrichment of animal manures economically feasible? What procedures are most effective for manure enrichment?
Multiple waste sources	Should animal manure and other waste streams (municipal sludge, newsprint, etc.) be blended for a more complete fertilizer or soil amendment?
Impact of diet	What effect does the diet of the animal have on the nutrient composition of the manure? What impact will manure composition have on nutrient availability to crops?

Following this protocol will ensure that the solution, when developed, will be acceptable to all stakeholders.

Issues related to the use of resources, including the amount that producers, processors, consumers, and the general public will pay must be addressed. Environmental, social, and economic requirements and implications associated with all possible solutions must be addressed when using the systems engineering process to conceptualize and develop a solution for the real problem. Ultimately, tradeoffs among alternatives must be examined and decisions made.

Throughout the entire systems engineering process, documentation is extremely critical so that the rationale and basis for the eventual solution is known and recorded. This documentation will be invaluable when during the course of the life-cycle the system requires replacement, and persons responsible for the original design are no longer available for consultation regarding why an issue was addressed in some particular manner.

The systems engineering process was recently demonstrated as a potential tool for research planning and coordination by Karlen et al. (1994a). To illustrate how the process might be used to design programs and strategies for animal manure management, we will discuss our experiences in focusing research efforts on beef feedlot manure and the use of systems engineering principles to design an integrated animal-crop-soil management research program for the northern Corn Belt.

V. Beef Feedlot Manure Utilization Research

A. Problem Situation

A systems research plan for environmentally and economically acceptable use of beef feedlot manure was to be developed in cooperation with several research scientists at the USDA-ARS Meat Animal Research Center (MARC) near Clay Center, NE and at the Soil/Water Conservation Research Unit (SWCR) in Lincoln, NE. A systems approach was desired because the scientists within these research units had been given a congressional charge to develop production systems that integrated animal manure back into crop production.

Problems with animal manure near the Chesapeake Bay and in the Florida Coastal Plain were important factors leading to this legislative mandate. MARC was selected for this research activity because the facility has several types of cattle, various types of production facilities, land for crop production, and facilities for handling manure. Corn (*Zea mays* L.), which can efficiently utilize the nutrients in the manure, is grown at this research site. The SWCR scientists were involved because of their expertise in nutrient cycling and crop production.

Runoff control facilities for the feedlot manure at MARC were installed in the early 1970s for unpaved feedlots. Flushing gutters that were developed in the early 1980s are used for animals in housed confinement. MARC is also located near two active commercial feedlots, so one requirement placed on the systems engineering process was that the results of these research programs must have direct applicability to owners, operators, and managers of beef feedlots.

Eghball and Power (1994) indicated that a substantial amount of research has been conducted on the use and management of manure from beef feedlots. They concluded that several questions remain unanswered (Table 5) and that well-planned research is justified. To facilitate a systems research approach, several stakeholder meetings were held during 1992 to examine several aspects of beef feedlot manure management. The meetings involved a feedlot manager, representatives of the Nebraska Department of Environmental Quality, and scientists and engineers from MARC, SWRC, and the University of Nebraska.

The informational meetings enabled the researchers to identify additional constraints to the systems research program that was to be designed. For example, they concluded that large feedlots could not afford to spread raw manure directly on the land because: (1) weed seed proliferation and soil compaction problems have reduced the willingness of local producers to allow spreading of raw manure; (2) available time and existing crop cover make it difficult to match cleaning feedlots with a suitable location for direct application; and (3) manure spreading equipment is expensive to purchase and maintain. These constraints were among those given by the local feedlot manager.

One possible solution emerging from those meetings was to compost the feedlot manure. A primary reason for examining this alternative more closely was that by marketing the compost within a reasonable distance, feedlot operators were able to recover processing, hauling, and field spreading costs. The stakeholders concluded,

Table 5. Factors affecting direct land application of beef feedlot manure and the type of research needed

I. Operational factors
A. Equipment for uniform application rates
B. Techniques that minimize soil compaction
C. Application equipment that preserves crop residue for erosion control while minimizing odor
II. Agronomic factors
A. Optimum crop sequences for efficient use of nutrients from animal manure
B. Reliable and rapid techniques for manure analysis
C. Soil sampling and testing protocol for accurate assessment of nutrient availability from manure
D. Accurate assessment and understanding of weed problems associated with manure application
E. Criteria for assessing water quality impact of manure application
F. Criteria for assessing soil quality impact of manure application
G. Criteria for assessing air quality impact of manure application

however, that long-term effects of using compost on soil fertility, soil structure, and infiltration were unknown. The scientists also learned that investment costs associated with developing large-scale composting operations were a drawback because the feedlot operators lacked assurance from the Environmental Protection Agency (EPA) that composting would be considered an environmentally and socially acceptable management practice for handling beef feedlot manure. Other issues related to composting that remain unresolved were the ability to provide quantitative estimates for: (1) N leaching from the feedlots, composting areas, and storage basins, ponds or lagoons; (2) runoff from these areas; and (3) technical information with regard to the effects of composting on nutrient availability, weed seed sterilization, volume reduction, and nuisance factors such as flies, odors, and aesthetics.

The informational meetings emphasized that systematic planning was needed to identify all factors that may affect or be affected by any potential solution. The plan was essential to prevent potential conflicts between agencies focusing on goals such as soil erosion control versus development of best management practice (BMP) recommendations for utilizing animal manure. For example, the Natural Resources Conservation Service (NRCS) requires farmers to maintain surface residue cover to minimize soil erosion and remain in conservation compliance, while the University of Nebraska recommends incorporation of animal manure to minimize odor and runoff and to maximize nutrient use efficiencies. However, the equipment needed to inject animal manure generally buries a substantial portion of the crop residue, thus increasing the potential for both wind and water erosion and putting the farmers out of compliance.

Other critical animal manure management issues for which there are currently no definitive answers included being able to quantify the effects of salt accumulation, excessive P loading, uniformity of application, volatilization losses, and possible land use changes created by the construction of wetlands or riparian waterways. Developing cropping sequences that can make the most efficient use of nutrients from the manure and providing educational information to help feedlot owners and operators become aware of requirements and options for manure management were also identified as critical components. These should both be considered when developing a comprehensive research and education program that addresses the utilization of beef feedlot manure.

To address these complex and interrelated issues, it was suggested (personal communication, J.A. Nienaber, 1993) that a computerized decision aid be developed to determine optimum uses for beef feedlot manure on cropland and pastureland. To be most flexible, the decision aid should be useful as a management tool for feedlot owners and operators as well as a research product. He suggested that it should also be sensitive to management factors such as: (1) type of cattle, (2) ration composition, (3) cattle management, (4) manure collection, (5) manure handling, (6) application practices, (7) seepage losses, (8) runoff, (9) N form, (10) crop nutrient requirements, and (11) relative costs of operation versus the potential for pollution. With this background, a systems engineering process was used to help establish program priorities.

B. Systems Engineering Planning Approach

The large but incomplete body of knowledge and experience regarding the use of manure in agricultural systems (Eghball and Power, 1994) demands that a structured approach be used to help identify the most critical research needs. Numerous processes and interactions occur when animal manure is applied to soils under varying climates and with various methods of handling the manure before or after removal from the feedlot. This includes compaction if the soils are wet, release of odors if the manure is surface applied, or possible loss of residue cover if it is injected. Manure application also affects nitrification and denitrification processes; P adsorption, runoff, and leaching losses; changes in soil organic matter; as well as microbial and microfaunal activity. Many of these interactions can have either negative or positive impacts on the environment by affecting air, surface water, groundwater, and soil quality.

Individual experiments have generally focused on animal manure as a fertilizer nutrient source, soil building amendment, or environmental pollutant. Some efforts have been initiated, but in general, a systems approach toward integrating the results from these component studies has not been developed or applied. This lack of integration has resulted in information and products that in general, have not been directly applicable to owners, operators, or managers of beef feedlots in Nebraska and throughout the U.S. Great Plains.

To develop a comprehensive research and technology transfer program that addressed all factors affecting beef feedlot manure management, the participants agreed that the project should be evaluated in terms of its response to system inputs. This included how well and how appropriate the results of the research were focused on the problem, and how well and at what cost the research program responded to the needs identified by the congressional mandate.

The initial research program was expected to last approximately 5 yr. This relatively short duration was chosen because the program was both a research plan and a system being designed to develop the decision aid for environmentally and economically acceptable management and use of beef feedlot manure. It was anticipated that advancements in systems applications to agricultural research would bring about conditions for an early replacement of the decision aid with an improved system for managing beef feedlot manure based on the results from an integrated research program.

System inputs were determined to be: (1) beef feedlot waste; (2) public demand for economically sound, socially acceptable, and environmentally safe beef production practices with regard to manure management; (3) operator preferences and requests, which are influenced by public demands and economic realities; (4) funds or commitment, generally measured in terms of fiscal and human resources provided for conducting the research, developing technology transfer tools, and implementing the system; and (5) feedback reflecting specific measurements of economic, environmental, and social impact of existing or modified management practices.

System outputs were identified as: (1) plans that provide manure management strategies that could be implemented by owners, operators, and managers of beef feedlots throughout the U.S. Great Plains; (2) information identifying highest priority research issues and knowledge gaps that prevent development of beef feedlot waste management strategies; and (3) feedback which, through monitoring, provides essential economic, environmental, and social information needed for continual evaluation and improvement of the system.

A functional design (Wymore, 1993) was conceptualized to include three critical issues that the research program would be expected to address. These were to: (1) provide guidelines for beef feedlot manure management practices that have direct applicability to owners, operators, and managers of feedlots throughout the U.S. Great Plains; (2) provide a framework for an information system, designed to initially address beef feedlot waste management, but applicable for comprehensive planning of a national animal waste management initiative; and (3) provide information for a decision aid that feedlot operators could use to develop and implement environmentally and economically acceptable manure management practices.

Inputs and outputs for the overall research program were envisioned to have essentially the same format for each of these applications. The primary difference would reflect how the functions are implemented. For example, research and development functions would require scientists at MARC and SWRC to serve as the primary system managers and information compilers. When used as a physical model for developing manure management practices, the functions would generally be performed by feedlot owners. Ultimately each of the functions would be incorporated into a decision aid that could be used as an information system for improved animal manure management strategies.

The systems engineering process requires documentation of the specific measurements to be made and criteria to be used for their interpretation and evaluation. These criteria will differ according to who makes the final evaluation for each system function and/or technology. This emphasizes the need for thorough documentation, because that may be the only record available when the system is retired or replaced.

Based on prior research, much of it conducted by MARC and SWRC scientists, several technologies that may be useful for helping to resolve the feedlot manure management problem were identified. Potential technologies for utilizing the beef feedlot manure that were proposed included: (1) Direct Application, (2) Composting, (3) Aquaculture, (4) Fuel, (5) Anaerobic Digestion, (6) Enrichment, and (7) Separation with re-feeding.

Priorities among technologies were established by the scientists using their knowledge and research experience with beef feedlot manure. Potential technologies were ranked on a scale of 1 (highest) to 7 (lowest). The rank and total score (value in parentheses) were: Composting (18), Direct Application (19), Enrichment (32), Anaerobic Digestion (42), Separation and re-feeding (50), Aquaculture (57), and Fuel (62). All participants agreed that the rankings would have been different for other types of animal manure or geographic locations (i.e., poultry waste in the southern U.S.).

Factors affecting the four highest priority technologies were established in a similar manner. Any factor that could impact any of the four technologies selected for further evaluation was accepted as an input for discussion. The issues were prioritized and summarized. The research questions that emerged during this process for direct land application, composting, enrichment, and anaerobic digestion of the feedlot manure are summarized in Tables 5 to 8.

The scientists then identified general factors that could affect each technology. Their conclusions were that: (1) education, through extension and technology transfer, (2) management practices, including cattle type, ration, manure handling, and labor, (3) assessment of water, soil, and air quality effects, and (4) policies, for addressing nuisance factors such as odor, weeds, and flies were the primary issues affecting any animal waste management program.

C. Current Research Program

As a result of the systems engineering planning process, and prior discussions with potential stakeholders (Nebraska Department of Environmental Quality, local feedlot owners, and University of Nebraska scientists), research focusing on composting was initiated by MARC and SWRC scientists. An initial evaluation of nitrate movement beneath a local composting site was prepared by Nienaber and Ferguson (1992). They found that both NO_3 and Cl concentrations were greater beneath compost areas than in soil profiles in adjacent furrow irrigated corn fields. Electrical conductivity measurements for the surface soils (0 to 15 cm) at the

Table 6. Factors affecting composting technology for beef feedlot manure and the type of research needed

I.   Processing factors
     A. Manure characteristics (sources, water content, weed seed content, volume, nutrients, and pathogens)
     B. Nutrient loss through volatilization, leaching, and runoff

II. Agronomic factors
     A. Optimum crop sequences for efficient use of nutrients in composted manure
     B. Soil sampling and testing protocol for accurate assessment of nutrient availability from compost
     C. Water quality impact of compost relative to direct application of animal manure
     D. Weed problems associated with composted manure relative to direct application of animal manure
     E. Soil quality impact of composted manure compared to noncomposted manure
     F. Air quality impact of the manure composting process and product application to the land when compared to direct application of manure

III. Economic factors
     A. Quantitative procedures for determining the cost:benefit ratios for composting manure
     B. Assessment of the impact of regulations on the economics associated with composting manure
     C. Methods for reducing transportation costs associated with composting manure

composting site were high (3.0 ds m^{-1}) and could lead to possible soil fertility and salinity problems, if the site were returned to crop production without leaching.

Nutrient and mass loss during composting of beef feedlot manure were also determined in a study at the University of Nebraska Agricultural Research Center near Mead, NE. The manure pack was composted as it was removed from the feedlot without adding a supplemental C source. In 1992, there was a 20% loss of manure mass during 110 days of composting (Table 9). In 1993, mass loss averaged 15% (data not presented). Loss of mass was lower than the normal range of 35 to 50%, presumably because the C:N ratio of the manure was 12:1 rather than the usual 20:1 or wider ratio. Visual estimates suggested that approximately 30% of the volume was lost during composting. Nitrogen loss was 42.5% of the total manure N (Table 9), with 3.2% lost in runoff and 96.8% apparently volatilized. In 1993, 19.3% of the total N was lost with 8% in runoff, and volatilization apparently accounting for 92%. These data suggest that volatilization of N as ammonia is the major mechanism for N loss during the composting process. Some N loss, as a

Table 7. Factors affecting enrichment technology for beef feedlot manure and the type of research needed

I. Technology development
   A. Development of procedures for manure enrichment
   B. Development of equipment for manure enrichment

II. Product composition
   A. Determination of the impact of alternate manure sources
   B. Quantitative procedures for assessing effects of manure additives

III. Agronomic factors
   A. Soil sampling and testing protocol for determining nutrient availability from enriched manures
   B. Water quality impact of enriched manure relative to direct application of animal manure
   C. Weed problems associated with enriched manure
   D. Soil quality impact of enriched manure
   E. Air quality impact of the manure enrichment process and product

IV. Economic factors
   A. Quantitative procedures for determining the cost:benefit ratios for manure enrichment
   B. Assessment of the impact of regulations on the economics associated with manure enrichment
   C. Marketing costs and strategies required to develop a demand for enriched animal manure
   D. Methods for reducing transportation costs associated with enriched animal manure

result of denitrification during composting, has also been reported. However, that process requires the manure to be saturated. The composting manure in this study was kept at 40-60% moisture throughout the period, so denitrification was probably minimal. Phosphorus loss during composting in 1992 was 0.8% of total manure P, with all of it in runoff. In 1993, 12% of the total P was lost. Approximately 17% of this loss was accounted for in the leachate, but the rest (83%) was unaccounted for. Unlike N, runoff appeared to be the primary mechanism for P loss during composting.

Temperatures at all depths within the compost pile reached 60°C (140°F) within 24 hours after starting the composting process and were not affected by the time of day. They remained constant, even at the 0.25 m depth, averaging approximately 55 to 65°C until day 65. At day 68, the temperature decreased to 40°C (104°F) indicating the end of thermophilic process. After this point, the pile was no longer turned, and the material was allowed to cure for an additional 43 days. At this time, the temperature of the composting material was approximately ambient and the

Table 8. Factors affecting anaerobic digestion of beef feedlot manure and the type of research needed

I. Processing factors
    A. Improvements in equipment from that developed during the 1970s
    B. Reduction in labor and maintenance requirements
    C. Criteria for by-product disposal

II. Product composition
    A. Determination of the impact of alternate manure sources
    B. Quantitative procedures for assessing effects of manure additives

III. Agronomic factors
    A. Soil sampling and testing to determine nutrient availability from anaerobically digested manure
    B. Water quality impact of anaerobically digested manure relative to direct application of manure
    C. Weed problems associated with anaerobically digested manure relative to direct application of manure
    D. Soil quality impact of anaerobically digested manure relative to direct application of manure
    E. Air quality impact of the anaerobic digestion process and product relative to direct application of manure

IV. Economic factors
    A. Quantitative procedures for determining the cost:benefit ratios for anaerobically digested manure
    B. Assessment of how regulations impact economics associated with anaerobically digested manure
    C. Marketing and transportation costs associated with anaerobically digested animal manure

Table 9. Mass and nutrient balance of composted beef feedlot manure

Variable	Mass	Nitrogen	Phosphorus	EC[a]	Ash
		----------kg----------		S m^{-1}	- % -
Manure	7002	37.5		1.21	58.7
Compost	5575	61	37.2	0.74	80.8
Amount lost	1427	45	0.3	------	------
Runoff loss	50	1.5	0.3	------	------

[a]Electrical conductivity, measured on a 2:1 manure:water mixture; S m^{-1} equals 10 mmho cm^{-1}.
(Unpublished data, B. Eghball and J. F. Power, Lincoln, NE.)

material was applied to the field where crop, soil, and other environmental responses are being measured.

When these and other studies have been completed to provide a more complete data base, the information will be used to help develop a decision aid product as identified by the systems planning. This will be an iterative process as stated by Karlen et al. (1994a), but by using a systems engineering planning process to identify the ultimate application for research information before it is collected, a more useful and user-friendly product can be developed.

VI. Integrated Manure Nutrient Management Research

Systems engineering was also used to help plan an integrated soil-crop-animal management research program that would focus on using swine manure. One goal for this project was to conceptualize systems that were scale neutral and acceptable to conventional and alternative agriculture practitioners. Another was to promote enterprise and landscape diversity by conceptualizing animal and crop production enterprises that would increase midwestern land use options.

A. Problem Situation

Compared to cash-grain farming, agricultural production systems that integrate livestock and crop production enterprises can reduce costs of purchased inputs such as fertilizer (NRC, 1989), reduce economic risk through diversification of farm enterprises (Kliebenstein and Ryan, 1991), and increase economic return to rural communities by providing jobs in sales or meat processing. Integrating ruminant animals and the manure they produce into farming systems may also decrease soil erosion and surface runoff, by increasing the potential for including forages in the crop rotation.

Integrated animal and crop production enterprises that use manure as the primary nutrient source for crop production are not without problems. Nutrient loading on a farm with an animal enterprise may exceed crop nutrient needs (Lanyon and Beegle, 1989). This can occur because the amount of nutrients being removed in products from the diversified enterprise is often lower from livestock operations than from cash-grain operations. Problems with nutrient loading may be exacerbated by timing of manure nutrient availability, which is often not synchronized with potential crop utilization (Lanyon, 1991). Excess nutrient loading and asynchronous timing with crop needs are two manure management problems that can result in contamination of ground and surface waters with N and P (NRC, 1989).

Nutrient management practices that minimize losses from the entire soil-plant-animal continuum are needed for economically feasible, environmentally safe, and socially acceptable agricultural systems (Karlen and Sharpley, 1994). Components of systems that include animal and crop production will include site selection, size and type of animal production unit, manure storage facilities, and appropriate manure application rates. Factors influencing application rate include manure analysis, physical, chemical and biological soil characteristics, crop species, yield goals, seasonal precipitation and temperature patterns, soil drainage, groundwater depth, and geological characteristics (Sweeten, 1991). This complexity suggests risk assessments related to manure use should be made for specific regions, rather than attempting to adopt generic standards across widespread geographical or geopolitical regions (NRC, 1993).

Compared to cash-grain enterprises, measurement and management of nutrient flow within integrated crop-livestock systems is difficult because there are several internal nutrient transfers and multiple input and output positions. To manage nutrients in an integrated crop-livestock system, including those from the manure, a systems approach is needed to monitor and control nutrient flow at the boundaries of each management unit. Critical factors affecting flow at the various boundaries are summarized in the following sections.

1. Nutrient Losses from Manure

Nitrogen loss from animal manure is a function of management. In beef feedlots, as much as 50% of the N excreted by the cattle can be lost before it is ever removed (Eghball and Power, 1994). Vanderholm (1975) reported 30 to 90% N losses to the atmosphere with surface applied manure. Warm, dry conditions at the time of spreading causes rapid ammonia volatilization (Lauer et al., 1976; Sommer et al., 1991; Whitehead and Raistrick, 1991) and can significantly increase volatile N loss. Ammonia-N losses can be reduced by injecting or immediately incorporating manure into the soil. Winter manure applications and surface runoff effects have been studied (Steenhuis et al., 1975; Young and Mutchler, 1976), but much is yet to be learned about manure application in cold weather. A recent Canadian study showed increased nitrate-N levels in field drain water following cold weather manure application (Foran et al., 1993).

Edwards and Daniel (1992) stated that surface water quality impact of land-applied poultry manure is affected primarily by factors influencing runoff and erosion. This includes the type of soil, rainfall intensity and duration, roughness characteristics of the surface, and topography. Loading rate and application timing are two other factors that can also have a significant impact. The negative impacts, especially when excessive quantities of pollutants such as sediment, N, P, and microorganisms including coliforms are transported to surface waters, include adverse effects on aesthetics, human and animal health, and aquatic wildlife. The impact on groundwater will be dependent upon subsurface transport and is thus dependent on the hydraulic characteristics of the soil-waste system as well as the amounts and forms of potential pollutants present. Several of these issues have been thoroughly discussed by other authors contributing to this publication.

Numerous studies have focused on optimizing the use of animal manure to supply crop N needs and on N losses from manure to groundwater, but few have addressed the problem of excessive P accumulation. Applications of manure (especially from swine and poultry) to soils at rates that supply adequate N for crops almost always result in P accumulations that exceed crop needs (Christie, 1987; Sharpley et al., 1984). Loading rates of P in excess of crop needs may result in nutrient imbalances and increased eutrophication of surface waters if P is lost through runoff. This problem is compounded in the Corn Belt because the majority of manured or unmanured agricultural soils already have soil-test P levels in excess of what is needed to maximize crop yields (Mallarino et al., 1991). Research focusing on efficient management of manure-P inputs to prevent the buildup of excessive P levels in soils has been identified as a fundamental need for manure management programs designed to reduce P loading in surface waters (NRC, 1993).

During manure storage and handling, P losses as high as 80% can also occur when solids accumulate in storage pits or lagoons (Midwest Plan Service, 1985). Phosphorus may be lost from soils as sediment-bound P or soluble P. Losses of sediment-bound P usually are large for cropland, while soluble P losses usually are large for pastureland (Young et al., 1993). Conservation practices that tend to reduce losses of sediment-bound P may not affect or may increase losses of soluble P (Sharpley and Menzel, 1987). The level of soluble P at or near the soil surface is the critical factor determining losses of soluble P to surface waters. The buffer capacity of the soil for P strongly influences the equilibrium between soluble P and adsorbed P in soils. This capacity varies among soil types because of differences in several mineralogical and chemical properties. Factors such as amounts, chemical form, and application method of P also influence the retention by soils.

Although numerous studies have addressed the capacity of soils to retain fertilizer P, few have addressed retention of manure P. Some reports (Abbott and Tucker, 1973; Pratt and Laag, 1981; Reddy et al., 1978) suggest that applying manure P to soils may result in higher soluble-P concentrations, more prolonged periods of high soluble-P concentrations, and greater downward movement of P, perhaps in soluble organic forms. Although it is generally assumed that P supplied by manure will be readily available for crops and independent of the form and level of P in livestock diets, this is not true. The form and proportion of organic and inorganic P of manures are affected by diet, and not all of the P applied through the manure will be available for crops. Some of the available P will be converted to unavailable forms at rates that vary with the adsorption capacity of the soil and other factors (Campbell et al., 1986; Sharpley et al., 1989).

2. Nutrient Use by Cropping Systems

Crop yield is strongly influenced by crop rotation (Karlen et al., 1994b). As yield increases, the quantity of nutrients moved from producing areas either off the farm or to another component within the farming system also increases. Loading rates and removal in harvested products largely determine the amounts of P that accumulate in soils and the potential for losses through soil erosion and runoff. Removal of P from soils by crops depends primarily on yield, capacities for P uptake, partitioning of absorbed P within plants, and whether grain or the entire plant is harvested. The capacity for P uptake by crops usually is largely dependent on biomass production, but with high soil P levels, luxury uptake of P may also occur.

An ideal cropping system that is designed for recycling N and P from swine manure would be able to remove large amounts of these nutrients from the soil profile relatively quickly, allow multiple applications during the growing season, and return some economic value to the overall enterprise. Reed canarygrass (*Phalaris arundinacea* L.), a cool-season species that produces most of its growth in the spring and fall (Marten, 1985), and switchgrass (*Panicum virgatum* L.), a warm-season species that produces most of its growth in the summer (Voight and Maclauchlan, 1985) are two species that might be integrated into a nutrient recycling system. Both species have very high N and P uptake potentials and because of their extensive root systems are very effective at scavenging nutrients from the soil. Preliminary data indicate that those species can remove up to 180 kg N and 30 kg P ha^{-1} yr^{-1} in the northern Corn Belt.

Nutrient uptake by both reed canarygrass and switchgrass is greatest during periods of active growth. By capitalizing on the differences in growth habit for these two plant species, it may be possible to develop cropping systems, including strip intercropping, that increase the temporal and spatial diversity of the landscape and provide an opportunity for season-long application of manure. Furthermore, when grown intensively as a forage, several cuttings of reed canarygrass and switchgrass can be made during each growing season, thus providing the opportunity for multiple applications of manure. Intensive management systems for these species could yield large quantities of good quality forage that can in turn be recycled through livestock. Alternatively, the biomass harvested could be used to produce ethanol (Cherney et al., 1988).

3. Forage Use and Nutrient Retention by Stocker Cattle

Raising beef cattle with a low input winter feeding program followed by subsequent summer and fall grazing and a short finishing period can provide positive economic return even at low breakeven prices (Viselmeyer et al., 1994). This enterprise can also provide an effective use for forages produced on land receiving manure from swine, poultry, or feedlot operations. The key to profitability for a forage-based beef production is the wintering system. In the northern Corn Belt, corn residues provide a feed resource that fits economically and logistically in beef production systems (Klopfenstein et al., 1987). To further reduce costs for supplemental feed and forage harvest, winter grazing of stock-piled hay can be used for either growing cattle (Allen et al., 1992) or mature cows (Bransby, 1989). Strip intercropping would facilitate such management by providing an opportunity to graze high-quality forage from stock-piled legume species grown in close proximity to the corn residues (personal communication, R.M. Cruse, 1994).

Grazing summer pastures has reduced the needs for grain required to finish cattle to a comparable grade by 15 to 45% (Russell et al., 1983). This decrease in grain feeding would not only reduce the cost of beef production, but also reduce the amounts of N and P imported onto the farm. Compared to grazing, however, nutrient removal from a field should be greater when forages are mechanically harvested, because forage harvest efficiency is greater (Rotz and Abrams, 1988; Rotz et al., 1993). Mechanical harvesting will increase the amount of manure nutrients that can be harvested, but if the forages are fed on the same farm, overall nutrient balance may not be changed (Lanyon and Beegle, 1989).

4. Nutrient Retention and Loss from Feedlot Cattle

Performance of well-managed cattle during the finishing phase has improved dramatically during the past 10 years, as the percentage of cattle with 'Continental' cross has increased (Baltz et al., 1992). However, the types of cattle being produced in the U.S. still vary immensely in size and this, in conjunction with their diet, creates significant variation in manure quantity and composition. Therefore, before an integrated livestock-crop production system can be developed for optimum nutrient flow and economic return, techniques for rapid and accurate manure analysis must be developed and research information will be needed to predict and model the effects of diet on both cattle performance and manure composition. For example, anabolic implants are known to increase retention of N and P in cattle (Preston, 1975). Including trenbolone acetate in the anabolic program causes a further increase in N retention (Hayden et al., 1992). No data are available on the effects of trenbolone acetate on P retention, but an increase in N and P retention would reduce the amount of these elements in the manure. The composition of milk is well established, and further research is not needed to accurately estimate N and P removal from the production system in this product.

Economic returns from cattle are dependent on performance and value of meat produced. Feeding to maximize gains and feed efficiency of cattle by feeding supplements usually results in optimum returns to the cattle feeding enterprise. In an integrated livestock-crop production system, maximum performance with major inputs of supplemental nutrients may not result in optimum economic returns to the system. Furthermore, because retention of supplemental nutrients by cattle is low (Lanyon and Beegle, 1989), the amount of nutrients lost as manure and returned to the soil is highly related to the amount of supplemental nutrients being fed. Cattle feeding studies at Iowa State University indicated that "Choice" beef with minimal animal waste fat could be produced with minimal quantities of grain (Trenkle, 1985). Large-framed steers produced a greater percentage of live weight as trimmed retail beef compared to smaller-framed animals. This production strategy would have greatest economic returns in a market compensating for yield of retail beef. Production of beef from different types of cattle, managed at different levels of intensity, needs to be evaluated at the whole farm level to optimize all aspects of the enterprise, including utilization of the manure as a resource.

5. Use of Biomass for Nutrient Removal

Farming systems designed to optimize the use of nutrients contained in animal manure may benefit by including a short rotation of woody plant species such as fast-growing hybrid poplars (*Populus x euramericana* 'Eugenei'), 'Austree' willow (*Salix matsudana x alba*), black willow (*Salix nigra*), silver maple (*Acer saccharinum* L.), black locust (*Robinia pseudoacaris*), sycamore (*Platanus occidentalis*), or green ash (*Fraxinu pennsylvanica* Marsh.). These would be planted at relatively close spacings in rotations lasting from 1 to 10 yr (Meridian Corp., 1986).

Research on short rotation of woody plant species for Iowa and the midwestern agricultural region has focused on species trials, cultural methods, spacing, and rotation length interactions; genetic selection and improvement; and integrated pest management and biotechnology since the early 1980s (Colletti et al., 1991; Schultz et al., 1991). Those studies show that it is possible to produce biomass for energy in 5 to 10 yr and fiber/timber products in 15 to 20 yr. Without fertilization, many of these woody species can produce approximately 9.0 to 15.7 dry Mg ha^{-1} yr^{-1} of biomass. These tree species reproduce vegetatively by stump or root sprouts, so one planting will produce 3 to 4 harvests. The large root systems allow rapid regrowth that provides continuity in water and nutrient uptake and physical stability of the soil throughout the life of the stand.

Current research in Iowa is focusing on the use of these woody species as a treatment for utilizing nutrients from municipal sewage sludge. Cottonwood (*Populus deltoides*) hybrids are being grown in an alley-cropping system with switchgrass and annual biomass crops. Sewage sludge is either not applied or applied by surface application in the summer and fall at rates providing 168 kg N ha^{-1} yr^{-1} (the apparent annual uptake of N by both the herbaceous and tree crops) or 336 kg N ha^{-1} yr^{-1}. The results suggest that tree yield may be doubled by the highest application rate compared to application of no sludge. Preliminary data also show that the wood of poplar cottonwood hybrids contains 0.65% N and 0.1% P under high fertility. This means that a typical plantation growing at a rate of 12 Mg ha^{-1} yr^{-1} would accumulate 78 kg N ha^{-1} yr^{-1}. If this plantation was harvested at 5 years of age, then 60 Mg ha^{-1} of wood would be harvested and would remove 392 kg N ha^{-1} from the soil. If manure application can double wood production, as observed from municipal sludge, then as much as 120 Mg ha^{-1} of wood could be harvested at the end of 5 years with a N content of 785 kg ha^{-1}.

B. Systems Engineering Design

The principles of systems engineering were used to conceptualize four integrated nutrient flow models for soil-crop-animal enterprises that could efficiently recycle nutrients contained in animal manure. Figures 2 and 3 show our initial attempt to capture the critical economic, social, environmental, and physical features by combining component information into various potential management systems. Each enterprise within any of the individual systems has its own nutrient inputs and

SUGGESTED SYSTEMS

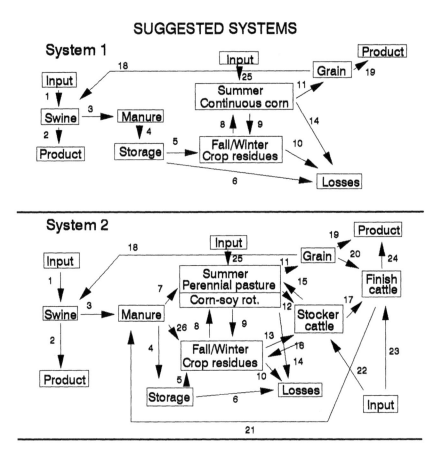

Figure 2. Conceptualized farming systems for utilizing animal manure with continuous corn (System 1) or mixed cropping (System 2) practices (numbers represent N and P transport vectors where existing or new information is used to alter management practices).

outputs, but interacts with other enterprises through nutrient flow in the grain, forage, or manure. Nutrient balance for the entire soil-crop-animal management system can be optimized by reducing external inputs, increasing the effectiveness of internal nutrient transfer between the enterprises, and increasing nutrient output in the form of meat, grain, or biomass products.

The primary goal for these systems is to maximize the use of N and P in the manure. For example, swine or poultry manure could be applied to corn, soybean, and forage in various spatial and temporal rotations across a landscape. Nutrients accumulated by these crops would be removed from or recycled within the system in various ways. A portion of the nutrients would be recycled in the grain produced using the manure as the primary nutrient source and used as an internal input for the swine feeding operation. Beef cattle could be integrated into the enterprise to utilize

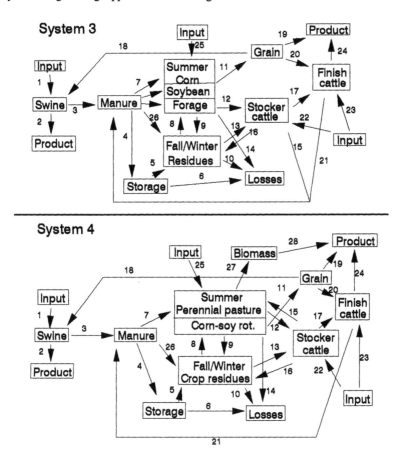

Figure 3. Conceptualized farming systems for utilizing animal manure with mixed cropping and animal practices (System 3) or mixed cropping, animal, and biomass enterprises (System 4) (numbers represent N and P transport vectors where existing or new information is used to alter management practices).

the forages and other crop residues. Woody or herbaceous biomass crops could be integrated as an outlet for excess manure and the nutrients it contains.

In assessing nutrient flow within farms or regions, operational boundaries for management units need to be established and flow measurements made at those boundaries (Lanyon and Beegle, 1989). These flow measurements may then be integrated into the total system. This approach thus allows data from reductionist or component research to be incorporated into evaluation of nutrient movement within the entire management system.

1. System Operation

System 1 consists of a farrow-to-finish swine enterprise with a continuous corn cropping system. Manure from the swine enterprise would be applied only during the fall and winter to land used for corn production. Grain would either be fed to the pigs or marketed.

System 2 consists of a farrow-to-finish swine enterprise, corn-soybean-meadow-meadow-meadow cropping sequence, and a growing-finishing beef enterprise. Manure from the swine and cattle enterprises would be applied to the row crop acres during the fall and winter and to the meadow during the summer. Stocker steers would graze crop residues supplemented with hay from the meadow during the winter and would graze the meadow during the summer. Corn grain and soybean would either be fed to the pigs and cattle or marketed.

System 3 consists of a farrow-to-finish swine enterprise, a corn-soybean-oat/legume strip intercropping system and a growing-finishing beef enterprise. Using the forage strips for wheel traffic control, manure from swine and cattle enterprises would be applied to the entire cropping area throughout the year, with the exception of 2 to 3 months between seeding and first harvest. Growing cattle would graze crop residues and stock-piled residual legume forage during the winter and would be fed fresh-chopped small grain and legume silage during the summer. Similar to system 2, corn grain and soybean would either be fed to the pigs and cattle or marketed.

System 4 would be identical to system 2 except that woody and herbaceous biomass would also be grown on a portion of the meadow to remove excess N and P from the overall system, and to provide a place for manure to be spread when it could not be applied to the row crops.

The arrows or vectors shown between individual components within the four conceptual soil-plant-animal systems (Figures 2 and 3) represent the anticipated direction for N and P flow or transport. For consistency, the same numbering sequence was used for all four figures. Therefore, as the conceptual systems became more complex, additional vectors were added. Several factors can affect the specific processes occurring at any one of the vectors, but by consensus among stakeholders who proposed the management systems, it was decided whether the primary factor(s) would be "decision points" where the owner/operator would currently have adequate information to make a decision, or a "knowledge gap" where information was unavailable or inadequate to be considered a simple decision. If there was a knowledge gap, additional process or component research would be required to fully evaluate and implement the system. The type of vector (decision point or knowledge gap) is listed in Table 10. As stated, these vectors show the direction of N and P transport, and therefore, they may represent several research questions (knowledge gaps) or management choices (decision points).

At each vector, systems analysis techniques (Bird et al., 1990) can be used to help reveal where nutrient balance might be improved by changing the management practices or obtaining new information. For example, nutrient losses from animal manure that is being stored in bunkers, lagoons, slurry tanks, piles, etc. (vector 6) are highly affected by the specific storage conditions. Similarly, nutrient losses from soil-applied manure (vectors 10 and 14) are highly affected by the method and timing of application. The requirement for external nutrient inputs such as fertilizer (vector 25) will be affected by the amount of manure that was applied (vectors 5 and 7), the type of cropping system (vectors 8 and 9), and the amount of nutrient loss (vectors 10 and 14). Nutrient removal as forage for the cattle enterprise (vectors 12 and 13) may be altered by the harvest efficiency (i.e., mechanical harvesting would transfer more nutrients from the field to the feeding area than grazing). External nutrients input during cattle feeding operations (vectors 22 and 23) would be highly dependent upon crop yield, composition, and quantity provided directly from areas receiving manure.

In addition to revealing areas where nutrient balance could be improved by changing management practices, systems analysis along each vector in the conceptual models (Figures 2 and 3) could help identify knowledge gaps where information is very limited or nonexistent. For example, to calculate the amount of N and P available from grazed pastures (vector 12) or crop residues (vector 13), the yield, forage composition, and harvest efficiency are needed. Yield and N concentration data on many forages are available, but few data are available on the P concentration of forages, and information on grazing efficiency is almost nonexistent. Furthermore, most grazing efficiency data that are available focus on forages that were not fertilized with manure, and therefore, would be of limited value for evaluating N and P transport in the conceptual systems.

Similarly, to calculate N and P transport from cattle grazing either summer pasture (vectors 12 and 15) or crop residues (vectors 13 and 16), it is necessary to know the amount and composition of forage consumed, the amount and composition of weight gain by the animals, and the proximity to and type of water supply. Of those variables, only bodyweight gain for different grazing systems is well-documented. Little information on the amount and composition of forage selected during grazing by stocker steers is available. Gradual changes in the type of cattle being grown further reduce the value of some data because the information used to estimate the composition of bodyweight gain is not likely to relate to the types of cattle that are currently being produced.

Other vectors that the stakeholders identified as having a high need for research to improve the accuracy of the conceptual models (Figs. 2 and 3) include nutrient losses from manure applied during the winter (vector 10) or summer (vector 14), total nutrients available for crop production (vector 8), uptake of manure nutrients by crops (vectors 11 and 12), nutrient retention (vector 24) and loss (vector 21) from finishing cattle, and methods of altering external inputs into crop (vector 25) and cattle (vectors 22 and 23) production.

The vectors in Figs. 2 and 3 that were not identified as knowledge gaps in these systems (Table 10) have been studied more thoroughly and generally have more available information that can be used for decision making. However, processes occurring at these vectors can have profound effects of N and P transfer and balance in soil-plant-animal systems that are designed to optimize the use of nutrients from manure. For example, the availability of P in feeds for monogastric animals is generally well established and ranges from 15 to 50% (Jongbloed, 1987; NRC, 1988). However, it has been reported that P availability may be increased by 18%

and the amount of fecal P decreased by 17% by feeding a microbial phytase with a corn-soybean meal diet to young pigs (Young et al., 1993). This type of interaction between vectors 1 and 18 thus creates a knowledge gap for which there is currently an insufficient amount of information to be considered a simple management decision.

2. System Evaluation

A final step when applying the principles of systems engineering to the design of soil-plant-animal management systems such as those shown in Figs. 2 and 3 is the need to determine and document how the information collected at each N and P transport vector will be evaluated. After the critical processes occurring at each vector have been determined and documented, we recommend developing criteria for standardized scoring functions (Wymore, 1993) as demonstrated for an integrated farm management systems (IFMS) research program by Karlen et al. (1994a). The specific criteria for evaluating N and P transport along each vector within the integrated crop and livestock systems discussed (Figs. 2 and 3) are currently being developed. This type of research is new and beyond the scope of this chapter, but it must be done and well documented to fulfill the requirements for a project that has been designed using the principles of systems engineering (Sage, 1992; Wymore 1993).

VII. Summary and Conclusions

Animal manure is one of the oldest resources applied to the soil to cycle plant nutrients, but in recent years, this potential "resource" has been considered a "waste" for many animal feeding operations. One reason for this change has been the separation of animal and crop production enterprises. This separation has created animal waste management problems for the swine, poultry, and beef industries, and at the same time, has decreased soil quality by reducing C input to many soils used for production of feed for these livestock enterprises. Our objective for this chapter was to show how principles of systems engineering could be used to develop management strategies that would change the general perception of animal manure from an agricultural waste to a resource when the environmental, economic, and social factors are considered.

The use of systems engineering requires a holistic approach that is supported by the mechanistic and component research. It often uses tools and techniques associated with systems science, which focuses on constructing both conceptual and mathematical models. Systems engineering also requires a clear definition of the problem to be solved. Therefore, we have briefly reviewed problems and critical issues associated with manure management for the swine, poultry, and beef feedlot enterprises.

Experiences with the use of systems engineering to focus research efforts and to design integrated soil-plant-animal management strategies are reviewed. The two examples used show how the systems engineering process can be applied to the issue of managing animal manure by seeking to: (1) define the problem that must be addressed, (2) determine how well the system must perform, (3) select criteria that will be used to measure performance, (4) identify technologies that must or can not be used, and (5) to document any economic, environmental, social, or other resource tradeoffs that must be considered.

Systems approaches are not new, but the complexity of agriculture has often resulted in emphasis on single- or limited-factor research. This is not acceptable for issues such as efficient utilization of nutrients in animal manure. Hopefully, the ideas shared in this chapter will stimulate communication among agriculturalists, legislators, policy makers, and the general public, and stimulate an increased emphasis on research planning and technology development to define and understand the interactions associated with issues such as manure management.

References

Abbott, J.L. and T.C. Tucker. 1973. Persistence of manure-phosphorus availability in calcareous soil. *Soil Sci. Soc. Am. J.* 37:60-63.

Allen, V.G., J.P. Fontenot, and D.R. Notter. 1992. Forage systems for beef production from conception to slaughter: II. Stocker systems. *J. Anim. Sci.* 70:588-596.

Baltz, T.C., J.W. Goodwin, and A.H. Brown, Jr. 1992. A review and analysis of beef carcass weight increases: Economic, biological, and industry relationships. *Prof. Anim. Sci.* 8:46-52.

Bird, G.W., T. Edens, F. Drummond, and E. Groden. 1990. Design of pest management systems for sustainable agriculture. pp. 55-110. In: C.A. Francis, C.B. Flora, and L.D. King (ed.), *Sustainable Agriculture in Temperate Zones.* John Wiley & Sons, Inc., N.Y.

Bransby, D.I. 1989. Compromises in the design and conduct of grazing experiments. pp. 53-67. In: G.C. Marten (ed.), *Grazing Research: Design, Methodology, and Analysis.* Crop Sci. Soc. Am., Inc., Madison, WI.

Campbell, C.A., M. Schnitzer, J.W.B. Stewart, V.O. Bierderbeck, and F. Selles. 1986. Effect of manure and P fertilizer on properties of a Black Chernozem in southern Saskatchewan. *Can. J. Soil Sci.* 66:601-613.

Cherney, J.H., K.D. Johnson, J.J. Volenec, and K.S. Anliker. 1988. Chemical composition of herbaceous grass and legume species grown for maximum biomass production. *Biomass* 17:215-238.

Christie, P. 1987. Long term effects of slurry on grassland. p. 301-304 In: *Animal Manure on Grassland and Fodder Crops: Fertilizer or Waste?* H.G. Van Der Meer et al. (eds.) Martinus Nijhoff. The Hague, The Netherlands.

Colletti, J.P., R.B. Hall, and R.C. Schultz. 1991. Hickory Grove, an Iowa example of short-rotation woody crops in agroforestry. *Forestry Chron.* 67:258-262.

Edwards, D.R. and T.C. Daniel. 1992. Environmental impacts of on-farm poultry waste disposal -- A review. *Bioresource Tech.* 41:9-33.

Eghball, B. and J.F. Power. 1994. Beef cattle feedlot manure management. *J. Soil Water Conserv.* 49:113-122.

Foran, M.E., D.M. Dean, and H.E. Taylor. 1993. The land application of liquid manure and its effect on tile water and ground water quality. pp. 279-280. In: *Proceedings of the Agricultural Research to Protect Water Quality Conference.* Soil Conservation Society, Ankeny, IA.

Hatfield, J.C., M.C. Brumm, and S.W. Melvin. 1994. Swine manure management. pp. 2-40 to 2-56. In: R.J. Wright (ed.), *Agricultural Utilization of Municipal, Animal and Industrial Wastes.* USDA-ARS. Washington, D.C.

Hayden, J.M., W.G. Bergen, and R.A. Merkel. 1992. Skeletal muscle protein metabolism and serum growth hormone, insulin and cortisol concentrations in growing steers implanted with estradiol-17b, trenbolone acetate, or estradiol-17b plus trenbolone acetate. *J. Anim. Sci.* 70:2109-2119.

Hileman, L.H. 1973. Response of orchardgrass to broiler litter and commercial fertilizer. Report Ser. 207, AR Agric. Expt. Stn., Univ. AR, Fayetteville.

Jongbloed, A.W. 1987. Phosphorus in the feeding of pigs: Effect of diet on the absorption and retention of phosphorus by growing pigs. Drukkeri, DeBoer, Leystad, Netherlands.

Karlen, D.L. and J.W. Doran. 1993. Agroecosystem responses to alternative crop and soil management systems in the U.S. corn-soybean belt. pp. 55-61. In: D.R. Buxton, R. Shibles, R.A. Forsberg, B.L. Blad, K.H. Asay, G.M. Paulsen, and R.F. Wilson (ed.), *International Crop Science*, I. Crop Sci. Soc. Am., Inc., Madison, WI.

Karlen, D.L. and A.N. Sharpley. 1994. Management strategies for sustainable soil fertility. pp. 47-108. In: J.L. Hatfield and D.L. Karlen (ed.), *Sustainable Agriculture Systems.* Lewis Publ., CRC Press, Inc., Boca Raton, FL.

Karlen, D.L., M.C. Shannon, S.M. Schneider, and C.R. Amerman. 1994a. Using systems engineering and reductionist approaches to design integrated farm management research programs. *J. Prod. Agric.* 7:144-150.

Karlen, D.L., G.E. Varvel, D.G. Bullock, and R.M. Cruse. 1994b. Crop rotations for the 21st Century. *Adv. Agron.* 53:1-45.

Kliebenstein, J.B. and V.D. Ryan. 1991. Integrating livestock industry and community development strategies. pp. 113-130. In: *The Livestock Industry and the Environment Conference Proceedings.* Iowa State Univ., Ames, IA.

Klopfenstein, T., L. Roth, S. Fernandez-Rivera, and M. Lewis. 1987. Corn residues in beef production systems. *J. Anim. Sci.* 65:1139-1148.

Lanyon, L.E. 1991. Livestock waste-nutrient source and crop utilization. pp. 46-50. In: *The Livestock Industry and the Environment Conference Proceedings.* Iowa State Univ., Ames, IA.

Lanyon, L.E. and D.B. Beegle. 1989. The role of on-farm nutrient balance assessments in an integrated approach to nutrient management. *J. Soil Water Conserv.* 44:164-168.

Lauer, D.A., D.R. Boulden, and S.P. Klausner. 1976. Ammonia volatilization from dairy manure spread on the soil surface. *J. Environ. Qual.* 5(2):134-141.

Mallarino, A.P., J.R. Webb, and A.M. Blackmer. 1991. Corn and soybean yields during 11 years of phosphorus and potassium fertilization on a high-testing soil. *J. Prod. Agric.* 4:312-317.

Marten, G.C. 1985. Reed canarygrass. pp. 207-216. In: M.E. Heath, R.F. Barnes, and D.S. Metcalfe (ed.), *Forages: The Science of Grassland Agriculture.* 4th ed., Iowa State University Press, Ames, IA.

Meridian Corp. 1986. Short-rotation intensive culture of woody crops for energy. Meridian Corp., Falls Church, VA. 85 pp.

Midwest Plan Service. 1985. Animal waste utilization. Livestock Waste Facilities Handbook MWPS-8. Iowa State Univ., Ames, IA.

Moore, P.A. Jr., T.C. Daniel, A.N. Sharpley, and C.W. Wood. 1994. Poultry manure management. pp. 2-18 to 2-39. In: R.J. Wright (ed.), *Agricultural Utilization of Municipal, Animal and Industrial Wastes.* USDA-ARS. Washington, D.C.

National Research Council (NRC). 1988. Nutrient Requirements of Swine. National Academy Press, Washington, D.C.

National Research Council (NRC). 1989. Alternative Agriculture. National Academy Press, Washington, D.C.

National Research Council (NRC). 1993. Soil and Water Quality: An Agenda for Agriculture. National Academy Press, Washington, D.C.

Nienaber, J.A. and R.B. Ferguson. 1992. Nitrate movement beneath a beef cattle manure composting site. Paper No. 922619. Am. Soc. Agric. Eng., St. Joseph, MI.

Pratt, P.F. and A.E. Laag. 1981. Effect of manure and irrigation on sodium bicarbonate-extractable phosphorus. *Soil Sci. Soc. Am. J.* 45:887-888.

Preston, R.L. 1975. Biological responses to estrogen additives in meat producing cattle and lambs. *J. Anim. Sci.* 41:1414-1430.

Reddy, G.Y., E.O. McLean, G.D. Hoyt, and T.J. Logan. 1978. Effects of soil, cover crop, and nutrient source on amounts and forms of phosphorus movement under simulated rainfall conditions. *J. Environ. Qual.* 7:50-54.

Rotz, C.A. and S.M. Abrams. 1988. Losses and quality changes during alfalfa hay harvest and storage. *Trans. ASAE* 31(2):350-355.

Rotz, C.A., R.E. Pitt, R.E. Muck, M.S. Allen, and D.R. Buckmuster. 1993. Direct-cut harvest and storage of alfalfa on the dairy farm. *Trans. ASAE* 36(3):621-628.

Russell, J.R., K.A. Albrecht, W.F. Wedin, and K.L. Driftmier. 1983. Effect of length of grazing period and feedlot grain level on the performance of growing and finishing steers. 1983 Annual Progress Report. Shelby-Grundy Research Center. Iowa State University, Ames, IA.

Sage, A.P. 1992. *Systems Engineering.* John Wiley & Sons, Inc., N.Y.

Schulte, D.D. and E.J. Kroeker. 1976. The role of systems analysis in the use of agricultural wastes. *J. Environ. Qual.* 5:221-227.

Schultz, R.C., J.P. Colletti, and R.B. Hall. 1991. Uses of short-rotation woody crops in agroforestry: An Iowa perspective. pp. 88-100. In: *Proc. First Conf. on Agroforestry in North America.* University of Guelph, Guelph, Canada.

Sharpley, A.N. and R.G. Menzel. 1987. The impact of soil and fertilizer phosphorus on the environment. *Adv. Agron.* 41:297-324.

Sharpley, A.N., I. Singh, G. Uehara, and J. Kimble. 1989. Modeling soil and plant phosphorus dynamics in calcareous and highly weathered soils. *Soil Sci. Soc. Am. J.* 53:153-158.

Sharpley, A.N., S.J. Smith, B.A. Stewart, and A.C. Mathers. 1984. Form of phosphorus in soils receiving cattle feedlot waste. *J. Environ. Qual.* 13:211-215.

Smith, A.M. 1952. *Manures and Fertilizers.* Thomas Nelson and Sons Ltd., London. p. 31.

Sommer, S.G., J.E. Oleson, and B.T. Christensen. 1991. Effects of temperature, wind speed and air humidity on ammonia volatilization from surface applied cattle slurry. *J. Ag. Sci.*, Camb. 117:91-100.

Steenhuis, T.S., G.D. Bubenzer, and J.C. Converse. 1975. Nutrient losses from manure under simulated winter conditions. *Managing Livestock Wastes*, The Proceedings of the 3rd International Symposium on Livestock Wastes, ASAE, St. Joseph, MI.

Stephenson, A.H., T.A. McCaskey, and B.G. Ruffin. 1990. A survey of broiler litter composition and potential value as a nutrient resource. *Biol. Wastes* 34:1-9.

Sweeten, J.M. 1991. Environmental management practices for cattle feedlots. pp. 71-81. In: *The Livestock Industry and the Environment Conference Proceedings.* Iowa State Univ., Ames, IA.

Trenkle, A. 1985. Feedlot performance and carcass quality of medium and large framed steers fed three levels of energy. pp. 47-51. 1985 Beef Cattle Research Report. Iowa State University, Ames, IA.

Vanderholm, D.H. 1975. Nutrient losses from livestock waste during storage, treatment, and handling. In: *Managing Livestock Wastes*, Proceedings of Third International Symposium on Livestock Wastes, pp. 282-285. ASAE, St. Joseph, MI.

VanDyne, D.L. and C.B. Gilbertson. 1978. Estimating U.S. livestock and poultry manure and nutrient production. USDA, ESCS-72. Government Printing Office. Washington, D.C.

Viselmeyer, B., T. Klopfenstein, R. Stock, and R. Huffman. 1994. Extensive beef production systems: Forage combinations managed as one unit. pp. 20-22. 1994 Nebraska Beef Cattle Report. University of Nebraska, Lincoln, NE.

Voight, P.W. and R.S. Maclauchlan. 1985. Native and other western grasses. pp. 177-187. In: M.E. Heath, R.F. Barnes, and D.S. Metcalfe (ed.), *Forages: The Science of Grassland Agriculture.* 4th ed., Iowa State University Press, Ames, IA.

Walter, M.F., T.L. Richard, P.D. Robillard, and R. Muck. 1987. Manure management with conservation tillage. pp. 253-270. In: T.J. Logan, J.M. Davidson, J.L. Baker, and M.R. Overcash (ed.), *Effects of Conservation Tillage on Groundwater Quality: Nitrates and Pesticides.* Lewis Publishers, CRC Press, Boca Raton, FL.

Whitehead, D.C. and N. Raistrick. 1991. Effects of some environmental factors on ammonia volatilization from simulated livestock urine applied to soil. *Biol. Fert. Soils* 11:279-284.

Wymore, A.W. 1993. *Model-Based Systems Engineering*. CRC Press Inc., Boca Raton, FL.

Young, R.A. and C.K. Mutchler. 1976. Pollution potential of manure spread on frozen ground. *J. Environ. Qual.* 5:174-179.

Young, L.G., M. Leunissen, and J.L. Atkinson. 1993. Addition of microbial phytase to diets of young pigs. *J. Anim. Sci.* 71:2147-2151.

Index

acidification 42, 65, 76

acid deposition 102, 115, 116, 186

acid rain 98, 101, 115, 116, 278

aerobic treatment 243, 265, 266, 272, 278

algae 66, 174, 265, 266, 277

alum 96, 98, 100, 101, 105, 106, 112, 113, 115, 116, 211, 213, 214, 274

aluminum sulfate 100, 101, 105

amino acids 50, 51

ammonia volatilization 76, 77, 81, 89, 93-97, 99, 107, 108, 115, 288, 301

anaerobic treatment 243, 256, 260, 277

aquatic weeds 174

aquifer 79, 132, 141, 175, 176

bedding materials 92, 210

beef cattle 4, 127, 128, 142, 143, 147, 160, 161, 165, 255, 283, 288, 303, 307

beef production 34, 45, 294, 303, 304

behavior 2, 3, 5, 11, 12, 29, 55, 73, 165

belief 5, 10, 17, 19, 29, 30

blue-baby syndrome 102

BMPs 107, 116, 130, 175, 177, 178, 197, 200, 212, 231

box spreader 23, 37

broiler litter 92, 95, 100, 105, 188, 211, 287

buffer strips 105, 107, 111, 116

cattle spacing 125, 134, 147

chlorophyll tests 201

chromium 49, 54, 55

climate 66, 70, 79, 134, 136, 158, 159, 165, 197, 220, 256, 288

compost 139, 140, 145, 210, 215, 245, 271, 278, 292, 296, 297, 299, 300

composting 58, 93, 125, 132, 145, 215, 243, 256, 269, 271-273, 277, 278, 292, 293, 295-298, 300

constraints 1, 19, 26, 28, 30, 31, 65, 79, 80, 230, 285, 289, 292

copper toxicity 89, 94

copper 49, 53, 54, 58, 66, 82, 89, 93, 94, 105

corn production 5, 19-21, 308

cost 5, 9, 21, 25, 30, 31, 34, 37-41, 44, 51, 71, 72, 79, 93, 100, 101, 109, 111, 137, 140, 144, 173, 175, 177, 178, 180, 210, 214-217, 229, 231, 260, 269, 275, 276, 278, 284, 285, 294, 296-299, 304

crediting 3, 8, 9, 11, 12, 23-25, 29, 30

cropland 9, 23, 24, 44, 126, 145, 146, 158-162, 244, 276, 288, 293, 302

daily haul 10-12, 14-19, 23, 25-27, 30

dairy manure 66, 71, 74, 75, 82, 131, 139, 174, 182, 184, 185, 192, 193, 198, 250, 275

decay series 73, 74, 205, 206

Delphi 289

denitrification 70, 72, 73, 78-81, 84, 107, 176, 179, 180, 182, 184, 186, 204, 206-209, 290, 294, 298

deserts 158

dissolved P 66, 187, 191, 219, 230

economic 2, 10, 15, 16, 21, 23, 25, 28-31, 33, 35, 36, 38, 40-43, 45, 46, 51, 66, 75, 79, 81, 82, 101, 111, 112, 178, 202, 203, 210, 212, 214-216, 218, 222, 230, 231, 279, 284, 289, 294-300, 303-306, 310, 311

educational 2, 26, 27, 178, 205, 211, 215-217, 231, 293

electricity generation 144, 278

environmental 1, 2, 15, 21, 25, 28-31, 38-40, 43, 65-67, 75, 76, 79-81, 83, 84, 89, 92-94, 102, 103, 106, 109, 111, 113, 115, 116, 125, 127, 142, 157, 158, 173, 175-179, 186, 188, 196, 202, 206, 210-213, 217-220, 227, 230, 231, 278, 279,